HANDBOOK
OF
LOGISTICS AND
SUPPLY-CHAIN MANAGEMENT

HANDBOOKS IN TRANSPORT

2

Series Editors

DAVID A. HENSHER

KENNETH J. BUTTON

HANDBOOK
OF
LOGISTICS AND
SUPPLY-CHAIN
MANAGEMENT

Edited by

ANN M. BREWER
Institute of Transport Studies,
University of Sydney

KENNETH J. BUTTON
The School of Public Policy,
George Mason University

DAVID A. HENSHER
Institute of Transport Studies,
University of Sydney

2001
PERGAMON
An Imprint of Elsevier Science
Amsterdam – London – New York – Oxford – Paris – Shannon – Tokyo

ELSEVIER SCIENCE Ltd
The Boulevard, Langford Lane
Kidlington, Oxford OX5 1GB, UK

First edition 2001

Library of Congress Cataloging in Publication Data
A catalog record from the Library of Congress has been applied for.

British Library Cataloguing in Publication Data
A catalogue record from the British Library has been applied for.

ISBN: 0 08 043593 9
Series ISSN: 1472-7889

∞ The paper used in this publication meets the requirements of ANSI/NISO Z39.48-1992 (Permanence of Paper).
Printed in The Netherlands.

INTRODUCTION TO THE SERIES

Transportation and logistics research has now reached maturity, with a solid foundation of established methodology for professionals to turn to and for future researchers and practitioners to build on. Elsevier is marking this stage in the life of the subject by launching a landmark series of reference works: *Elsevier's Handbooks in Transport*. Comprising specially commissioned chapters from the leading experts of their topics, each title in the series will encapsulate the essential knowledge of a major area within transportation and logistics. To practitioners, researchers and students alike, these books will be authoritative, accessible and invaluable.

David A. Hensher
Kenneth J. Button

**The Institute of
Logistics and Transport** ™

This Handbook is supported by The Institute of Logistics and Transport (ILT).

The Institute is the professional body for individuals and organizations involved in logistics, transport and supply-chain management. The Institute supports its 22 000 members and each sector, by actively promoting professional excellence and best practice and encourages the adoption of efficient and sustainable logistics and transport policies.

The Institute of Logistics and Transport
11/12 Buckingham Gate
London
SW1E 6LB
UK

Tel: +44 (0)1536 740100
Fax: +44 (0)20 7592 3111
Email: enquiry@iolt.org.uk
http://www.iolt.org.uk

CONTENTS

Chapter 8

The Concept of Value: Symbolic Artifact or Useful Tool?
ANN M. BREWER

Chapter 9

Intermodal Transportation
BRIAN SLACK

Chapter 1

INTRODUCTION

ANN M. BREWER
University of Sydney

KENNETH J. BUTTON
George Mason University, Fairfax

DAVID A. HENSHER
University of Sydney

1. Introduction

Freight transportation is both a major industry in its own right and a core input into the production process. It is therefore important to understand how commodities, including raw materials and semi-finished products, are moved about in a modern economy in a manner that serves the needs of business more generally. Important changes in technology, markets, institutional structures, and management theory have led to new ways of thinking about this process and to the development of ways of more closely tying transportation into the larger production process.

Logistics was initially a military activity concerned with getting men and munitions to the battlefront in time for the fight. It was seen as important but not vital for military prowess; Alexander the Great excelled at it, while Napoleon was dismal. But it is now appreciated as being an integral part of the modern production process. While only a few years ago it was virtually unheard of for companies to have logistics experts on boards of large manufacturing companies (homage being paid to financial and production management), this has now changed considerably. There are now also many more specialized companies that offer a full range of logistics services to meet the needs of companies which find it more efficient to out-source that part of their needs. The quality of transport logistics is now recognized as often being vital to the success of many organizations. Nationally, the ability to transport goods quickly, safely, economically, and reliably is now seen as vital to a nation's prosperity and capacity to compete in a rapidly globalizing economy.

Handbook of Logistics and Supply Chain Management, Edited by A.M. Brewer et al.
© 2001, *Elsevier Science Ltd*

What has brought this change about? Certainly there have been technology changes both in the strict transport hardware that is used and in the supporting hardware such as packaging (modern logistics would not take its current form without the advent of the container). There have also been massive changes in the information technology field which, in simple terms, allow a closer matching of the needs of transport users with suppliers, both within businesses and outside. But there have also been institutional changes, often neglected in the management literature, such as the deregulation of transportation, financial, and communications markets and the liberalization of international trade regimes. Perhaps of most importance, however, have been the speed and the ingenuity shown by management in taking advantage of these developments to restructure and reformulate the notion of logistics.

We see many of these changes taking place around us and, indeed, everyone is part of the process. Ignoring for one moment those professionally concerned with the new logistics, how are changes in the way logistics is being conducted affecting the man in the street? There are more trucks about because customers are demanding (at least in the light of current cost structures and final demands) more frequent deliveries and also deliveries that are distant from older terminals such as railheads. The size of trucks is often smaller, especially in cities, as the nature of goods carried has changed and warehousing strategies have been modified. Where large trucks are used they tend to have containers on them. There are more (although not vastly more) goods being delivered to homes as a result of e-mail transactions. Moving around the countryside it is increasingly common to come upon a large warehouse or distribution center. At airports, the traveler sees more freight planes being loaded, much of it in international carriers. The changes within businesses are even more dramatic.

Upon scanning this volume for the first time, the reader may gain an impression that what is driving the so-called "new economy in logistics and supply chain management" is technological change and the globalization of markets. For years, businesses have utilized new physical technologies coupled with the upgrading of communications infrastructure to compress supply chains and get commodities and freight moving more rapidly. While these trends seem likely to continue, there are additional challenges remaining, involving procurement as well as managing and co-ordinating warehousing and transportation processes. It should also be noted that these developments in logistics extend beyond what is often called the "new economy" (so-called "high-tech" companies) and extend to the rethinking of ways of enhancing the performance of many very traditional sectors.

One recent tactic has been the use of supply chain management and information technology to plan procurement and fulfillment as well as to monitor directly products through the shipment and billing processes. On the one hand, with the introduction of e-business the nature of competition has changed from a narrow business focus to a broader, supply chain one. On the other hand, e-business is leading to a transformation in supply chain management. Retailers, distributors, and manufacturers want to share logistics information by giving suppliers, buyers, and consumers instant access to their supply chain management systems.

In addition, increasingly, industries are starting to use shared online supply chains in business-to-business exchanges, providing specialized services to out-source key points in the chain, such as freight management. More and more, business processes are concentrating on some form of collaboration to source new customers and suppliers, as well as to service existing ones. For example, Ford, Daimler/Chrysler, and General Motors are transferring all their purchases to the web over the next few years via a single e-hub exchange for auto parts, and General Electric has been rapidly moving in the same direction. The savings in doing business can be considerable, with estimated savings in the U.S.A. of up to 39% for electronic components, 20% for computing, 25% for forestry products, and 22% for machining. The distinction with technology in the new economy is that it enables stakeholders, customers, suppliers, and buyers, to communicate and process knowledge without regard for spatial and temporal boundaries. In knowledge terms, the only boundary that remains today is political or proprietary. Otherwise, it is achievable for everyone to have the same information, regardless of position, organization, or national context.

2. How is the new economy different from the old view?

To understand the role of logistics in the modern economy it is important to appreciate some of the background changes that have been occurring in the economy more generally. In particular, there have been major adjustments in both the way many economies are now structured and in the ways that we think about how economic growth takes place. The importance of some of the changes have perhaps been exaggerated, and equally some of the new ideas may not be as powerful, and ultimately as enduring, as their advocates would like to think, but nevertheless changes there are.

To be useful, any focused set of papers on logistics must begin by providing a definition of what is really meant by the new, as opposed to the old, economy. While there have been some exceptions, theoretical economic models of the "old economy" view development as a result of input accumulation and technological development in a world of nearly steady returns to scale. However, how to measure inputs and how to define technological development has always been a contentious issue. There is, however, considerable agreement that economic development concerns trade-offs – there are opportunity costs involved. For example, increasing capital may mean more investment and savings but at the expense of declining consumption, while increasing labor input may require education expenditure and diminishing leisure time. Any residual, or "unexplained," growth after taking into account increases in factor supply, is attributed to total factor productivity growth. This is a measure of overall cost efficiency after accounting for the role of inputs. Productivity growth is important in contributing to profitability and, ultimately, the proficiency of supply chains and businesses. Specifically, total factor productivity growth models look at changing relationships between the cost of outputs and inputs of labor, energy, and capital investment in materials and technology. One of the factors which is usually not taken into account in the total factor productivity growth equation, at least in a quantitative sense, is strategic capacity. Such capacity, however, is vital for achieving high levels of co-ordination and integration across supply chains.

The new economy assumes a large contribution from total factor productivity growth, with significant industry sectors (e.g., information technology and biotechnology) gaining from increasing returns, positive externalities, standards, and network economies. In a network economy, more is leveraged off more, and ultimately value is created. For example, technology transfer triggers innovation. While these assumptions are not necessarily inconsistent with views of the old economy, proponents of the new economy attribute them with greater significance and see information technology and globalization as the principal drivers of economic development. This view also has important implications for wider economic thinking because it means that the cumulative effects of these various economies makes it difficult for nations without a critical threshold of appropriate industries to prosper, while those with the critical mass, through internal synergies, expand at an accelerating pace. This has macroeconomic and public policy implications beyond the scope of this volume, but is one that should be borne in mind.

3. What is knowledge in the economy?

This volume is about transport logistics, but in the modern world it is vital to understand how the efficient and effective movement of raw materials, intermediate goods, and final products relies on information and knowledge. There is a need to know if transportation is needed and, if so, what is required by users of transport, when is the service needed, and what can be offered and at what cost, not only in money terms but also in terms of taking transport hardware from other uses. There is the further issue of whether transport should be done by the producing company or be out-sourced. Tied to this is the need to appreciate where transportation fits into the modern logistics chain.

The knowledge base of a nation, industry, supply chain and business has to be transformed into value. "Value" in this context may be thought of largely in monetary terms, but in some cases there are wider considerations concerning the environment (e.g., "green logistics"). Businesses have always been in the knowledge game. That is why Japan so carefully protected, on pain of death, how silk worms are cultivated, why fishermen have always been shy to say where they get their best catches, and why tulip growers in The Netherlands in the 17th century were secretive about their cultivation techniques. The situation today is no different – all businesses are in the knowledge business as they move to source products for customers and to provide solutions to their needs, adding value derived from the quality of advice and broad knowledge. Perhaps the key difference is the amount of knowledge that is now available, how it can be acquired, and the speed at which it can be transmitted. Where businesses are directly involved in producing commodities, these are highly customized and/or have a strong focus towards application to specific customer demands in some way. The process for capturing and using knowledge has to be continually upgraded and accessible to everyone, otherwise value-creation will not ensue.

Paving the way towards a new economy is the structural change taking place in corporate management. Industries have entered an era in which commercial organizations alone can no longer guarantee their competitive advantage in large national, let alone the global, marketplaces. Management structures and systems need to be flexible enough to capture and harness not only the know-how but also the knowledge of the new economy. The paradox of knowledge is that it is impermanent and inherently unmanageable. If ways to capture knowledge are not realized, there can be a failure to transform supply chain management to value networks. One approach to this involves refining and creating new concepts, structures, and markets embodying transportation within a wider knowledge-based framework.

It is helpful to clarify the terms "technology," "know-how," and "knowledge." These terms often appear abstract, but have important practical connotations. Technology is the accrued "know-how" and modus operandi built into the production processes. The design of technology incorporates and signifies the knowledge and skill base of a nation. Successful technology transfer requires that an adopter has ample absorptive capacity to integrate new technologies within its production processes. It can be inferred that the real driver of change is not technology but the quest for knowledge to create value principally by addressing customer demands. As globalization of competition has affected many regional markets, many product and service life cycles have been so condensed that industries focus mainly on market-share competition. Increasing returns to scale and scope are progressively more significant in the competition stakes. Businesses have an even greater incentive to enter the market early because they have only a very short time to recover their development investments. The transport input, by necessity, must be flexible and responsive, whether supplied by specialists or in-house.

The transformation of supply chain management resides in stakeholders' openness to ideas from the outside world as much as to changes in technology.

Knowledge takes many forms, from the strategic use of management and allocation of investment, through to research and development to capitalize on the use of technology and technology transfer. To add weight to the argument that knowledge, not technology, is the real propellant of change, management technologies such as lean, continuous improvement and total quality, strive to create and leverage the collective knowledge of employees, contractors, and stakeholders through restructuring conventional processes and relationships. An important part of knowledge management is learning by doing, as well as improving the proficiency of processes through experience. This applies to the transportation component of the process as to all others.

Taking the concept of knowledge channels further, alliances, in every sense, are having a profound impact on the nature of international relations and global business. They abound in transport, with strategic airline alliances such as the STAR Alliance and railway alliances, and are spreading in international shipping, but are even more pervasive in other sectors. It is often in industries where product life cycles are decreasing most rapidly that alliance formation is most common. The very perception of economic prospect depends on knowledge of markets, competitors, new technology, government policy, and general economic and political conditions. This is what is referred to as "strategic intellectual capital." Relationships with customers, buyers, and suppliers are an important source of learning, as well as being part of our overall intellectual capital. This may be one reason why users and suppliers of transport services often develop long-term relationships.

Management today is struggling to move beyond models wedded to the old economy. Strategically, there is often seen to be a need to integrate logistics in corporate planning with accountability for it throughout every layer of the business. Accounting, for example, will have to move from cost accounting towards supply chain, performance-based outcomes (i.e., activity-based accounting). Unlike cost accounting, which asks how much something costs on a component basis, activity-based accounting enquires whether an activity must be done and, if so, where. For example, for freight forwarders, which activities are best handled by fleet and which are more efficiently out-sourced? The main impediment to the new value economy is to find ways of rethinking the control models of management. Attempting to think beyond control is often constrained by institutions and structures that do not support the transition to knowledge creation demanded by the new value economy.

The proficiency of supply chains depends heavily on timely, reliable information channels and flows. Logistics and supply chain management need to be based on concepts that are extremely flexible and yet rigorous enough to support new enterprise models. A fundamentally changed business model that integrates multifarious stakeholders underpins e-business, or the "e-supply chain." e-Business is an enabler, making certain things feasible that would be impossible otherwise. Specifically, collaborative mechanisms such as demand forecasting, purchase order management, shipment scheduling, export/import compliance checks, multiple owners of inventory, and requisition generation, review, and acceptance are some real-time processes to minimize transportation search costs, accelerate logistics decision-making, and improve communications among all the stakeholders in a supply chain. As a result, brand owners, knowledge brokers, program managers, global distribution dealers, capital and third-party providers, to name a few, will need to be increasingly incorporated within internal as well as external supply chains. Consequently, an integrated view of supply chain management, at both the micro and macro level, is essential.

Moving away from control and ownership, the nature of management takes on a different meaning in the new economy. Human resources will have to move away from an asset to a process focus. For example, the organizational chart does not define the organization, and intangible assets such as intellectual capital easily classified by inventories. The structures and systems associated with control models of management virtually lead to asset stripping, not enhancement. Indeed, equating people to assets becomes meaningless in the knowledge-based economy. Business cannot "own" assets as assumed under the old economic regime; rather, shaping, promoting, attracting, and sharing assets will be the raison d'être of the new economy. A common, interactive knowledge base needs to be established, along with clear accountabilities for the management of it. Furthermore, the balance of knowledge and skills needs to be transformed by building a small but highly capable group of *knowledge brokers*, who can identify processes which are used by competitors, partners, suppliers, and customers. Stakeholders need to know about progress in creating and maintaining a knowledge organization, mainly because focused and consistent internal signals are essential if culture change is to be achieved.

The transition from the old frontier to the leading edge is an enormous challenge. The major test ahead would seem to be deciding not what will be required for the future, but how strategic actions construct the new value economy and ensuring that the capabilities necessary for the future are being designed. Projections from the old economy simply extend the current mindsets and do not move beyond them. The new focus is customer-dominated: who will be the customers, what will customers expect, and how can a business source and utilize its knowledge to create a real competitive advantage?

4. The Handbook

This book is the second volume in the series *Handbooks in Transport*. As with the other volumes in the series, the coverage is meant to be neither comprehensive nor always excessively deep; it is neither a textbook nor a research monograph.

Some topics not covered here can be found in a companion volume (e.g., freight modeling topics are covered in Volume 1 of this series), simply because they had as much right to be there as to be in this collection. The material has been organized with papers grouped to reflect the important elements of modern transport logistics.

The papers are all original, and the international collection of authors was selected for both their knowledge of a subject area and, of equal importance, their ability to put on paper in few words the core of that subject. Not all the authors have gone about their task in exactly the same way, but this adds to the richness of the material and reflects the diversity of approaches that can be adopted. It also shows the individuality of the various contributors and their ways of thinking.

The common denominator extending across the contributions is that they try to be up to date in their treatment of topics. The topics themselves were selected to embrace most elements of modern transport logistics. These extend beyond domestic issues, and the coverage embraces many matters important for efficient international trade. The authors have approached their task using a variety of tools, including synthesis and case studies, but aimed at ensuring that the reader is provided with contemporary material. Equally, some of the topics and their coverage are more abstract, while others take a much more pragmatic bent. Again this is by design. Furthermore, there is a degree of interlinkage and overlap between some contributions, which reflects the artificial boundaries that often have to be drawn in our quest for understanding. Humans are creatures who habitually compartmentalize things, but the resultant divisions should be seen as porous.

The audience that the authors were asked to target is not an easy one. Handbooks serve the purpose of keeping informed and up-dating those already working in a field while also offering an opening for those who wish to gain an introductory acquaintance with it. This has by necessity limited the technical components of some papers, but equally has drawn forth some very articulate verbal accounts of topics. Overall, we think that the contributors have done an admirable job in meeting the challenge posed them.

Part 1

GLOBALIZATION AND INTERNATIONAL PERSPECTIVES

Chapter 2

PERSPECTIVES ON GLOBAL PERFORMANCE ISSUES*

TREVOR D. HEAVER
University of British Columbia, Vancouver

1. Introduction

Two aspects of logistics performance are the focus of this chapter: how to achieve a high level of performance, and how to measure the level of performance. These two aspects are related, because without measurement good performance is difficult to achieve. However, this linkage is only one reason for them being treated together. Another important reason is that both are important at the level of the company (the micro level) and at the level of the economy (the macro level). There is value in a global perspective on logistics performance at both levels.

The concept of "performance" merits some clarification. It embraces the process of measurement, but the approaches to and methods of measurement differ depending on the purposes of the measurement and on the characteristics of the logistics services. For example, companies may wish to know the value of resources used in logistics activities and how the costs are changing. They may be interested in the effectiveness with which the resources are being used by assessing the value of the tangible and intangible outputs of logistics activities. The same interests in inputs and outputs may be shown by members of a supply chain or may exist at a national or regional level.

The nature of logistics raises issues for the assessment of performance. While the measures of logistics performance at the level of the company are taken up elsewhere in this book, the general characteristics of the difficulties of measuring performance are relevant here. They include the multifaceted nature of the objectives and characteristics of logistics services. The challenges are aggravated because the measurement of performance of individual companies and of the supply chains of which they are a part must be considered in the context of the economy within which companies operate. The interdependence of micro and macro conditions has particular importance in logistics, as the services are more

*The author gratefully acknowledges the helpful comments of Lennart E. Henriksson of the University of British Columbia on a draft of this paper. All errors and omissions remain the responsibility of the author.

influenced by public investments and regulatory policies than are many sectors of the economy.

Logistics systems utilize a wide range of resources and services within and external to individual companies. External relationships are important, in part, because of the importance of transportation and communications with their significant investments in infrastructure. However, a wide use of external services is also likely because of the wide range of functions that logistics must encompass in procurement and distribution. Furthermore, the essential spatial dimension of logistics activities makes their provision and control by a single enterprise unlikely. Two results have particular relevance here. The first result is the importance to good performance of managing the interrelationships well. In current terminology, how well the supply chain is managed. The second result is that enterprises in a region or nation are likely to experience similar challenges and opportunities because of the extent of common conditions. These may arise from the same geographical circumstances, from characteristics of shared infrastructure (such as information systems and transportation), or from widely held cultural, political, or economic attitudes to which individual enterprises are largely captive.

Achieving a high level of performance in logistics is important for the profitability of companies and for the efficiency of national economies and the global economy. As international trade increases as a percentage of national domestic activity, so the interactive effects of the productivity of national and international logistics increases. It is understandable, then, that corporations and nations are interested in measures of performance at the macro level.

The rest of this chapter is in four sections. Section 2 deals with the nature of logistics and the implications that this has for the pursuit and measurement of logistics performance. This is followed in Section 3 by a description of issues in the pursuit of good logistics performance at both the micro and the macro level. An issue that is given separate treatment (in Section 4) is the measurement of logistics performance. This is done by considering the types of performance measurement as influenced by data availability and the purposes for measurement. Section 5 concludes the chapter by examining the application performance measures at the micro and macro levels in the economy.

2. Logistics service and performance measurement

Logistics is an aspect of management that has developed as a recognized field in business only in the last 40 years. The use of the term "logistics management" in manufacturing and distribution organizations has been popular in Europe and North America only since the mid-1980s. Its recent and slow recognition is in part because it delivers a service that is multifaceted and largely intangible. The importance of inter-organizational relationships in logistics has led to use of the term "supply chain management" to ensure attention to the challenges that arise from complex organizational structures. These characteristics and the recent development of logistics have implications for approaches to logistics performance.

2.1. The aspects of logistics in an economy

For a long time, the integrated contribution of logistics activities was not recognized, let alone measured. Measurement was of individual components such as transportation costs, the value of inventory, and warehouse costs. The individual components were viewed as cost centres, with the performance of the activities judged on the basis of cost levels. Interpretation of cost-based results had to be made taking into account required changes in service levels and the scale and geographical extent of the services, to mention just a few variables affecting costs.

Recognition of the greater value associated with the management of logistics as an integrated function has been associated with recognition of the contributions of logistics to creating value through enhanced customer service. Logistics has shifted from a purely cost centre to a centre creating value. This makes the range of service attributes required to meet logistics requirements greater and their measurement more difficult.

As use of the term "logistics management" has gained acceptance, and as various factors have contributed to the use of out-sourcing, so a market has grown up for "logistics services." Some "third-party logistics service providers" (3PLs) are freight forwarders who have traditionally provided a mix of services designed to meet a variety of logistics needs. Previously, their services were not structured and marketed as logistics services. The main other source of 3PLs is companies who have had assets and related management capabilities in a part of logistics to which they have added other capabilities. They have done this particularly through the addition of information-system assets and knowledge-based expertise. A good American example is Ryder System Inc., which has incorporated additional information technologies and logistics management skills in its fleet management expertise to become a major logistics service provider. Many transportation companies have operating divisions or subsidiaries providing logistics services to complement their carrier activities.

Thus, logistics in the economy has two dimensions: logistics management in manufacturing and distribution organizations, and logistics organizations providing services to the manufacturing and distribution companies. While the distinction is blurred to the extent that some for-hire logistics services are provided by manufacturing and logistics companies, the characteristics and performance of the two sectors are generally treated quite separately. However, both are affected by the varied activities that make up logistics.

2.2. *The multifaceted and intangible nature of logistics*

Logistics delivers a service that is multifaceted and largely intangible. "Good service" can be hard to define and difficult to measure, while the magnitude of its effects on the corporate supply chain and national objectives are even harder to quantify.

Logistics contributes to the creation of time, place, and even form utility through the management of processes that enable companies to get the right goods to the right place at the right time in the right condition and at the right cost. These basic utilities have been recognized for many years. Of course, the changing costs of resource inputs and changing technology and concepts have meant that the characteristics of logistics systems have changed radically in physical and organizational terms. The levels of service expected of logistics systems have increased as response times have had to be reduced and made more reliable and flexible, while the pressures to reduce costs have persisted. The developments are reflected in new practices such as cross-docking and efficient customer response and in new organizational structures evident in the term "supply chain management" and in the growth of 3PLs. The changes in concepts, technologies, practices, and organization structures pose new challenges for governments trying to foster conditions favorable to efficient logistics and for companies trying to improve the contributions of logistics to corporate performance.

The objectives of logistics must be aligned with corporate goals, within the influences of national interests and regulations. The goals of companies are diverse, reflecting different market conditions and corporate priorities. They are affected by the various economic, political, social, and environmental conditions under which companies operate. Goals can include return on capital, customer satisfaction, employee satisfaction, growth, cost reduction, environmental quality, and others. It is not appropriate here to explore the extent to which these objectives have interdependencies and conflicts (for a review, see Chow *et al.*, 1994). The reality is that multiple objectives co-exist and must be reflected in the structure of logistics services and in the measures of performance. The existence of diverse objectives precludes the possibility of "one best way" of defining performance. It is not surprising, then, that there is no universally-accepted measure of logistics performance.

3. Types of performance measure

The notion that if an activity cannot be measured it cannot be properly managed is well known. The need to manage activities is a key reason for their measurement. Well-designed measures serve to focus attention on the key performance objectives that they monitor and, thereby, promote fast response to issues affecting them (Atkinson *et al.*, 1997). However, while this provides a fundamental and simple reason for measurement, it leaves open the various uses and reasons of decision-makers for measures. Different types of measures are used in different circumstances.

Different approaches are used in activity measurement depending on the availability of data and the purpose of the study. Three aspects of performance measurement are highlighted here: the use of input and output data, the source and nature of the data, and the purposes of the measurement.

3.1. The use of input and output data

It is easier to measure the inputs to logistics than it is to measure the outputs. Inputs are the costs that show up in tangible monetary form. These may be freight rates, the total transport cost, the value of inventory held, the capital invested in logistics assets, or the amount/cost of labor employed in certain logistics activities. These may be expressed on a total basis or an amount per unit of output, such as per tonne or as a percentage of the value of goods sold. However, such input measures do not indicate the extent to which on-time delivery has been achieved, the amount of flexibility of delivery to suit user needs, the contributions of logistics to customer satisfaction, or the achievement of corporate internal goals such as return on capital and employee satisfaction. The "consumption" of the environment is a cost which is both difficult to measure and to shift from an externality of companies to an internal monetary cost.

Input data may be used when private or public sector organizations want to know how large an activity is. For example, estimates of the amount of fuel used by a mode, not a usual measure of performance, may be used as a proxy for a the effects of a mode on pollution. Changes in the size of a mode of transport may be used as an indicator of its competitive success. Input or output data may be used in this case, but the output data are at best partial. For example, the tonne-kilometer is a measure of the physical work done but not of its value. Furthermore, the value of a transport mode may be viewed as an input to the broader function of logistics in providing a multidimensional service.

Decision-makers are interested in knowing how well resources are being used. This is indicated most easily by the use of numerous partial productivity measures. Some measure particular activity patterns (e.g., inventory turns), while others compare a level of activity with an input (e.g., tonne-kilometers per train hour or warehouse throughput per employee). Key productivity indicators can provide well-informed managers with an up-to-the-minute insight into changes taking place in operations. However, partial productivity measures can only be interpreted properly with a thorough knowledge of all the changes in the output mix and the other inputs that affect productivity. The most common context in which this is important is when explaining changes in labor productivity per hour. Often an increase in labor productivity is enabled by a substitution of capital for labor. If too much capital is employed, it is possible that total productivity will decrease, even though labor productivity is still increasing. Unfortunately, the breadth of data and the econometric methods required for accurate total factor productivity measurement preclude its use as a short-run source of information for decision-makers.

Intuitively, the best application of performance measurement is when the results are compared with the objectives. What service value has been provided to customers? What has been the contribution to the return on capital? Are employees "satisfied?" Such measures are difficult to define and to quantify, and have to be weighted. Deliberate efforts are needed to collect information on logistics outputs. Dimensions of value may be the extent to which orders are filled fully and on time, the accuracy of billing, or the level of flexibility in meeting customer requirements. To get such measures a company may go beyond developing databases from its own activities to soliciting specific performance data from customers. Both data derived from the customers' operating files and subjective views are of value.

3.2. The sources of data

The source of data is an important issue, whether the assessment is carried out within a company or externally (e.g., by governments or academics). The most readily available data, especially for broad international comparisons, are often based on financial accounting data (see Chapter 33).

For reporting purposes, organizations are required to maintain financial records according to accepted accounting practices. This has the advantage of promoting consistency within accounting jurisdictions, but it does not provide consistency across jurisdictions (e.g., between countries or groups of countries). Nor does it necessarily provide management with the best measures for decision-making. For example, depreciation amounts and book values may be out of step with market realities. For decision-making purposes, organizations use measures of costs that are more consistent with economic realities, such as those derived from activity-based costing. However, financial measures may have a role to play in measuring logistics inputs at a macro level, as they may be the only data available on a consistent basis in a country.

The multiple dimensions of logistics performance are a real source of data problems. Within a company, where the confidentiality of data is not a major problem, data may be taken from transaction files (e.g., order fill rates from shipping files) or gathered through special reports (e.g., satisfaction surveys of employees). However, reliance on internally generated data would create self-assessment for customer service. Use of such data is much less desirable than are evaluations using "hard" data (e.g., product availability) and "soft" data (e.g., personal contact experience and overall satisfaction ratings) provided by customers. The richness of data for internal studies provides companies with considerable opportunity to link variables affecting output with performance (e.g., breaking services down by customer, product, market, and other attributes). However, cause–effect relationships may still be a problem. For example, the net contributions of logistics to the corporate rate of return on capital may be ambiguous, but that may not stop the overall rate of return being used as an element in assessing logistics performance.

There are greater data problems associated with external assessments of performance, which are generally conducted by academics and consultants using surveys. The use of surveys creates many problems. There are response biases; that is, the companies responding may not be representative and the responses may be "conditioned" by the questionnaire. Self-assessment of customer service performance is usual. Finally, making comparisons among surveys is made difficult by the lack of standard definitions and measures. This is unfortunate, because comparison is usually an important part of performance assessment exercises.

3.3. The purposes of performance measurement

Measurement may be undertaken simply to determine the size of an activity. This is most frequently the case in the collection of public data. It is expected that the data can be used in various ways, some of which will be performance related. For example, time series on the size of the modes of transport, the prices that they charge, and the profitability of companies may all be inputs to assessing the effectiveness of public policies. Agencies collecting public data must anticipate the uses to which they may be put. Privately collected data are usually tied more directly to assessment objectives.

Measurement may be undertaken to enable comparison of an activity against a standard or goal, over time, between different activities at one time or a combination of these. For example, it may be desirable to know whether the tonnage of rail freight carried in Europe has increased in a period and how this experience compares with that elsewhere. Companies may wish to know how their share of activity compares with that of other enterprises, whether in the context of market share or to understand their importance in a supply chain. The relative level of logistics costs is commonly used as a measure of their importance at a national, industry, or corporate level.

Productivity measures are a particular type of performance measure. Whether a rate of return on capital, a partial productivity measure, or a total factor productivity measurement, the data are useful in comparison with either standards, past data, or the data of other organizations. Benchmarking is a particular form of comparison deserving special attention.

3.4. Benchmarking

Benchmarking has been adopted widely as a method by which organizations can improve their performance (see Chapter 20). It has grown out of the experience of Xerox in the early 1980s when a team of managers was established to identify internal standards of performance for processes, such as order fulfillment, distribution, and production costs, and to compare these with the standards of market rivals. The process of identifying activities, measuring standards, and comparing with the experience of other companies proved so beneficial that Xerox expanded the process to all areas of its business. It gained the greatest success when learning from companies that were perceived as the best at a function in a similar or related industry. Benchmarking became a methodology of comparing internal performance and processes against organizations with the "best practices." In 1992, Robert Camp, Xerox manager of benchmarking competency, emphasized that the objective is "to understand the processes that will provide a competitive advantage ... Target setting is secondary" (Bonney, 1992).

Benchmarking is now a specific, widely recognized business-improvement process. Unfortunately, as with many successful practices, the term "benchmarking" is now used in situations when output measures are compared even though conditions are heterogeneous and no analysis of the processes is involved. An example is the *International Benchmarking of the Australian Waterfront* (Australian Productivity Commission, 1998) which compares productivities and concludes that better performance is possible on the Australian ports but does not attempt to understand the processes of port operations. Consequently, the actual nature of benchmarking studies is now uncertain. The popularity of the term has made it a buzzword of uncertain meaning.

The utility of the different measures of logistics depends on the context and purpose for which the measures are to be used. The development of more output measurement in recent years is a reflection of the growing maturity of logistics management. Exploration of issues associated with measures is done in the context of the different scales on which performance measurement is conducted.

4. Performance measurement at different levels

Logistics performance is of interest at different levels in the economy. The levels may be grouped as micro and macro. Micro measures focus on the performance of the individual company or enterprise, but may extend to the supply chain of which individual enterprises are a part. Micro measures are based on a vertical approach to value creation in a part or the whole of a supply chain. A supply chain is all those activities, including intercorporate relationships, that contribute to the translation of basic resources into the products and services of value to end customers. A supply chain may fall only in one country, may be international or, increasingly, may be "global" in extent. Macro measures focus on the performance of sectors of the economy defined by an industry sector or a geographic area such as a city, region, nation, or trading area. Macro studies follow a horizontal approach to an industry or industries. Micro and macro studies serve different purposes.

4.1. Performance measurement at the micro level

The reasons for an individual company or other organization to be interested in data on its performance are readily appreciated. The measurement of performance is integral to the management of its resources to achieve its objectives. However, it is at the level of the individual company that there is likely to be the greatest variety of objectives that are meaningful to the assessment of performance. For example, a company may include an objective of adding value for employees as well as for customers in a multiple-objective evaluation. Such internal objectives are not usually evident in measures attempted on wider bases.

There are three characteristics of performance measurement by the company that warrant recognition. First, assessments of logistics performance are still becoming more sophisticated as increasing use is made of multiple objectives and of more sophisticated measures. The greater complexity of measures is required as the value creation roles of logistics and the complexity of customer service requirements gain recognition. Second, the measures of performance are affected by the difficulty of attributions to functional areas as greater integration of decision-making is achieved and cross-functional teams play a greater role. Third, companies are affected by their need to consider performance in the context of the supply chain of which they are a part. Core performance measures may not change, but companies need to assess that they are better off with their piece of a larger pie than if all partners were making suboptimal decisions.

Recognition of the wide interdependencies among companies (and public sector organizations) in bringing value to the end customers has led to an interest in broadening trade-offs from within organizations to trade-offs among organizations in the supply chain. Assessment of the performance of the parts and of the whole supply chain can be instrumental in advancing positive co-operation in relationships that might otherwise be subject to suboptimization as a result of excessive self-interest and competition. Simulations can provide good insight into the consequences for the performance of the supply chain of different actions by individual enterprises. They are one way for supply chain members to develop attitudes supportive of the exchange of information and of reward mechanisms consistent with optimizing supply chain performance. Simulations across a whole supply chain are easier to conduct in the natural resource sector (e.g., grain or coal) because the range of products and organizations involved is smaller than for consumer goods, such as grocery products.

The level of leadership in a chain influences the extent to which supply chain optimization is practiced. The greatest integration is likely when a single interest controls many elements of a supply chain. Examples of the greatest span of control are in resource industries where the complexity of the product mix and the number of supply points and destinations are relatively limited. One example is the ability of Methanex, the world's largest producer of methanol, to produce and deliver product to customers globally. Buyers of coal for power plants or steel may control their sourcing with the total costs in mind, but they generally do not own the international producers. For consumer products, large producers such as Nike or retail chains such as Marks & Spencer and Wal-Mart may order products which they have designed on ex works terms from the supplier, often located in another country. The buyers may not perform the logistics function but they control and assess performance on a supply chain basis. Increasingly, a 3PL performs the logistics service. Under ex works terms, the break in logistics responsibility occurs at the point of manufacture, which may be inland in a different country from the customer. However, frequently in international trade, the terms of shipment used result in the break in responsibility occurring at the export port (Heaver, 1981). For example, coal, grain, sulfur, or other raw products move from the production point to the port under the sellers' control. Their sellers' objective is to meet customer delivery times, as indicated by ship arrival times, at least cost. Although some market competition may exist among the producers, co-operation between them and others in the port supply chain is desirable to reduce costs. In such cases, methods of performance evaluation can be crucial to demonstrating and realizing the benefits of co-operation in planning and operations control in order to achieve optimal efficiency.

4.2. Performance at the macro level

Companies or governments may be interested in measures of logistics performance at the industry level in order to assess the effects of government policies on logistics service suppliers or on the logistics performance of manufacturing and related companies. Companies within an industry may also find that industry data are helpful in rating their performance against the industry's experience. Such comparison is not formal benchmarking, but it may be a step in the process of benchmarking. Indeed, some industry studies, or at least studies by groups of similar companies, have been undertaken in the expectation that they would facilitate benchmarking activities.

Concern for the lack of consistent definitions and methods across studies led the U.S. Supply-Chain Council to establish a program to develop the supply chain operations reference (SCOR) model (Saccomano, 1996.) The purpose of the SCOR model is to establish standards of process definition for supply chain management on a par with the use of "generally accepted accounting principles." The SCOR model is intended to enable companies to compare their own supply chain processes with "best practices" and to strive to achieve the latter by using recommended software.

Studies of logistics performance in manufacturing or in logistics services may be carried out at various levels for policy appraisal and performance comparison. A community aspiring to serve as a national or regional gateway may wish to compare its performance with others or to assess changes in its performance over time. International comparisons, when available among countries or, increasingly, among trade areas, may be of value.

Measurements of logistics at the macro level are useful. In many countries, data show that logistics performance is improving (Waters, 1999; Delaney, 2000). Recognition that economies are functioning with a smaller and more responsive level of inventory has important implications for economic forecasting. Expectations are for less cyclical volatility caused by inventory adjustments. Insights can also be gained by making comparisons between the logistics performance of different countries. For example, it is beneficial to have data on the response of transport and other logistics service providers to public regulatory and investment policies. However, the limitations of national data and of international comparisons must be recognized.

The measurement of logistics at the macro level frequently focuses on inputs, because of data availability. Relevant data may be compiled from various public sources, such as those of the U.K. Office of National Statistics, as used by Waters (1999). Interpretation of such national data must reflect the various geographic, economic, and cultural factors affecting productivity and the various objectives set for logistics services.

National data commonly include input and partial output data on major components of the logistics service industry (e.g., for-hire transport companies and warehousing companies). However, there are many sectors that they do not include, such as information services and private activities in transport and other logistics activities. The inclusion and categorization of data are dependent on the specifics of definitions, which vary among countries. The definitions normally result in data being categorized according to the principal business of a company. For example, all activities of a trucking company will be so listed even though it engages in a widening range of logistics services such as more storage and distribution services.

The national data that are most frequently available in continuous time series deal with the value of stock held. Good time-series records by industry and for the country generally show decreasing levels of stock in relation to gross domestic product. This is a partial measure of improved logistics performance, as it does not indicate non-inventory costs such as transportation.

International comparison of logistics performance is difficult for many reasons. As already noted, definitions differ between the countries, so that the data are not strictly comparable. However, more importantly, logistics costs are affected by a wide range of geographic, economic, and cultural conditions that are intimately associated with the way of life in national cultures. Product mixes vary greatly between countries. Logistics costs are likely to be higher where distances are large, where densities of economic activity and population are low, and where economic and cultural conditions limit opportunities for large-scale distribution activities.

Finally, the use of national input data to measure performance has the obvious shortcoming of not indicating the values created. For example, such data do not measure the extent to which on-time delivery has been achieved, the amount of flexibility of delivery to suit user needs, the contributions of logistics to customer satisfaction, the achievement of corporate internal goals (e.g., return on capital and employee satisfaction), or the achievement of national goals (e.g., reductions in air pollution).

There is increasing concern about the relationships of private logistics strategies and public interests and policies. Examples include the effects of greater use of frequent deliveries by trucks on road congestion and on air pollution. In order to determine optimal public policies to guide logistics strategies appropriately, it is necessary to understand fully the complex responses of logistics systems to policies affecting the elements of logistics costs and service qualities. The responses include the location of facilities, the use of transport modes and the competitiveness of industries nationally and globally. The level of complexity of the relationship between public policies and logistics performance is a result of the increasing importance and complexity of the roles of logistics management.

5. Issues in the pursuit of good logistics performance

In a field as wide as logistics and with diverse objectives, a wide array of issues is to be expected. Three are discussed here: the roles of public policy in logistics, the roles of co-operation and competition among companies, and the challenges in matching human resource capabilities to the information technology requirements of the new environment.

5.1. The roles of public policy in logistics

Governments have various roles to play in logistics performance (see Chapter 23). Some of these are associated with traditional roles in the provision of transport infrastructure. Others are new, arising from recognition of new concerns, such as the environment. Finally, the government is faced with the need to make decisions in the light of international competitive forces with which improved logistics is associated.

The regulation of logistics services

Until recently, public policies relevant to logistics have focused primarily on transport services, for various reasons. Transport was seen as an important cost center by companies and by governments. Also, transport was a sector of the economy in which public investment and regulation were important in most countries. Variations among countries have been matters of the form and extent of public intervention. The evolution of corporate transport management to logistics management has gone hand in hand with deregulation of the transport industry, more by accident than design. The deregulation of transport was needed for different reasons by mode and country but, generally, the objectives were to improve the efficiency of the modes and to cause the investment in the sector to reflect economic potential and not be constrained by the limitations of public funds. What was not anticipated was the effect of deregulation on shippers, who needed improved decision-making in order to optimize the use of new services available from carriers. The deregulation of carriers has coincided with improvements in information systems and the adoption of new and more integrated approaches to logistics, such as just-in-time replenishment management. The new environment was a great stimulus to improved management and performance by carriers and shippers.

The success of the move to deregulate carriers does not mean that regulatory issues have disappeared or will disappear. Two types of issue persist and are likely to do so. First, there remain pockets of regulation and government monopolies the continuation of which should be challenged. They exist in the telecommunications industry, which is undergoing rapid technological and organizational change, as well as in transport. Government regulations and licensing arrangements affect the level of competition in many countries. Examples include telephone and other information technologies, postal-type services, rail freight, and international liner shipping. Second, issues will continue to arise over the preservation of efficient market structures in the face of merger and other consolidation initiatives. Such initiatives have conflicting potential effects through the benefits of economies of scale and scope and the disadvantages of the diminished effectiveness of competition.

Investment in and pricing of infrastructure

Governments also face logistics-related issues in the provision and pricing of infrastructure and related services. The efficiencies in logistics through faster and more reliable transport have increased truck use, which aggravates the pressures for road investments, and heightens environmental concerns associated with trucking and highway congestion. The alternatives of greater rail and water transport are caught up with a range of issues associated with the role of public support in the modes. The pressures on governments for optimal investment policies among the modes come when governments face constraints on expenditures and international pressures for efficient and fair public policies. The issues are evident in the efforts of the European Union to implement a common transport policy. The investment and pricing decisions will affect the character and performance of logistics. Whether the outcome will be optimal depends on the ability to make decisions accurately reflecting the logistical and other consequences.

Environmental protection

The role of environmental concerns associated with logistics is a relatively new influence on industry structure (see Chapter 21). The issues of air pollution associated with the use of fossil fuels are greater now than formerly, but the direct effects on logistics of measures such as tax rates on fuels and requirements for cleaner fuels and more efficient vehicles are difficult to identify at this time. Nevertheless, they are present and are likely to become more significant. Currently, the more easily identified effect of environmental concerns on logistics has been the problem of waste disposal. It has given rise to the concept of "reverse logistics." Requirements for recycling have not just given rise to new recycling logistics businesses in many countries, but have also led to new product life cycle strategies that affect the basic design of products and the location of plants. The opportunity to gain value through metal recycling has long been a factor in the steel and other metal industries, but the process of recycling is having new effects in other sectors. For example, places with lots of trees used to be the only location for paper plants. Now, cities can be an attractive alternative, by transporting a small quantity of wood product to be added to recycled paper in an urban location.

The appreciation of logistics by governments

The improvements that have taken place in logistics performance for over a decade have had a material effect on the performance of most economies. Resources are used more efficiently, as evidenced by the reduced needs for inventory to support a given level of economic activity, and inventory adjustments to changes in demand have become less dramatic so that the economies are less volatile. The adoption of new concepts such as supply chain management and of new technologies such as business-to-business e-commerce are having material effects benefiting most economies. It is important that the contributions of logistics to a country's economic performance are seen as wider and more complex than when the focus was largely on transport.

There are two reasons why industries are interested in governments' appreciation of the new efficiency of logistics internationally. First, government departments are often a part of supply chains. An example is the role of customs in international trade. Customs have numerous functions to perform for imports and exports in most countries. However, they need to approach these tasks as service suppliers in international supply chains. In Europe, for example, where the efficiency of international logistics services affects not only the local manufacturing industries but also can affect the gateway role of the economy, customs in the U.K. and The Netherlands have demonstrated a high level of innovation. They have provided model roles in the adoption of new technologies and approaches, and have promoted change (rather than lagging or delaying change to suit slow companies and industries).

Companies have a second, more general, interest in the appreciation of logistics conditions by governments. They seek to ensure that governments fully appreciate that the new efficiency of logistics contributes significantly to the increased mobility of industry, maybe across oceans and certainly across borders. Consequently, the whole range of government policies and the manner of their implementation need to be aligned with the competitive realities. Often, the concern is not with the nature of a decision or with its effect when made, but with the protracted and uncertain nature of the decision-making process. In a competitive environment where change happens quickly, time is important, and first movers gain momentum. This can be important in the development of gateway services for which volumes affect service and cost levels.

5.2. The roles of co-operation and rivalry among companies

There are a vast number of particular technical issues affecting the logistics performance of companies. What software and information technologies should be used? How much out-sourcing should be practiced? What should be the roles of vertical and horizontal decision structures within organizations or of functional management along traditional departmental lines and process management requiring inputs from across departments? Such issues are important and affect businesses generally (Chow *et al.*, 1995.) The wide issue with particular importance for logistics relates to the amount of co-operation and rivalry to be expected in supply chains.

Co-operation and rivalry are present in most business relationships. Exchange is beneficial to both parties, but each has selfish interests in the size of its share of the benefits. In western societies, greater weight has been given to rivalry and competition among independent companies than in Asia. For buyers, the threat of using alternative suppliers was seen as the best way to ensure effective price and service competition among suppliers. However, as the service dimensions of logistics have gained more recognition and the opportunities for better integrated operations have been enhanced by new information technologies, so the benefits of dealing more co-operatively with fewer suppliers has come to the fore. Shifts from short-term transactional relationships to long-term "partnerships" have become common. The notion of supply chain management promotes close relationships between the participants in the supply chain. It is not uncommon for the supplier of a logistics service to have one or more of its employees working out of its customer's facilities. The expected benefits of partnership relationships include better and faster product or service innovation, greater reliability of supply, and lower costs (through the joint optimization, rather than suboptimization, and through the benefits of specialization, scale and scope).

In spite of the publicity given to supply chain management concepts, a recent book has found that the majority of relationships are bilateral, not multilateral (Keebler, 1999.) The findings of this study suggest that there are still very many logistics systems that can benefit from greater integration within the supply chain. It may also be that the challenges of sharing information and managing responses in a multi-echelon system have been given too little attention. There are uncertainties about the number of organizations in a supply chain that can be managed effectively. It is also true that a diversity of relationships should be expected to be rational.

Technology is allowing markets to function more effectively. e-Commerce is facilitating markets where goods and services with prescribed characteristics can be traded. Truckload movements may be traded as a commodity. While specialized service requirements preclude the use of such transactional relationships in all situations, they may still have a role in well-managed supply chains. Furthermore, integration in some supply chains has been mandated by a large organization with buying power. Some just-in-time contract requirements may essentially push inventory down the supply chain. It may be difficult to judge in a dynamic environment whether the dominant company is optimizing or suboptimizing. Certainly a case can be made that there are opportunities for companies to benefit from leveraging power to get suppliers to eradicate waste rather to rely on incentives in a "partnership" relationship (Watson *et al.*, 1999). If companies are able, they will make and capture rents. So, the interplay of co-operation and rivalry can be expected to persist in the pursuit of improved performance.

5.3. *Human resource capabilities and information technology*

Information technologies have had radical effects on the organization structure of supply chains and on the ability of managers to make improved choices between logistical alternatives. However, the superior performance and sustained competitive advantage linked with the use of new technologies are dependent on the strategic role of people (Manheim and Medina, 1999). There is, therefore, a challenge for organizations to seek out individuals with the vision to translate the potential for supply chain integration into dynamic reality and to continually push the applications of new technologies. This also creates a challenge, for educational institutions have to adjust existing programs or develop new ones to meet the new needs in executive development and degree programs targeted at entry-level management positions.

Briefly, the challenges for education are to provide mixes of education in technical expertise in areas such as information technology, inventory management, and transport management, while ensuring cross-functional knowledge and skills across business areas and within a supply chain context. Evidence of the change is to be found in programs in transport or operations management shifting to logistics or supply chain management. The challenges are difficult to meet within the tightening budget and time constraints of education. Perhaps a more effective "supply chain approach" in which the buyers (employers) are more involved with the suppliers (educators) is required.

6. Conclusion

The progression of the art and science of logistics management has witnessed the application of more sophisticated measures of logistics performance and increased attention to the systems of which logistics is a part. Within companies, logistics is now commonly an integral part of corporate strategy. The strategy of companies is often based on their place in the supply chains of which they are a part. Consequently, the measurement of logistics performance is increasing in importance, but the correct interpretation of the results is becoming more difficult as a company's logistics decisions reflect trade-offs with other factors in the company and supply chains. In this corporate environment, the role of logistics/supply chain managers is gaining greater recognition and appreciation.

The role of logistics in the broad competitiveness of economies is also gaining recognition. It is leading governments to change their perception of distance from one based on transport to one based on logistics. They are also recognizing that improved logistics performance is having identifiable effects on the dynamics of their economies. These developments, too, are making measures of the performance of logistics more important, but also more difficult, because of their integration in national and global supply chains.

References

Atkinson, A.A., J.H. Waterhouse and R.B. Wells (1997) "A stakeholder approach to strategic performance measurement", *Sloan Management Review*, 38(3):23–38.

Australian Productivity Commission (1998) *International of the Australian waterfront*. Melborne: Productivity Commission.

Bonney, J. (1992) "What benchmarking is – and isn't", *American Shipper*, 34(5):6/13.

Chow, G., T.D. Heaver and L.E. Henriksson (1994) "Logistics performance, definition and measurement", *International Journal of Physical Distribution and Logistics Management*, 24(1):17–28.

Chow, G., T.D. Heaver and L.E. Henriksson (1995) "Strategy, structure and performance: A framework for logistics research", *The Logistics and Transportation Review*, 31(4):285–307.

Delaney, R. (2000) "11th annual state of logistics report", Cass Information Systems (http://www.cassinfo.com/bob_press_conf_2000.html).

Heaver, T.D. (1981) "Terms of shipment and efficiency in overseas trade", *Journal of Maritime Policy and Management*, 8(4):235–252.

Keebler, J.S. (1999) *Keeping score: Measuring the business value of logistics in the supply chain*. Oak Brook: Council of Logistics Management.

Manheim, M.L. and B.B.V. Medina (1999) "Beyond supply chain integration: Opportunities for competitive advantage", in: D. Waters, ed., *Global logistics and distribution planning*, 3rd edn., pp. 85–124. Dover: Kogan Page/CRC Press.

Saccomano, A. (1996) "Order out of chaos", *Traffic World*, November 25:38–39.

Waters, D. (1999) "Judging the performance of supply chain management", in: D. Waters, ed., *Global logistics and distribution planning*, 3rd edn., pp. 141–154. Dover: Kogan Page/CRC Press.

Watson, G., A. Cox and J. Sanderson (1999) "Thinking strategically about supply chain management", in: D. Waters, ed., *Global logistics and distribution planning*, 3rd edn., pp. 125–137. Dover: Kogan Page/CRC Press.

Chapter 3

EUROPEAN TRANSPORT: INSIGHTS AND CHALLENGES

CEES RUIJGROK
TNO-INRO, Delft

1. Introduction

European transport has been subject to a number of structural changes in the last 10 years, which have been induced by both political and technological circumstances. This chapter reviews these circumstances briefly and describes how these factors have had and are expected to have an influence on the structure of the demand for freight transport. From this description the challenges that can be identified, both for politicians and for entrepreneurs, are indicated. The objective of this chapter is to give a brief overview of the main developments that influence the demand for freight transport in Europe. Readers can find more detailed description in the references mentioned.

Since the beginning of the European unification process, which started with the Treaty of Rome in 1956, one of the aims of the process has been to create an open market in which, through a free flow of information, people, and goods, economic and social developments would occur. This process began rather modestly with the involvement of six Europe countries (France, Germany, Italy, and the Benelux countries (Belgium, Luxembourg and The Netherlands)). It was not easy to convince people that the benefits of this unification process outweighed the disadvantages. Many people, regions, and countries had the idea that the best situation could be reached by protecting the home market from foreign competition.

In the so-called Cecchini report (1988), named after the chairman of the commission that investigated the negative effects of an incomplete market with many of trade barriers and border problems, it was indicated that Europe could benefit tremendously by lifting these trade barriers. In 1985, a decision by the European Court of Justice, which blamed the European Commission of being too passive in these matters, initiated a period of fast liberalization. This liberalization was, firstly, mainly focused on the road freight market. Later, other markets were involved in the so-called "single market program" (SMP) aimed at the elimination

Handbook of Logistics and Supply Chain Management, Edited by A.M. Brewer et al.

of non-tariff barriers (particularly technical and administrative barriers) to trade and investment, and the free movement of goods and persons.

For the transport sector this meant the abolishment of:

(1) bilateral quota restrictions,
(2) price restrictions,
(3) discriminatory licensing restrictions,
(4) cabotage restrictions (i.e., restrictions on the free transport within an area by an outside (foreign) party),
(5) different rules for the weights and dimensions of vehicles,
(6) different rules for road safety and working hours, and
(7) border formalities.

As a result, the road freight market is now widely liberalized in most EU member states, although differences in tax levels and charges still exist. Also, transport on inland waterways has been liberalized considerably, which has helped to reduce the subsidies that mainly helped to maintain this economically weak sector. It is only in the railway sector, although many initiatives for deregulation and liberalization have been taken, that the opening up of the European market has not yet been very successful. A remaining problem that has not yet been solved is the enforcement of social regulations, especially with regard to working conditions in the transport sector.

In a report from the European Commission to the Council and the European Parliament, *The 1996 Single Market Review* (1996), it was concluded that, overall, the EU has benefited tremendously from the SMP process. One of the conclusions reached was that it was difficult to separate these effects from other developments that affected the economy, such as the globalization of the economy, cyclical effects, technological advances, and political changes such as the demise of the former U.S.S.R., the reunification of Germany, and the liberalization of world trade.

It was concluded that the reorganization of the distribution process generated by the SMP led to logistic cost reductions between 1987 and 1992 of as much as 29% (thus leading to a fall in the share of total revenue from an average of 14.3% to 10.1%). The largest cost reductions observed were in the transport sector, where cost reductions of up to 50% were realized in some subsectors. Idle time at borders declined and service quality improvements in the form of reduced service failures were reported. It was estimated that the yearly savings in direct costs, time savings, and other indirect cost savings would amount to ECU 130 billion per year in 2005, when the European market was anticipated to be completed. The estimated yearly cost of formalities and border controls is presented in Table 1. This indicates that, although, European integration has been successful, a lot needs still to be done to overcome the final burdens.

Thus it can be concluded that the unification process has not been finalized yet. A big step forward was the signing of the Treaty of Maastricht in 1996, which led to

Table 1
Estimated yearly cost of administrative formalities and border controls

	$ U.S. billion (1988)
Administration	9
Delays	1
Public spending on custom operations	1
Lost business opportunities	5–19
Total	16–30

a common monetary policy. As of 1 January 2001, the Euro will be the main monetary unit. At the same time the geographical scope of the EU has widened, from the initial six countries that began the process to 15, and further growth (up to approximately 24 countries) is expected. In the Common Transport Policy Action Program 1995–2000, the main targets for this period have been identified as:

(1) the development of integrated transport systems based on advanced technologies that contribute to environmental safety and economic objectives in order to achieve improved quality;
(2) the promotion of more efficient and user-friendly transport services, while safeguarding social standards and choice by improving the functioning of the single market;
(3) the broadening of the external dimension by improving transport links with third countries and fostering the access of EU operators to other markets.

It is clear that, although structural changes in the transport sector have taken place in the last decade, further significant changes will occur. In the next two sections the trends in transport and logistics are identified, with a focus on Europe. The chapter concludes with a list of challenges for policy-makers and entrepreneurs.

2. Trends in transport and logistics

The focus in this section is on the current developments in transport and logistics.

2.1. Major European trends and their impact on logistics and transport

Economic growth

Transport is a second-order activity which is generated by other economic activities. As such, the demand for transport depends heavily on economic activities and consumption, and changes in both of these. When the economy is growing, both production and consumption will grow, hence leading to an increase in the demand for transport, and vice versa. Figure 1 shows the growth in freight transport and in GDP for EU countries. Although these two measures show the same pattern, freight transport has been growing faster than GDP in recent years.

Economic developments and the single European market are the main drivers of the growth in freight transport. It is expected that the growth in freight transport will continue, due to the further unification of Europe (the integration of Eastern European countries in the EU), globalization of trade relations, and the potential of developments in information and communication technology

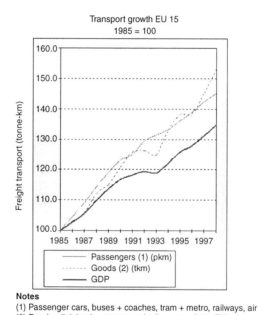

Notes
(1) Passenger cars, buses + coaches, tram + metro, railways, air
(2) Road, rail, inland waterways, pipelines, sea (intra-EU)

Annual growth rates EU 15 (% change)

	1980–90	1990–96	1997	1998
GDP	2.4	1.8	2.5	2.7
Industrial production	1.8	0.2	3.8	3.4
Passenger transport, pkm (5 modes)	3.2	2.0	2.5	2.0
Freight transport, tkm (5 modes)	1.9	2.4	5.0	4.0

Figure 1. Growth in freight transport and GDP(source:
http://europa.eu.int/comm/transport/tif/).

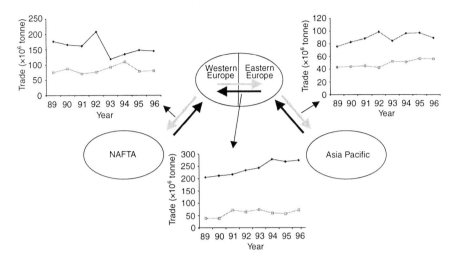

Figure 2. Time series of the volume of trade between Europe and NAFTA, between Europe and the Asian Pacific, and between the EU countries and Eastern Europe.

(ICT) (the "new economy"). In particular, road freight transport is expected to grow because of the quick, flexible, and reliable door-to-door services it can offer.

Increase in international trade relations

The flows of trade, capital, goods, and information have an increasingly international character. This is largely caused by liberalization of world trade and geographical specialization of production. Internationalization can be found both at the global level and at the European level. As a consequence, the flows of goods increase. Compared to domestic transport, international flows have shown a higher growth rate since the mid-1980s, and this trend is expected to continue as national economies converge. Recent model calculations indicate that international freight, even within Western Europe, could grow at twice the rate of domestic transport (Tavasszy, 1996).

Figures 2 and 3 show the expansion in trade volumes between 1980 and 1995 for the three key regions of the world economy (North America; Asia and the Middle East; and Europe). At a global level, the growth in the value of international trade since 1989 (up 190%) has substantially outstripped the growth in production (up by 80%) (Holland International Distribution Council, 1998; Trilog Consortium, 1999).

It can be seen from Figure 2 that a relatively stable quantity of goods is being traded between the trade blocks, certainly if only the years 1994–1996 are taken into account. For all the trade relations of Western Europe shown, imports exceed exports. The amount of trade with NAFTA is higher than that with Asian Pacific (225 million tonne as compared with 140 million tonne). The trade with Eastern Europe has shown a steady rise since the beginning of the 1990s. Figure 3 shows the same trade flows in terms of value.

One general conclusion is that the value per tonne is rising, both for goods coming from and going to Eastern Europe. In other words, the EU is importing and exporting higher value goods from and to Eastern Europe. A second conclusion is that there is a huge difference in the value of goods imported to and exported from the EU – the EU is exporting far more valuable goods compared with the imports.

Forecasts of the change in the number of tonnes of goods transferred between Western Europe and Eastern Europe are given in Table 2. The expected growth in trade between Eastern and Western Europe far exceeds the expected growth within Western Europe, for both short and long distances.

Limits to growth

An observation regarding Europe is that concentrations of population, especially along the London–Milan axis, are becoming increasingly congested, mainly along the north–south transport axes, reaching from The Netherlands, along the Ruhr/Rhine industrial area, down to the Alps, and continuing into northern Italy. Freight traffic will suffer increasingly from congestion along this corridor. The

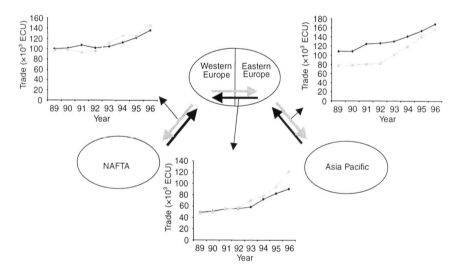

Figure 3. Time series of the value of trade between Europe and NAFTA, between Europe and the Asian Pacific, and between the EU countries and Eastern Europe.

Table 2
Forecasts of development of tonnes East/West 1995–2020 (NEA, 1999)

	Total transferred (tonne)
Within Western Europe, short distance	+35%
Within Western Europe, long distance	+78%
Western Europe to CEEC	+168%
CEEC to Western Europe	+124%

latest forecasts of passenger and freight traffic on the European infrastructure network (IWW/MkMetric, 1999; NEA, 1999; TNO, 1999) indicate that, if the integration between Western and Central European economies continues as planned, the total length of network links that now reach traffic volumes between 60 000 and 120 000 vehicles/day (the two directions added together) will be doubled by 2010. Approximately three-quarters of these high-intensity network links now lie (and will remain) within the London–Milan corridor. The economic damage of lost travel time in Germany, the main road freight transit country in Europe, is estimated to be about DM 550 million per year (approximately U.S. \$275 million). With congestion increasing, many countries in Europe are preparing plans for tolls and taxes. The German government aims to introduce a specific road freight oriented tax by 2002 of Pfennig 25 (approximately U.S. \$0.12) per vehicle-kilometer. The impact on traffic volumes is expected to be very limited. The income of these levies will, however, generate funds for capacity improvements on 20 major road network junctions, as well as improvements for other modes of transport.

2.2. Logistics trends in Europe

In this section logistics trends derived from a review of the recent literature on trends in logistics and supply chain management, and from consulting the European Redefine project (Redefine Consortium, 1998) and the European part of the Trilog project (Trilog Consortium, 1999), are presented. Both Redefine and Trilog addressed the link between logistics practices and the demand for road freight transport. The Redefine study provides a taxonomy of logistics and supply chain trends, already synthesized from existing commentary and research, to place trends according to the level at which they affect logistics decision-making for companies. Decisions concerning logistics can be divided into those concerned with:

(1) the *structure* of the supply chain (i.e., the location and size of production or processing plants, storage sites);

(2) the *alignment* of the supply chain (i.e., the breakdown of the chain into different processing segments, the number and location of supplies, and the ultimate destination of the product);

(3) the *scheduling* of the product flow (i.e., the frequency of delivery, the mode of ordering, and delivery); and

(4) the *management of logistics resources* (i.e., the size of vehicles used, types of handling and storage system, and their effectiveness of use).

For each of these logistics decisions, several examples of trends are given in the following subsections.

The structure of the supply chain

Decisions on the structure of the supply chain cover many important developments:

(1) the centralization of inventory has been one of the most pronounced trends in logistics over the past 30 years (Trilog Consortium, 1999);

(2) the stockholding and break-bulk operations which have traditionally been performed in the same locations have now been geographically separated, with the former becoming more centralized while the latter remain decentralized (McKinnon, 1989);

(3) having reached an advanced stage within individual countries, inventory centralization is now occurring on a larger geographical scale within the Single European Market as companies take advantage of the removal of border controls, the deregulation of international road freight, and improvements in road and rail infrastructure.

The above trends all lead to an increase in transport distances.

Centralization has also occurred in parcel and mail delivery systems. Unlike break bulk/trans-shipment systems that are for single users, in parcel and mail delivery companies deal with consignments that have already been created and are working for multiple users. The adoption of hub-satellite systems has increased the volume of parcel movement (in tonne-kilometers), thus amplifying the effect of the steep growth in the parcel business on the transport system. It is likely to have had a much smaller effect on the number of vehicle-kilometers, as the load factors on trunk vehicles have been increased substantially. Despite the fact that individual consignments now follow much more circuitous routes, the net effect of the restructuring of parcel delivery networks on traffic levels may have been quite modest.

The alignment of the supply chain

The transformation of raw materials into the final product usually consists of different steps executed by different parties, and involves a number of transport

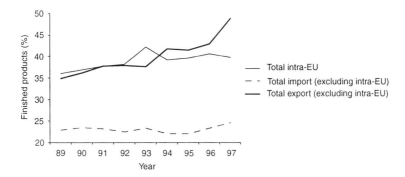

Figure 4. Share of the volume of finished products in various trade flows.

links in the chain. In many sectors, firms have been concentrating on core competences while subcontracting non-core, ancillary activities to outside contractors. This process of vertical disintegration, which reverses the prevailing trend during the 1960s and 1970s, adds extra links to the supply chain and increases the transport intensity of the production process. In other words, geographical change may lead to an increase in transport intensity and to handling (although if vertical disintegration occurs within a plant because of a change of ownership there may be no change in transport intensity). The share of final products in the imports, exports, and intra-EU trade is shown in Figure 4.

As can be seen from Figure 4, for exports, imports, and intra-EU trade the percentage of finished products was higher in 1997 than in 1989, the trend being most pronounced for exports. For imports, the share rose again and the increase stabilized in the mid-1990s. There was a greater flow of components over a larger geographical scale after 1993.

Companies have steadily reduced the number of suppliers used to provide a particular product or part. In so doing, companies have reduced the transaction cost and strengthened their negotiating position with respect to the chosen supplier. At the same time they have exposed themselves to greater vulnerability should their chosen supplier fail to deliver. This rationalization of the supply base applies to the purchase of logistical services as well as to material goods. Several surveys have confirmed that the average number of logistics service providers that companies use has been declining (PE Consulting, 1996; Holland International Distribution Council, 1998; Pellew, 1998). As a consequence, logistics service providers are repositioning themselves, and new forms of logistics service provision have appeared on the horizon (Figure 5).

Economies of scale in terminal and vehicle operation have led to the concentration of international trade through a smaller number of hub ports and airports. The latest generation of deep-sea vessels, for example, can only be accommodated at a few major gateway ports, such as Rotterdam, Antwerp, and Le Havre (see Chapter 27).

The growth in the volume of air freight handled by all-cargo aircraft (rather than in the belly holds of passenger aircraft) is creating the potential for the development of all-cargo airports away from established airport locations, such as at Vatry in France. Such airports can also become nuclei for the development of high-tech manufacturing. For instance, the first of a proposed network of "global transparks" has been established at Raleigh, North Carolina. This network will link manufacturing complexes around the world on a just-in-time basis by dedicated air freight services. Boeing (1998) predicts that, worldwide, air freight traffic will grow at a rate of around 6.5% per annum over the next 20 years. Recently, Peters and Wright (1999) predicted a steady increase in the proportion of air freight handled by integrated express carriers, all of which concentrate their sorting operations on hub airports.

Scheduling of the product flow

A series of new management principles and approaches, such as just-in-time (see Chapter 13; Hutchins, 1988), quick response (Fernie, 1994), lead-time management (Christopher and Braithwaite 1989), lean logistics (see Chapter 11; Jones et al., 1997), agile logistics (Christopher, 1998), and efficient consumer response (Kurt Salmon Associates, 1993), have been developed over the past 20 years to help firms accelerate their logistical operations. Process and pipeline mapping techniques

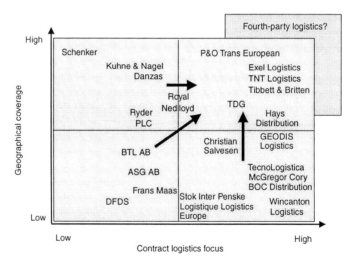

Figure 5. Strategic positioning of leading European logistics operators, 1998 (Peters, 1999).

Table 3
Customer service requirements

Order cycle time	Delivery time window			
	0 h	1 h	1 day	1 week
<1 day	0%	4%	9%	0%
1–5 days	0%	13%	36%	10%
>1 week	1%	3%	8%	18%

have been developed to analyze the expenditure of time in the supply chain and assess the opportunity for eliminating slack time and non-value-adding activities (Jones et al., 1997).

This mounting pressure to "time-compress" logistical systems may seem at odds with the lengthening of supply chain links outlined earlier, particularly where these links cross international frontiers. Long transit times from distant factories can, after all, make the international supply chain unresponsive to short-term variations in demand. For example, the abrupt decline in retail spending that occurred in several European countries in 1991 left some large retailers which source much of their merchandise from the Far East exposed, with several weeks worth of inventory already en route to their outlets. By restructuring their global supply chains, however, multinational retailers can drastically reduce delivery times. Marks and Spencer, for example, have been able to reduce delivery times from suppliers in the U.K. to their shops in Hong Kong from around 25 days to 11 days.

Survey work, conducted by CELO in 1993 and 1999, reveals that customer service requirements seem not to have changed very much. CELO measured this along two dimensions: order delivery time (i.e., the time between the moment a customer places an order and the time he receives it) and delivery time window (i.e., the time period during which the product on order can actually be delivered to the end customer). As can be seen from Table 3, in 72% of cases both the order delivery time and the delivery time window were 1 day or more (this was 70% in the previous survey done in 1993). In fact, these service requirements are rather modest and not very challenging.

Firms operating a nominated-day delivery system achieve much higher levels of transport efficiency by forcing customers to adhere to an ordering and delivery timetable. Customers are informed that a vehicle will be visiting their area on a "nominated" day, and that to receive a delivery on that day they must submit their order a certain period in advance. The advertised order lead time is thus conditional on the customer complying with the order schedule. By concentrating deliveries in particular areas on particular days, suppliers can achieve higher levels of load consolidation, drop density, and vehicle utilization. The resulting reductions in traffic levels can be significant.

The nominated-day principle is now widely applied to distribution within countries. It is less common in international distribution, although it is also becoming more widely adopted at this level. In many firms, the introduction of this principle has been resisted by sales and marketing staff who fear that the resulting of loss of customer service might jeopardize sales. It clearly conflicts with the trend towards more flexible, quick-response distribution in many industrial sectors. However, by accelerating the transmission and processing of orders many firms operating the nominated-day system have managed to reduce the lead time between the receipt of an order and delivery day. In some cases this lead time is now 24 hours or less. This has relieved some of the earlier worries about customer service and widened the acceptability of this practice.

Management of logistics resources

The performance of different modes of transport relative to other modes changes over time due to a variety of factors, such as changes in the cost structure (e.g., fuel prices) and improvements in technology. All modes of transport have enjoyed reductions in unit costs due to improvements in vehicle design, vehicle production processes, lower maintenance requirements, better fuel consumption, etc. On balance, these improvements have favored road transport, as reflected in the increasing share of the freight market. This is reflected and confirmed by looking at the change in modal split over time (Figure 6). From Trilog we learn that experts

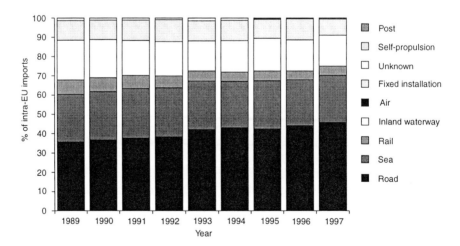

Figure 6. Modal split of the volume of intra-EU imports. The key corresponds to the vertical order of the modes in the chart

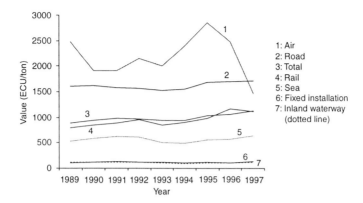

Figure 7. Change in value per tonne of product carried within the EU.

indicate that road transport will continue to take market share up to the year 2005 (Trilog Consortium, 1999). They suggest that there will be only small changes in market share for other transport modes. This should be viewed against their prediction of a steady growth in freight transport overall.

However, looking at tonne-kilometers disguises a dramatic change in the use of air freight. Figure 7 shows how the value of a ton of material moved by particular modes within Europe has changed since 1989. Care should be taken to read the air freight line against the right-hand axis. Since 1994 there has been a marked reduction in the value of product moved by air freight. This is almost certainly explained by a decrease in the cost of air freight. The value of 1 ton moved by other modes has remained relatively static over the same period, although sea and rail are showing a slight increase.

ICT developments are of major importance to the way in which information flows can be handled, both within and between companies. In general, the following advantages of ICT-supported information exchange are recognized: increase of accuracy, increase of storage capacity, reduction in transaction costs, and increase of worldwide coverage.

Probably the most eye-catching developments are related to the internet. The world wide web, the global computer network of information databases, has shown phenomenal growth, in terms of both the amount of information available and the total number of users accessing that information. In fact, in many European countries, large internet providers (and others, such as banks) are now offering free access to the internet, accelerating the use of the net even further.

Forrester Research (quoted in the *Economist* 26 June 1999) predict that intercompany trade in goods over the internet will double every year for the next 5 years, increasing its global value from around U.S. $50 billion in 1998 to between U.S. $1800 and U.S. $3200 billion in 2003. This growth in e-commerce between

Table 4
Predicted entry of e-commerce in countries and sectors into phase of hypergrowth
(source: Forrester Research)

Country		Sector	
U.S.A.	Early 1998	Computing/electronics	Mid-1999
Canada	Late 1999	Aerospace/defense	Late 1999
U.K.	Early 2000	Utilities	Late 2000
Germany	Mid-2000	Automotive	Early 2002
Japan	Late 2001	Shipping/warehousing	Mid-2002
Italy	Early 2002	Petrochemicals	Mid-2002
		Pharmaceuticals/medical	Mid-2003

companies will greatly exceed business-to-consumer e-commerce and will extend the globalization of trade. It is forecasted in the U.S.A., for example, that the value of internet sales to consumers will decline from 18% of the value of e-commerce in 1998 to 8% in 2003. This exponential growth in e-commerce will be attributable to so-called "network effects," as firms at different levels in the supply chain multiply their internet links. Forrester Research argues that, once a critical mass of companies are trading by this means, the volume of e-commerce will enter a phase of "hypergrowth." Countries and sectors will enter this growth phase at different times (Table 4).

The rapid development of ICT is one of the key factors influencing the structure and performance of supply chains. Point-of-sale scanners allow companies to capture the customer's voice. Electronic data interchange enables all stages of the supply chain hear that voice and react to it by using flexible manufacturing, automated warehousing, and rapid logistics. Furthermore, new concepts, such as quick response, efficient consumer response, accurate response, mass customization, lean manufacturing, and agile manufacturing, offer models for applying the new technology to improve performance (see also Fisher, 1997).

3. Expectations and challenges

Based on the overview of developments given above, we give here the following expectations and challenges for transport in Europe.

3.1. Enlargement of the European Union

A lot of progress has been made with the adoption of the acquis (the legal framework for EU membership) by the accession countries in Central and

Eastern Europe (CEEC). Key challenges that remain for the near future are the following (adapted from ESTO, 1999):

(1) Infrastructure development, particularly road links between CEEC and the current EU member states, to reduce traffic congestion at border crossings. The main limitations to progress lie in the availability of financial resources.
(2) The rate of progress in the reduction of state influence in the national railways companies, the introduction of market forces in the transport sector, and the harmonization of taxation systems will determine the extent to which rail transport can regain its share of the market.
(3) The harmonization of price and wage levels with markets within the present EU countries is progressing, but in the short-term is expected to have a large influence on the competition conditions within the EU market. Existing standards for material and working conditions tend to be less strict in the CEEC countries, while no effective control systems exist.

3.2. *Infrastructure, regional economies, and the environment*

As barriers for the international exchange of goods and services are being lowered, international traffic is increasing at a much higher rate than is domestic traffic. This is partly due to the reorientation of trade flows, but is also a result of the restructuring of logistics networks. The EU infrastructure policy, expressed by the Trans European Nework (TEN) Guidelines, aims to complete missing links in the European networks, including border links and terminals of European interest. The TEN policy is a necessary, although not sufficient, condition to respond to the long-term challenge of an efficient transport system. Beside the problems of implementation in the multi-colored political context (for which the TEN policy offers no solutions), the following necessary conceptual extensions of these guidelines can be mentioned (see also the excellent monograph on European infrastructure development by Turro (1999)):

(1) A clear cost–benefit analysis of potential projects that allow prioritization on welfare economics grounds. A recent study has shown that two-thirds of the TEN projects benefit countries other than the ones where these projects are realized (Bröcker, 1999). Mainly on the north–south axis, and in particular within Germany, transit traffic causes huge congestion problems, which affect both domestic and international economic relations.
(2) Completion of the networks in the direction of the CEEC countries; the TINA program (Transport Infrastructure Needs Assessment) was devoted to extending the TEN network along the 14 corridors towards these countries, but should now be followed up with the implementation of the defined projects. Public funds will not be sufficient to build up these

networks. The experiences with public–private co-operation in these countries are mixed, as they have resulted in high costs for the user (through tolling) or for the national governments.

(3) An overall strategy to develop the multimodal dimension of the TEN, including views on the complementarity between different modes of transport as well as priorities for ports and inland trans-shipment terminals, within different socio-economic scenarios. The development of the Trans European Rail Freight Freeways (TERFFS) is an important first step, but has not yet resulted in a visible growth in the share of rail transport, or of any other alternative to road transport.

3.3. The implementation of EU policy

The White Book on transport policy sketches the framework for policy measures that are expected to improve the efficiency and sustainability (in environmental and economic terms) of the EU transport system. Two types of challenge can be identified: those associated with the operational form of the measures (i.e., the specific points of intervention in the system (infrastructural, organizational, or market related)) and the instruments (fiscal and regulatory) to be applied; and those pertaining to the road map towards implementation.

(1) As an example, the internalization of external costs is generally accepted as a means of making the transport system more sustainable. Even if levies and tolls do not result in the desired behavioral change (e.g., a shift to more environmentally friendly modes of transport), users would be paying for the external costs of transport. Major research studies are being conducted to define as clearly as possible which modes and which parts of transport networks should receive which levels of pricing, and under what principles.

(2) The subsidiarity principle provides the member countries with varying degrees of freedom (including time) for the ratification and implementation of measures. As it is expected that both infrastructural and pricing measures will influence the competitive position of regions and sectors within Europe, the rate of adoption and the extent of compensation measures will be determined by the political processes within each member country.

3.4. Development of markets

The heterogeneous character of the market for logistics services within the EU, and the political forces that shape the framework conditions which aim at a fast convergence, lead to the following expectations and challenges:

(1) *Mergers and acquisitions:* the rationalization of the market typically leads to a decrease in the number of logistics players. In particular, international efforts at co-operation through joint ventures, mergers, and acquisitions have not all been successful. Foreign markets, even within the same continent, are often difficult to maintain. Subtle differences in business (e.g., speed of payment) or culture tended to frustrate the efforts made in the 1990s to establish Europe-wide networks for formerly national players.

(2) *Differences in framework conditions for logistics services:* financial, physical and ICT infrastructure, logistics education, and regulatory conditions are not harmonized between different countries. In the study presenting these findings (TNO, 1998), logistics benchmarking was proposed as a means to obtain insight into differences between regions in terms of framework conditions.

(3) *Balancing inequalities:* the question of how to balance these inequalities has, among other ways, been answered by appointing "disadvantaged regions", which have relatively easy access to EU funds for the support of economic growth. One may argue that poor framework conditions may explain poor logistics services, although examples exist of the opposite, where companies have learned to cope with the specific disadvantages of a region (McKinnon, 1989).

(4) *Common standards:* within Europe there is a wide range of standards for transport materials including, for example, pallet, container, and trailer measurements for trucks, and gauge width and power systems for rail. The harmonization of taxation systems and social regulations is ongoing. However, due to the influence on the competitive position of national markets, the progress is slow.

References

Boeing (1998) *World air cargo forecast 1998/1999*. Seattle, WA: Boeing.

Bröcker, J. (1999) *Trans-European effects of trans-European networks: results from a spatial CGE analysis*. Dresden: University of Technology of Dresden.

Cecchini, P. (1988) *The European challenge 1992: the benefits of a single market*. Aldershot: Wildwood House.

Christopher, M. (1998) *Logistics and supply chain management*. London: Pitman.

Christopher, M. and A. Braithwaite (1989) "Managing strategic lead times", *Logistics Information Management*, 2(4).

European Commission (1996) *The 1996 single market review*. London: Office for Official Publications of the European Communities/Kogan Page.

ESTO (1999) *Deregulation and transport in an enlarged European Union*. Seville: ESTO.

Fernie, J. (1994) "Quick response: an international perspective", *International Journal of Physical Distribution and Logistics Management*, 24(6):38–46.

Fisher, M.L. (1997) "What is the right supply chain for your product", *Harvard Business Review*, March–April, 105–116.

Holland International Distribution Council (1998) *Worldwide logistics, the future of supply chain services*. The Hague: HIDC.

Hutchins, D. (1988) *Just in time*. Aldershot: Gower Press.

IWW/MkMetric (1999) *Transport forecasting goods and passengers 2020. Passenger transport and assignment results*. Brussels: DG TREN, European Commission.

Jones, D., P. Hines and N. Rich (1997), "Lean logistics", *International Journal of Physical Distribution and Logistics Management*, 27(4):153–713.

Kurt Salmon Associates (1993) *Efficient consumer response: enhancing consumer value in the grocery industry*. Atlanta, GA: Kurt Salmon Associates.

McKinnon, A.C. (1989) *Physical distribution systems*. London: Routledge.

NEA (1999) *Transport forecasting goods and passengers 2020, Freight transport results*. DG TREN, Brussels: European Commission.

PE Consulting (1996) *The changing role of third-party logistics – Can the customer ever be satisfied?* Corby: Institute of Logistics.

Pellew, M. (ed.) (1998) "Pan-European logistics". London: Financial Times, Management Report.

Peters, M. and W. Wright (1999) "A research report on international air express distribution", Cranfield University.

Peters, M. (1999) "The logistics market in the UK and Europe, a service provider perspective", in: *Proceedings of the Conference 'Shaping the Pan-European Supply Chains of the Future*. Brussels: Pricewaterhouse Coopers.

Redefine Consortium (1998) *REDEFINE: relationship between demand for freight-transport and industrial effects*. Rotterdam: NEI.

Tavasszy, L.A. (1996) *Modelling European freight transport flows*. Delft: Delft University of Technology.

TNO (1998) *Benchmarking logistics services and their framework conditions in Europe*. Delft: TNO INRO.

TNO (1999) *Transport forecasting goods and passengers 2020. Design of scenarios*. Brussels: DG TREN, European Commission.

Trilog Consortium (1999) *TRILOG – Europe end report*. Delft: TNO INRO.

Turro, M. (1999) *Going trans-European*. Oxford: Pergamon.

Chapter 4

NORTH AMERICA: INSIGHTS AND CHALLENGES

THOMAS M. CORSI and SANDOR BOYSON
University of Maryland, College Park

1. Evolving logistics best practices

As we enter the new millennium, we are witnessing the emergence of the net-centric logistics model. Net-centricity is the unprecedented ability to connect, communicate, and share information with suppliers, customers, and business partners in "real" time (Boyson and Corsi, 2000). North American manufacturers and retailers have been at the center of this new model, with several major corporations, in particular, Federal Express, Sun Microsystems, and Cisco, having made great strides toward implementation of a fully integrated net-centric supply chain.

The structural foundation of net-centricity is the convergence of networking and communications technologies – driven by the internet – combined with common data formats and open systems standards. The old proprietary, mainframe-based computing platforms and applications, or legacy systems, are giving way to a unified information technology (IT) model that enables different systems (and people) to easily "talk to one another." This proliferation of connectivity has virtually eliminated transaction-processing lag times and enabled businesses to automate many management decisions and processes using dynamic, rules-based applications. The result is a more cost-effective and nimble organization, able to respond quickly to market shifts, customer needs, and supplier inquiries.

In the 1990s, manufacturing firms began to discover the benefits of net-centric operations. Greater business volatility and risk have given rise to a new generation of "high velocity enterprises" that manage global manufacturing supply chains on a real-time basis. Manufacturers are now embracing applications that previously disconnected elements of the electronic supply chain. This is giving corporations "forward visibility" over their own global processes and those of their customers, distributors, and suppliers. For example, an order query can be recorded, worldwide multi-enterprise inventories scanned for component availability, and an order delivery date given to the customer – all within seconds. Such rapid response capability is revolutionizing supply chain management.

Handbook of Logistics and Supply Chain Management, Edited by A.M. Brewer et al.
© 2001, Elsevier Science Ltd

In order to keep pace with the demands of a net-centric organization, many logistics and supply chain management executives in American firms are adopting a set of key practices (Boyson et al., 1999). These practices are summarized below.

(1) *Integrated supply chain orientations, cultures, and practices.* Managers view the entire set of activities stretching from supply points to production points to warehouse/distribution points to the customer as one cross-functional, integrated process. Their constant priority is to eliminate hand-off times and process disconnects across the supply chain (Boyson and Corsi, 1999).

(2) *A supply chain management champion.* Best practices companies usually charge a vice president of operations or similar function to inform senior managers of the tremendous cost, flexibility, and customer service advantages enjoyed by net-centric supply chain companies. This champion usually leads a cross-functional team that maps out the present physical distribution network to full implementation of a supply chain optimization model.

(3) *Distributing supply chain personnel throughout the organization.* Best practice companies are experimenting with a wide range of initiatives to better place supply chain practitioners within the organization to set directions and exert strategic control across the various business units. The most common mechanisms are highly empowered chief logistics officers or vice presidents of supply chain.

(4) *Re-centralized and shared logistics for all the operating units.* A core logistics/ supply chain management hub at the corporate center is established. This hub will have capabilities in customer order management, international transportation management, and information/data networking across with vendors and customers, and performance/metrics management.

(5) *Segmenting and focusing the supplier base.* Companies must establish strategic partnerships with: critical suppliers; transportation, warehousing, and distribution center service providers; and with a new breed of third-party companies who provide supply chain system modeling, optimization strategies, and turnkey management/information technology assistance to implement improvements.

(6) *Greater synchronization of production, distribution, and customer order activities.* This often brings dramatic gains. Work in process and finished goods inventory are slashed. Order to delivery cycle times are compressed and accelerated.

Table 1 illustrates the evolution of logistics management practices. The last column highlights the best practices that are emerging as organizations make the transition to a net-centric structure. Table 1 provides the fundamental contrasts between the old logistics model and the emerging net-centric model. As indicated, the old model suffered from a "stove-piping" of logistics functions into separate

Table 1
Evolving logistics management best practices

Old logistics	Current logistics	Emerging logistics
Prevailing wisdom:	**Prevailing wisdom:**	**Emerging wisdom:**
Decentralize for faster response	Centralize to gain scale economies	Enable a shared, real-time, seamless enterprise-wide system
Business model:	**Business model:**	**Business model:**
Logistics managers at business units	Chief logistics officer at headquarters	Vice president of supply chain management
Force multiple carriers/suppliers to compete	Create a logistics hub as a shared service for all business units	Create a multi-enterprise management team composed of headquarters, business unit, supplier, distributor, and customer executives
	Reduce the supplier/carrier base and focus on strategic partnerships	Focus on strategic partnerships, and inject competition and redundancy through online bidding among suppliers
Major problems with the model:	**Major problems with the model:**	**Major problems with the model:**
Lack of leveraging across business units and lack of scale in purchasing logistics services and managing vendors	Vulnerability to core carriers/suppliers on price and service Lack of nimbleness to adapt to customer requirements	Problem of multiple, "mirroring" supply chains Need to set management boundaries and define strategic control

activities that were poorly linked and integrated. This decentralized approach left the organization without the opportunity to leverage its buying power and to better manage its vendors. In the transition to the net-centric model, there is a move toward centralization of logistics functions in order to gain leverage and consolidated buying power for purchases of transportation services, raw materials, and components. Logistics expertise is consolidated and shared across the organization. The emerging net-centric logistics model builds on the base of the centralized logistics systems and employs real-time data sharing across the seamless extended enterprise. While the emerging net-centric model envisions close partnerships with vendors, distributors, and retailers, it also recognizes some inherent flexibility advantages in conducting online bidding among suppliers via electronic market exchanges.

2. North American supply chain leaders

At present, very few organizations in North America have succeeded in seamlessly integrating their supply chain and incorporating major aspects of a net-centric supply chain. The ones that have accomplished the emerging vision are reaping big competitive rewards and offer us a window into net-centric business processes into the future. In the following sections we present insights into the net-centric supply chains of Federal Express, Sun Microsystems, and Cisco Systems.

2.1. Federal Express

Today, Federal Express is a leading model of simultaneous processing across an extended supply chain. Federal Express employs real-time data transmission to assist in routing and tracking packages. Information recorded by portable bar-code scanners is transmitted to a central database and can be made available to all employees and customers, not just managers in traditional decision-making roles. The Federal Express corporate communications network is one of the world's most sophisticated and most reliable, each day processing nearly 400 000 customer-service calls and tracking the location, pickup time, and delivery time of 2.5 million packages.

The best example of how this is playing out today is the example of the alliance of Federal Express with Proflowers.com, an internet company that runs a portal for ordering fresh flowers. When Proflowers receives a web order, Proflowers' webserver simultaneously records the transaction and messages Federal Express. Based on this message, Federal Express generates both a shipping label (which is returned to Proflowers' webserver and downloaded to the grower), and a request to its fleet to pick up the order at the grower's site. When the pickup occurs, the Federal Express shipping label with all the requisite customer information is already on the carton of flowers to be shipped.

In essence, Federal Express and Proflowers.com are using a single shared trigger event, a customer order on the web, to generate multiple supply chain transactions. This is a completely new paradigm of supply chain, a webbed model (rather than a sequential model based on serial hand-offs) to attain quantum leaps in time savings and administrative processing costs.

2.2. Sun Microsystems

At Sun Microsystems, net-centric systems are changing the shape and culture of supply chain organizations of the future. Sun Microsystems is the industry leader in scalable servers to power PC networks and web sites, with U.S. $9.7 billion in revenue, 88% of which is hardware sales. Two of Sun Microsystems's products, JAVA (a cross-platform programming language) and JINI (software tools that enable devices to plug into and interoperate via public networks) as well as similar products from competitors, will affect the shape of the company of tomorrow. Sun Microsystems's major vision of the future is a net-centric one.

Sun Peak, a web-centric infrastructure (servers and JAVA workstations) links 50 000 employees, suppliers, and distributors. Sun Microsystems's last mainframe was unplugged in January 1999. The Sun Peak provides the foundation for future growth and flexibility to compete, and serves as a demonstration of the role of Sun Microsystems in networking. It will help Sun Microsystems transform itself to a net-centric way of doing business. The new type of networking will allow Sun Microsystems to reduce its planning and testing times and costs, and move to a collaborative product innovation system based on rapid experimentation and correction off real market/customer data.

Sun Peak infrastructure is being used to support a U.S. $250 million supply chain business process re-engineering project designed to cut 5 weeks from 14-week product cycle times and 25% of cost. This re-engineering of the supply chain is vital, given the speed and volatility of the high-tech marketplace.

The current management challenges that Sun Microsystems is grappling with are many. First, it must master better surge management techniques to cope with volatility in the marketplace, with 480% spikes in demand for a product in 1 year. The motto for this is: "Better to be flexible than right." It cannot simply be better at forecasting; the Sun Microsystems supply chain must be better at adapting to real-time demands on a chain-wide basis. One way to do this is via a shared extended enterprise-wide data network and more collaborative relationships across the chain that exploits this available real-time data.

Another way to do this is to implement Jini-enabled supply chains that automatically adjust networked assets, machines and inventories to real demand. Once plugged into a network, a Jini-enabled device, which has an address, can broadcast its capabilities to other devices in the network. With Jini-enabled devices, shared power, shared knowledge, equal access, expandability, and adaptability are possible. An important characteristic of Jini is its ability to work over existing networking software and protocols. So far, companies such as Canon, Epson, Ericsson, Mitsubishi, Quantum, Seagate, Sony, and Toshiba are onboard with Jini.

2.3. Cisco Systems

Cisco Systems has tripled in market value just in the past year to surpass Microsoft and to become the world's most valuable company. This valuation reflects not only the quality of its router products; not only the dynamism of its core market with internet traffic doubling every 3 months, but also its profound mastery of supply chain and e-business processes utilizing the internet.

Cisco Systems has employed the internet to re-engineer processes across its whole span of operations and has become a global networked organization. Every step of its supply chain, from customer order self-service/product configuration to supplier management, employs internet-based processes. This has worked so well that the per employee revenue at Cisco Systems is almost three times that of competitors such as Lucent. Cisco Systems has reaped huge benefits from successfully and seamlessly integrating the fundamental technologies and components of the internet on an enterprise-wide basis. There are many benefits from the pervasiveness of internet processes.

Cisco Systems customers can use a website to price, configure, validate, and order products. In addition, customers can employ the website to obtain copies of their invoice, review shipping schedules, and receive shipment notification or change orders. The information from the customer is linked to a centralized internal system that coordinates the entire supply chain of Cisco Systems. This co-ordination provides for information sharing among Cisco Systems' major suppliers. Overall, Cisco Systems has lowered the overall cost of taking orders as well as reducing errors in product configuration. In fact, the company has reduced, from 15% to 2%, the orders that require reworking. In addition, product delivery lead times have been reduced by 2–5 days. In total, the supply chain initiatives at Cisco Systems save it over U.S. $70 million on an annual basis.

3. Major gaps between leaders and the rest of corporate America

While the examples of Federal Express, Sun Microsystems, and Cisco Systems represent how far certain American firms have progressed in the process of adopting net-centric supply chains to take advantage of recent connectivity advances, the actions of these companies should not be viewed as the norm of American firms in general (Boyson et al., 1999). The Supply Chain Management Center at the Robert H. Smith School of Business conducted a comprehensive analysis of members of the National Industrial Transportation League (NITL) in the Spring of 1999 to establish what some of the largest firms in American industry were doing to re-evaluate their supply chain processes. The following paragraphs summarize key findings from that analysis.

There were over 212 responses to the NITL survey conducted by the Supply Chain Management Center. This represented a response rate of 36.9% (575 companies have NITL membership). A very large percentage of the respondent firms generated large annual sales. Indeed, 46% of the respondents represented firms generating at least U.S. $1 billion in sales volume during 1998. In sharp contrast to the overall size of firms, 46.2% of the respondents indicated that their firms projected spending on logistics systems for 1999 at U.S. $100 000 or less. This significant disconnect between firm size (as measured in overall annual size) and the perceived importance of logistics (as represented by spending on logistics systems) strongly suggests that many of the firms represented in the survey undervalue the central role that logistics/supply chain management plays in the overall profitability of the firm. These results suggests that many respondents need to acquire better skills and a better understanding of the strategic importance of supply chain/logistics in order to act as advocates in their own organizations for these activities.

The lack of focus by the respondents on the need for logistics restructuring within their organizations is also suggested by the low implementation level for the management action of re-engineering logistics business processes. Indeed, only 21.1% of the respondents said their firms had implemented a re-engineering of logistics business processes as a way to address the critical shipping factors in their business environment. Again, the results suggest the need for respondents to better communicate to their top management the manner in which a more holistic supply chain approach can add significant value to the firm.

Much recent attention has focused on the critical role that electronic commerce will play in shaping the more holistic supply chain approach mentioned in the preceding paragraph. However, when presented with questions about electronic commerce and establishing automated links with supply chain partners, survey respondents were surprisingly lukewarm in their response. Indeed, only 12.4% of the respondents said they had implemented actions to automate links with their extended enterprise partners. Also, when asked, the respondents said that the new opportunities to improve logistics processes through the internet and e-commerce would have an average impact (mid-range on a scale ranging from no impact to a significant impact). This overall lack of enthusiasm for the new supply chain paradigm and the strategic importance of e-commerce would suggest a need to develop enhanced skills and knowledge about the future role of logistics and supply chain among the respondents. The survey results substantiate a significant gap between the leaders in the American corporate community in adopting the net-centric approach to their supply chain and many other corporations, often some of the largest and most established companies.

4. Bridging the gap: diffusion of the net-centric supply chain

With an understanding of some of the advantages available to firms who re-engineer their supply chain processes to achieve better customer service, reduced costs, and better co-ordination between manufacturing processes and suppliers, the interesting question is how long will it take before larger sections of the

American corporate community adopt new supply chain management processes and procedures. There are, indeed, a series of forces working together that make the transition to a net-centric supply chain process a likely prospect on a widespread basis in the next 3–5 years.

4.1. Greater use of third-party logistics for out-sourcing

First, many firms recognize that they do not have the required skill sets internally to put in place a re-engineered net-centric supply chain that takes advantage of internet technology and links across all members of the firm's extended enterprise. Furthermore, many firms do not have the technology base or the software applications layer to initiate change processes. Trying to keep up with the pace of the changes in supply chain technology can be a huge challenge. Consequently, an entire industry of third-party logistics (3PL) organizations has emerged to offer solutions to these challenges (Boyson et al., 1999b). The 3PL organizations provide logistics services and access to the newest technology. Companies opting to go the out-sourcing route, then, benefit from the significant investments made by the 3PL in technology, information systems, people, and physical assets.

In 1997, researchers from the Robert H. Smith School of Business (Rabinovich et al., 1999) surveyed 300 companies about their out-sourcing experiences at the University of Maryland. Included in the results were responses from 114 Fortune 500 companies. Results showed that 75% of the respondent companies believed that out-sourcing all supply chain functions contributed to their competitive advantage. About 4% of the respondents believed that out-sourcing logistics functions contributed to a better level of customer service. Survey respondents indicated that the first-year logistics cost savings from total supply chain out-sourcing averaged 21.3%, with additional annual savings of 15.1% in years two, three, and four.

It is likely, then, that firms with serious questions about their ability to re-engineer their supply chains will increasingly turn to the 3PL. The greater reliance on 3PL will accelerate the diffusion of the net-centric supply chain management model. The 3PL will have access to technology, software, and business processes and can apply them to meet the needs of individual corporations.

4.2. Rise of the applications service providers

A second factor that will facilitate the diffusion of the net-centric supply chain approach is the rise of applications service providers (ASPs). Applications developers are moving from selling their product as static, installed software, to offering those same applications as a service delivered over the internet. This means that, rather than go through an expensive software-installation process, whereby the application resides on the company's internal IT infrastructure, companies can access these applications online in a pay-per-use business model. Virtually all the supply chain software applications will be available to companies in this format within the next several years.

At present, many of the tools of the net-centric supply chain (e.g., collaborative planning and integrated forecasting, planning, and manufacturing modules) are only available through the purchase of very expensive software acquisition. In addition to price tags that far exceed the capabilities of many small and medium-sized firms, there are impediments to implementation as a result of the very complexities of the software packages. However, within the next several years these packages will be widely available for delivery by ASPs. Furthermore, the 3PL can take advantage of these ASP services as well in putting together a package of supply chain tools for their clients.

4.3. Advances in middleware

One of the important barriers to a faster diffusion of the net-centric supply chain model is the fact that many companies have invested significant amounts of time and money in specific software applications to perform the various functions of the supply chain. Unfortunately, because these applications are discrete solutions, often produced by multiple vendors, they are neither integrated nor interconnected. The result is that multiple information systems cannot talk to one another, thereby causing massive disconnects both within companies and between trading partners. These disconnects create confusion and add cost and time to the supply chain.

No company can afford to throw away its investments in software. At the same time, the need for discrete information systems at companies to be able to talk to one another has never been more pressing. Middleware offers a solution to this challenge. Middleware works at many different levels, both within and between organizations. It acts as the glue that brings together the different technology platforms and different application layers that multiple companies along the supply chain might have in place at any given moment. It is a way to extract information from those diverse platforms and systems to produce a common shared pool of information – and to do so in a timely, responsive manner. Along with providing this glue and extracting key information required to manage the supply chain in real time, middleware can route messages to critical decision-makers. Decision-makers can then use this information to make decisions in real time based on timely, accurate data.

At the moment, there is a gap between what middleware programs are capable of doing and the reality of current business practice. It will take some time for the kind of real-time interchanges described above to become common practice. In fact, there are only a few organizations that now realize the full benefits of middleware, the NASDAQ stock exchange being the best known of these.

Using middleware, the NASDAQ has created a localized, real-time messaging network over which it can send financial information, stock quotes and the like in real-time to middleware network subscribers. These subscribers can receive real-time alerts on up-to-the-minute developments and use that information to make critical financial decisions. The financial industry has been a leader in adopting these sorts of middleware technologies. Manufacturing and other services are beginning to follow suit.

Middleware will make it easier for companies to link with their extended enterprise partners across the supply chain by removing the need for all partners to have exactly the same suite of software applications. It allows each partner to maintain their own software applications and extracts from each application the information that needs to be shared across the organizations in real time. The transition to a fully implemented middleware system is difficult and will require 3–5 years in order for widespread diffusion to occur. However, growth in middleware offerings and capabilities will facilitate the diffusion of the net-centric supply chain among American corporations, since it relies so intrinsically on the ability of partners (i.e., vendors, distributors, retailers, and manufacturers) to share data and communicate in real time.

4.4. Electronic marketplaces/exchanges

Within the past several years, industries have started to change the way in which they source and sell source materials. The internet has become a very important tool for matching supply and demand across global industries. Whole supply chains are collectively reaching out, through electronic exchanges, to find new partners and materials on an as-needed basis in real time. This virtual electronic marketplace has posed challenges for corporations that have formerly partnered with just a few core suppliers. A company with a set of core partnerships has to balance its commitment to these partners, while still taking advantage of the cost saving opportunities offered in the global marketplace.

Currently, the electronic marketplaces are matching suppliers with manufacturers in a virtual bulleting board arrangement. However, in the future, the electronic exchanges will provide a host of additional services (financing, quality control, etc.) and will link back directly into the enterprise resource planning systems of both the buyer and the seller. The example of such an exchange at the current status are given below to clarify the situation.

Ventro (formerly Chemdex) is a classic example of the logic of an exchange. It was founded in 1997 to meet the needs of researchers in the life sciences who need to obtain specialty chemicals. These products are very specialized, catalogs are largely out of date, and there are no standard descriptions of products. It typically takes 5 hours of search to locate a supplier for an item. Ventro provides a hub for the 300 000 buyers and 1500 suppliers. Whereas distributors in this fragmented supply command a fee of as much as 80% of the purchase price, Ventro takes a fee of just 5%. Furthermore, it provides additional services, linking the data systems of both the buyer and seller in order to ensure accurate billing and fast delivery.

Ventro represents only one of many electronic market exchanges that have been created in the past several years. Obviously, participants in these exchanges must have made a commitment to update their supply chain practices to incorporate the basic tenants of net-centricity. The exchanges require the participants to link their data systems with the information hub and to be able to share information across their supply chains in real time. Clearly, as these exchanges grow in importance and become the standard way for vendors and buyers to conduct business, more and more companies will feel increasing pressure to re-engineer their supply chain practices.

5. A view to the future

This review of developments in North American supply chain management has provided evidence that several large firms, most notably, Federal Express, Sun Microsystems, and Cisco Systems, have led a new wave by initiating net-centric real-time supply chains. The net-centric supply chain takes advantage of connectivity and bandwidth enhancements that facilitate real-time internet connections among all partners in an extended enterprise. Combined with real-time connectivity enhancements is the improved ability for supply chain partners to share information through rapidly developing middleware, which allows firms to keep legacy supply chain software systems but to share common data across applications.

The advantages of real-time net-centric supply chain management are compelling. All the various logistics functions are co-ordinated and linked together in real time across the organization. Thus, when customer orders are placed, either through the internet or by phone or fax, the information is shared across the organization and with extended enterprise partners in real time. The customer-initiated order results in an immediate search of inventories, manufacturing schedules, and vendor component inventories and schedules to arrive at an available date to promise notification to the customer. This entire searching process is automatic, based on pre-established business rules, and enables a response to the customer within seconds. Orders, once placed, are fed to the manufacturing group, shared with suppliers, and tracked throughout the process all the way to delivery. As noted above, the net-centric real-time supply chain results in fewer mistakes in orders, reduced lead times, shortened cycle times, reduced inventories, and better resource utilization across the supply chain.

However, despite these impressive potential gains, the net-centric real-time supply chain has not been widely adopted throughout American industry. As shown in the recent NITL survey, there are significant obstacles to overcome before the diffusion becomes widespread. Yet, the move toward the new model is inevitable. The forces for change are overpowering, and are led by the clear economic gains to be realized from the new approach.

The growth of 3PL with a sound understanding of the new technology and an ability to leverage ASPs for delivery of supply chain software is a particularly compelling reason to believe that diffusion will occur. Certainly, the Supply Chain Management Center's survey of logistics out-sourcing discussed above, concluded that logistics managers realize the potential advantages of out-sourcing. The 3PL will be a particularly effective way to diffuse new ideas, approaches, and technology to many medium and small-sized firms as well as to firms who have not been changing their logistics practices in any way to take advantage of new developments. These firms are stuck in the "old logistics" approaches.

Another important fact that will encourage firms to adopt net-centric real-time supply chain management approaches is the growth of electronic marketplaces. More and more industries are coming under the influence of these exchanges, which bring together buyers and sellers. Participation in these exchanges require firms to have a certain level of technology know-how and sophistication. Clearly, participation requires firms to take initial steps on the path toward a fully net-centric supply chain in order to participate in the marketplace. Those firms not changing any practices and staying outside the electronic marketplace will be impacted in a negative fashion.

The next 5–10 years will see the diffusion of net-centric real-time supply chain management practices to a much greater portion of American industry. The advantages of the new approach are compelling. The hurdles to adoption will be overcome particularly through the efforts of 3PL suppliers as well as the growth of ASPs, to deliver software competently on an as-used basis and without serious implementation barriers. Clearly, the next 5–10 years will be an exciting time as we watch the rapid diffusion of net-centric supply chain practices among American companies.

References

Boyson, S. and T.M. Corsi (1999a) "Emerging logistics: adopting the emerging best practices", *Executive Excellence*, 16:19–20.
Boyson, S. and T.M. Corsi (1999b) "Matching demand and supply in real time: the value of supply chain collaboration", *Global Purchasing and Supply Chain Management*, October, 130–133.

Boyson, S. and T.M. Corsi (2000) "The year 2010 netcentric supply chain: an exploratory technology forecast," Supply Chain Management Center Working Paper, Robert H. Smith School of Business, University of Maryland, College Park, MA.

Boyson, S., T.M. Corsi, M. Dresner and L. Harrington (1999a) *Logistics and the extended enterprise: benchmarks and best practices for the manufacturing professional*. New York: Wiley.

Boyson, S., T.M. Corsi, M. Dresner and E. Rabinovich (1999b) "Managing effective third party logistics partnerships: what does it take?", *Journal of Business Logistics*, 20(1):73–100.

Rabinovich, E., M. Dresner, R.J. Windle and T.M. Corsi (1999) "Outsourcing of integrated logistics functions: an examination of industry practices", *International Journal of Physical Distribution and Logistics Management*, 29(5):353–373.

Chapter 5

INTERNATIONAL LOGISTICS: A CONTINUOUS SEARCH FOR COMPETITIVENESS

HILDE MEERSMAN and EDDY VAN DE VOORDE
University of Antwerp (UFSIA/ITMMA)

1. Introduction

There is clearly growing interest from both the corporate and the scientific worlds for the concept of international logistics (see, e.g., Miyazaki et al., 1999). In part this is due to the fact that an increasing number of companies are getting involved in global markets, as greater emphasis is put on export and import trade.

International logistics distinguishes itself from domestic logistics in a number of ways. The dissimilarities are mostly due to the fact that goods are moved between countries that may differ from each other considerably. Young (1995), referring to Wood et al. (1995), rightly asserts in this context that "international logistics is not the same as domestic logistics with perhaps the addition of one or more international border crossings. International logistics certainly contains border crossing issues, but also must accommodate a significantly higher level of complexity comprising various combinations of cultural, political, technological, and economic variables."

A number of recent developments in the commercial environment in which companies operate have consequences in terms of international logistics. Globalization of production and trade generate substantial goods flows between countries, which must be dealt with as efficiently as possible. Furthermore, there is a growing tendency in international logistics towards supply chain management and time-based competition. Finally, developments in informatics, and especially electronic data interchange (EDI), have made the organization of international networks for goods and documents a lot more simple and efficient.

The complex nature of international logistics implies that many different parties may be involved. In order to retain and/or improve their competitive position, a number of these may benefit from forms of co-operation. This may lead to a better, more streamlined operations in different components of the logistics chain. The complex nature of international logistics brings with it that many different parties are involved. Their behavior is affected mostly by such

Handbook of Logistics and Supply Chain Management, Edited by A.M. Brewer et al.
© 2001, Elsevier Science Ltd

Table 1
Barriers to interoperability in European transport

	Road haulage		Railways		Inland navigation		Intermodal rail/road
	Infrastructure	Operations	Infrastructure	Operations	Infrastructure	Operations	
Austria		Technical standards	Technical (track, signaling), institutional Knowledge of track	Technical (vehicles)			
Belgium					Some canals too small		
Denmark			Different electric system to neighboring countries	Sea crossings			Relatively high reloading costs
Finland			Technical (on international operations)				Different container types
France				Incompatibility between HST and regional services			
Germany	Sometimes capacity problems			Technical (capacity problems depending on route and time)	Technical, financial/institutional (price system for using the network)		Kombiverkehr (own account operators are barred from it)
Ireland			Loading gauge prevents use of continental wagons				

Country						
The Netherlands			Technical and environmental norms	Some canals are too small	Different sizes of containers, hopper barges, etc.	
Norway	Different axle load capacity on state-owned roads and others					
Portugal	Technical, organizational (lack of uniformity of infrastructure)	Technical (gauge), capacity	Technical (lack of interfaces), cultural			Lack of an appropriate international network
Spain	Technical (speed, environment), maximum loads, physical (national road III, Madrid–Valencia) Strategic barriers	Technical (gauge, signalling, etc.), bottlenecks and pinchpoints	Broad gauge is a barrier to international traffic			
Sweden					Hull standards for chemicals and oil transports, container standards	Different container types
Switzerland	28-tonne limit	Capacity, technical and financial	Technical, staff education	Individual canals, variable water depth, limited network		
U.K.		Track gauge, loading gauge, signaling, electric supply, pinchpoints and bottlenecks	Institutional and organizational, central reservation systems, physical (estuaries and sea crossings)			Limited loading gauge

Source: based on Beaumont et al. (1996).

Table 2
Barriers to interconnectivity in European transport

	Road haulage		Railways		Inland navigation		Intermodal rail/road
	Infrastructure	Operations	Infrastructure	Operations	Infrastructure	Operations	
Austria	Technical, financial		Only a small rail network construction; financial	Agreements necessary			
Belgium						Connections with France	
Denmark				Not enough co-ordination in timetables and pricing systems			Different container types
Finland		Institutional (when crossing the Russian border)				No or insufficient interconnections	Different container types
France							
Germany		Bottlenecks Difficult network planning	Financial and institutional (co-ordination)	Missing links			
Ireland	Institutional, financial and environmental	Cultural (Northern Ireland)		Lack of integrated ticketing			

The Netherlands			No connections to other infrastructure Institutional	Organization (port of Rotterdam)	No co-operation with rail	
Norway	Closed roads in winter					
Portugal	Technical, organizational, no uniformity of infrastructure (road)	Gauge	Lack of interfaces and connections			
Spain	Product identification standards					
Sweden					Institutional barriers between land and ship	Different container types
Switzerland	Barriers against Alps transit	Financial and institutional				
U.K.				Limited and discontinuous network		Limited loading gauge

Source: based on Beaumont et al. (1996).

important developments as globalization, logistics chain management, and computerization.

2. International logistics: what's in a name?

Logistics is defined in many different ways, which is a consequence of its history, the specific sphere of activity, and the wide range of operations involved. However, many authors adopt Ballou's definition as their own starting point (Ballou, 1992). Ballou argues that logistics is all about guaranteeing that the right goods and services are delivered at the right time, in the right place, and in the right condition, and that, in making this happen, the company concerned should attain the highest possible yield. This definition applies to logistics in general, irrespective of whether it concerns a domestic or an international operation. Expanding operations from a domestic to the international arena brings with it very specific problems. Goods are not merely moved from one location to another; to cross a border often also means that the goods will enter an entirely different environment. Such differences may typically concern prevailing legal restrictions, culturally determined attitudes, the available infrastructure, and possibly even restrictive use of money. On this basis, one can distinguish between different channels in international logistics:

(1) the international transaction and payment channel,
(2) the international distribution channel through which the goods physically move, and
(3) the documentation–communications channel.

The resulting problems that may arise in terms of international goods flows can be illustrated quite clearly in a European context, where – even over relatively short distances – many frontiers may have to be crossed. One of the most significant problems is that national transport systems are often inadequately geared to each other. The SORT-IT study (Strategic Organization and Regulation in Transport – Interurban Travel), which was conducted at the request of the European Commission, shows that interconnection and interoperability are not merely impeded by technical differences. In fact, organizational and regulatory problems often constitute a much more serious barrier to an efficient goods traffic in Europe. Tables 1 and 2 offer an overview of some of the existing obstacles for selected European countries. The European Commission has introduced a number of measures aimed at facilitating international transport of goods (and passengers). These include: the development of trans-European networks (TENs), deregulation of the transport market by a number of directives such as 91/440 for rail transport, the implementation of the directive on the interoperability of the trans-European high-speed rail system, and the proposal in 1999 for a directive on the interoperability of the trans-European conventional rail system (see Chapter 3).

Besides a smooth physical movement of goods through an international network, sufficient attention needs to be paid to terms of sale and terms of payment, as these become much more complicated when the buyer and seller are from different countries and as more intermediaries are involved in the transaction. Indeed, differences in commercial and insurance law may be considerable. It may demand a lot of time and effort before all parties involved are able to reach consensus on the time and place for the transfer of ownership and responsibility for insurance of the goods. Moreover, goods may be in transit for some time, certainly in the case of maritime traffic, so that terms of payment can be crucially important.

By means of conventions, UNCITRAL (the United Nations Commission on International Trade Law) tries to harmonize legislation pertaining to international trade. The United Nations Convention on Contracts for the International Sale of Goods (Vienna, 1980, which came into force in January 1988) and the UN Convention on International Bills of Exchange and International Promissory Notes (New York, 1988) illustrate how much importance the United Nations attaches to a smooth globalization of international trade.

One of the consequences of the complexity of terms of sale and terms of payment is that an international commercial transaction requires a great many documents (e.g., a commercial invoice, a certificate of origin, insurance certificates, a letter of credit, an import licence, and an export declaration). It does not suffice that one possesses all the necessary documents, it is also crucially important that one is able to produce them at the appropriate moment. An adequate co-ordination of goods flow and document flow can prevent unnecessarily long and considerably costly delays. It is in this sense that EDI is able to make an increasingly important contribution to international logistics.

The complexity of international logistics is further enhanced by the fact that, often, a large number of parties are involved, each with their own objectives (e.g., the consignor/receiver of goods, shippers, forwarders, transport firms, legal experts, and insurance brokers). The changing economic and trade environment implies that international logistics must assure that international commerce can be conducted as smoothly as possible. Among other things, this presupposes an adequate streamlining of the supply chain (which may result in cost savings), cargo management (which encompasses both goods and documents), and minimizing financial and transaction costs, for example by avoiding import duties and barriers as much as possible. It is self-evident that if a company merely has sporadic or limited international business contacts, it will tend to call on external experts for such assignments. If, however, such international contacts are frequent and/or sizeable, the company may consider establishing its own international logistics department.

3. Altered environmental factors

The key to success in international trade is logistical support. A properly
functioning logistics system can, after all, cut costs considerably and may thus
contribute towards improving the competitive position of a firm. This explains
why so much significance is attached to studying the characteristics of
international logistics systems and the environment in which they operate.
International logistical activities clearly take place in an environment that has
changed a great deal and indeed continues to evolve all the time. Consignors must
take this into account, as such developments have a significant impact on the
relation between costs and services.

A number of important developments affecting the international flow of goods
and international logistics are: globalization of the production process; growing
competition in international trade; supply chain management strategies; time-
based competition; and growth of computerization, EDI, and global e-commerce.

3.1. Globalization of the production process

Essentially, any decision about the relocation of production or parts of the
production process to a country other than the country where the product is sold
depends on cost considerations. The following question is of central importance in
this respect: Do the economies of scale offered by factories specializing in the
production of certain components for a global market outweigh the economies of
scope offered by factories that produce more extensive packages for a local
market? The answer to this question depends on the following factors:

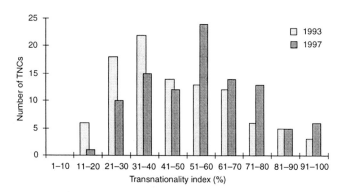

Figure 1. Distribution of the transnationality index of the top 100 transnational companies
(TNCs) (source: based on United Nations, 1995, 1998).

Table 3
Costs (U.S. $ 1990) of air transportation, telephone calls, and computer price deflator

Year	Average air transportation revenue per passenger mile	Costs of a 3-minute call, New York–London	U.S. computer price deflator*
1930	0.68	244.65	–
1940	0.46	188.51	–
1950	0.30	53.20	–
1960	0.24	45.86	125 000
1970	0.16	31.58	19 474
1980	0.10	4.80	3 620
1990	0.11	3.32	1 000

*1990 = 1000.
Source: Herring and Litan (1995).

(1) the extent of modulation and standardization of the production process;
(2) the evolution of the local consumption level;
(3) the possibility of spreading existing technologies geographically; and
(4) the share of the transport costs in the overall cost structure.

On the one hand, a rising local consumption, stimulated by economic growth, will lead to a sufficiently high demand to allow local production. As a consequence, companies will go international and invest a considerable amount of their financial assets abroad. Figure 1 offers an overview of the transnationality index of the top 100 transnational companies for 1993 and 1997. The transnationality index is calculated as the average of foreign assets to total assets, of foreign sales to total sales, and of foreign employment to total employment. Over the period 1993–1997, transnational companies became increasingly internationalized, selling more abroad and generating more employment abroad.

However, declining real transport costs, possibly enhanced by an increasing value of the goods transported and a declining ratio of weight against volume, is conducive to a concentration of production (or part of it) in specialized factories. Declining telecommunication and computer costs (Table 3) may contribute further to a smoother internationalization of the production process.

3.2. Growing international trade

Besides globalization of the production process, there is also a trend towards globalization of product markets. World trade has been growing at an average rate which is considerably greater than the growth rate of GDP (Figure 2). Factors that stimulate international trade play an important role in this respect. Policies aimed at the reduction of tariffs and quotas have led to an increase in the bilateral trade flows between countries.

This can be illustrated by the trade thickness indicator, which is a quantity-based measure for market integration (Knetter and Slaughter, 1999). It is the ratio between the number of actual bilateral trade flows in the sample of countries under consideration, and the total number of possible bilateral trade flows (for a sample of N countries this total is given by $N(N-1)$). It is clear from Figure 3 that, during the 1980s, there was a strong increase of the trade thickness indicator

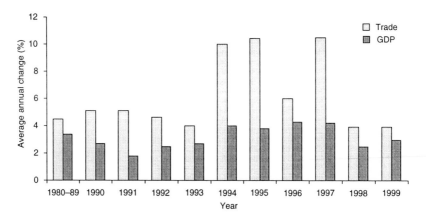

Figure 2. Rate of change in world GDP and world trade (source: International Monetary Fund, 1999).

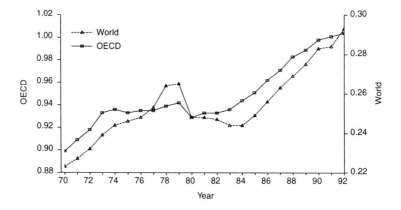

Figure 3. Trade thickness indicator (source: Knetter and Slaughter, 1999).

Table 4
Growth, revenue and number of employees of DHL

Year	Growth (%)	Revenue (U.S. $ billion)	Number of employees
1996–1997	17	4.7	53 222
1997–1998	12	4.8	59 211
1998–1999	8	4.4	60 486
1999–2000	18	5.2	63 552

Source: DHL.

within the Organisation for Economic Cooperation and Development and on the world level.

3.3. The emergence of supply chain management

Another important development is the growing importance of supply chain management strategies. This concept is based on establishing relationships with partners further up or down the logistics chain. The purpose is twofold. On the one hand, one aims for quality improvement of the logistics product. This implies, among other things, greater reliability, a smoother goods flow through the chain, and more efficient connections between the various links in the chain. On the other hand, one strives to realize this at the lowest possible cost for the chain as a whole. This can only be achieved if one has overall control over the logistics chain. From this perspective, it is quite clear that one should not concentrate exclusively on domestic logistics, as smooth international logistics also represent an important asset for a company to attain a more favorable and more competitive position.

3.4. Time-based competition (TBC)

Differentiation between products and services is often founded on so-called time-based competition. In essence, this concept boils down to developing strategies for production and delivery with the purpose of supplying customers with products in as little time as possible. Cost, value, and speed are no longer regarded as possible trade-offs for each other, but as objectives in their own right, to be realized through effective TBC programs (Hise, 1995). The significance of time-based competition is especially apparent in the growing success of companies such as DHL (Table 4), which provide a range number of services (including strategic supply management and return repair inventory) and specialize in speedy delivery.

3.5. Growing global e-commerce

The introduction of internet, e-commerce, and e-business has important consequences for international logistics. From 1997 to 1999 the number of internet connections per 10 000 people in the world almost tripled, but these resources are still a privilege of the high-income economies (Table 5).

In the U.S.A., the UPS Logistics Group accounted for approximately 55% of all purchases on the internet in 1998, followed by U.S. Postal Services with 32% and Federal Express with 10%. By the end of 1999, at least 10% of customers of the UPS Logistics Group were selling electronically, a proportion that continues to increase rapidly. According to the International Communication Union, e-commerce will increase in volume from U.S. $95 billion in 1999 to between U.S. $1000 and U.S. $3000 billion in 2005. Cable and Wireless expects a global turnover of U.S. $700 billion by 2002, while IDC Consulting envisages a turnover of between U.S. $800 and U.S. $1300 billion in 2003. Of this turnover, only 10% will be generated by business-to-consumer trade. The remaining 90% will result from business-to-business trade. As a consequence of this development, production for stock will decline, while production on order will increase. This will result in transportation of smaller batches and growing problems with regard to reverse logistics.

Each of the aforementioned developments clearly affects the international logistics process. While globalization of production and product markets enhance the significance of international logistics, the growing interest in supply-chain management and time-based competition tend to have an impact on the organization and structuring of the international logistics chain. The emergence of worldwide commerce on the internet not only leads to a greater need for international logistics systems, but also to important procedural changes. This puts increasing pressure on a provision of services that is aimed at speed,

Table 5
The number of computers with active AP-addresses connected to the internet, per 10 000 people

	July 1997	July 1998	July 1999
World	35.18	63.1	94.47
High income economies	203.46	374.89	607.55
Low- and middle-income economies	1.53	2.41	4.16
East Asia and Pacific	0.57	0.6	2.39
South Asia	0.06	0.11	0.17
Latin America and the Caribbean	3.48	7.65	14.78
Eastern Europe and Central Asia	6.53	10.55	15.47
Middle East and North Africa	0.2	0.23	0.37
Sub-Saharan Africa	2.03	2.32	2.32

Source: World Bank, 1998, 1999, 2000 (Table 5.11).

reliability, frequency, and low costs. Obviously, the repercussions will vary depending on the goods category, transport mode, and geographical area.

4. Consequences for market players

The globalization of the world economy not only increases the significance of international logistics, but also has an impact on the behavior of all those who are directly or indirectly involved in organizing this logistical process. For providers in particular the question arises of how they choose to react to this changing environment. Is their response limited to an ordinary adaptation to new circumstances, or is there evidence of real change in terms of market behavior and an altered market structure?

First and foremost, there is a growing trend towards providing an integral door-to-door service. At the same time, efforts are made to sell additional value-added logistics services. In addition, one obviously strives to an optimal control of logistical variables, such as speed, the number of routes to be served, the number of nodes to be called at (ports, terminals, etc.), the frequency of scheduling, the means to be deployed, and the question of whether or not to make use of hubs.

Providing an integral door-to-door service is a typical example of a logistics chain approach, whereby one considers the totality of costs and yields, from the point of production to the moment of consumption. This is where the question arises of control over the constituting parts of the logistics chain (e.g., forwarders, port terminals, and inland carriers) In practice, different strategies may be deployed, ranging from far-reaching integration to no integration at all (e.g., port-to-port shipping by Hapag Lloyd). Yet, a general trend towards the creation of DHL and UPS-type integrators is on the cards.

Cost control has led to business relationships involving fewer companies, inevitably coupled with negotiations about more substantial flows, often at tighter profit margins. For this reason, an increasing number of providers are looking for additional revenues from value-added services. The possibilities are quite diverse: storage of goods and related activities, such as groupage/consolidation; the division of large batches; sorting and other processing activities; planning and control of the production flow; and handling of logistics information systems, including invoicing and collection.

This raises a number of strategic issues (Heaver, 2000):

(1) How far should/can one go in spending scarce financial and other resources on additional value-added logistics?
(2) What is the appropriate organizational relationship between these kinds of services and other (core) activities?
(3) What should be the focus of logistical services?

Table 6
Co-operation agreements between various market players

Market players	Shipping companies	Stevedores	Hinterland transport	Port authorities
Shipping companies	Vessel-sharing agreements Joint ventures Conferences Consortia Strategic (global) alliances (e.g., Grand Alliance, New World Alliance) Cartel agreements (e.g., TAA) Mergers	–	–	–
Stevedores	Financial stake of shipping company in stevedore (e.g., CMB in Hessenatie, Nedlloyd in ECT) Joint ventures (e.g., MSC and Hessenatie in Antwerp) (Dedicated terminals) (e.g., ECT Maersk in Rotterdam)	Participation in capital (e.g., Hutchison Whampoa in ECT, PSA in Voltri Genova)	–	–
Hinterland transport modes	Block trains and capacity sharing (e.g., from Rotterdam to Italy) Alliances (e.g., CSX with DB and NS)	Joint ventures (e.g., between NMBS and Noordnatie for operating a terminal in Antwerp)	Takeover strategy of railway companies (e.g., DB and NS cargo, NMBS and THL)	–
Port authorities	(Dedicated terminals) (cf. land-use and concession policy)	Financial stakes port authorities (e.g., 30% ECT by Rotterdam, ECT in Trieste, Sea-ro in Zeebruges)	Antwerp in Rijn shipping terminal of Germersheim	Alliances (e.g., Rotterdam and Vlissingen, Antwerp and Zeebruges)

Source: Heaver et al.(2000).

The operational structure of many companies demonstrates that long-term co-operation agreements between buyers and sellers are increasingly common. Especially in service-oriented companies, this has led to a lower cost per transaction and greater customer satisfaction (Daugherty et al., 1996). These benefits have inspired producing companies to establish strategic relationships with external, third-party service providers, although preferably with a limited number of (large) competitors (Davis, 1987).

For this reason, the previously mentioned service providers themselves are aiming at scale increases, as can be observed quite clearly in maritime liner shipping. Consignors of goods prefer to work with a limited number of carriers offering a total service (i.e., an extensive network). This approach results in a number of specific supply characteristics: shorter transit times, thanks to faster vessels and fewer ports of call; more reliable services; adequate system capacity; and sufficiently high frequency (e.g., one departure per week). In other words, by combining these characteristics, one strives for network efficiency.

However, this is only possible if the service provider is sufficiently large. This (and cost reduction) largely explains the recent wave of mergers, slot chartering agreements, and various forms of strategic alliances. Besides economies of scale, international alliances also offer the advantages associated with increasing globalization.

In maritime liner shipping, these developments are not restricted to the shipping companies themselves; they are also occurring among the other market players involved in the logistics chain. It is quite clear in this respect that the interests and intentions of each of the market players tend to develop rapidly, especially as they are continuously striving towards gaining direct control over an increasingly large part of the logistics chain. We are clearly seeing a shift, then, towards a greater degree of vertical integration within the logistics chain.

The kinds of co-operation agreement that have been reached in recent years between the principal market players (shipping companies, stevedores, hinterland transport modes, and port authorities) are summarized briefly (with examples) in Table 6. Clearly, shipping companies remain the most actively involved in establishing co-operation agreements. As a consequence, they have acquired a much more solid negotiating position vis-à-vis the other players (e.g., port authorities, stevedores, and hinterland transport modes). It is self-evident that the other parties will need to respond. Goods handlers have reacted in various ways, ranging from the setting up of joint ventures for operating dedicated terminals to attracting fresh capital from international groups. The latter is illustrative of the growing importance of international logistics, but may also cause problems in terms of market dominance and conflicts of interest.

The parties involved in a co-operation agreement will invariably aim for a win–win situation. This may be attained through economies of scale by increasing the freight volume on a specific route, spreading the fixed costs over a more substantial service, and serving consignors more efficiently. An improved service may be achieved by extending the network (i.e., a supply that is even better geared to demand and to the expectations of consignors) and by closer integration of door-to-door services.

5. Conclusion

International logistics is very different from domestic logistics. As a consequence, it is subject to a number of developments that are less relevant to the domestic arena. The trend towards a greater degree of integration of product markets, a scattering of components of the production process over different countries, the further globalization of world trade, and the rapid emergence of e-commerce have all contributed to more substantial international goods flows. The organization of international logistics is, moreover, affected by the introduction of new concepts, such as supply chain management and time-based competition.

The success of any international operation depends partly on the ability to recognize and cope with such developments. It is important in this respect that one should take into account the entire logistics chain. Thinking in terms of a chain implies moving beyond the separate links in the logistics system and giving due consideration to interactions between these links (and the consequences in relation to costs). This requires each management to transcend a narrow, fragmentary approach to operations. Clearly, international logistics occupies an important place in this move towards globalization.

The growing potential of the international logistics sector has a strong impact on the behavior of providers. They are increasingly trying to gain control over the entire logistics chain, so that the market structures within which providers operate are subject to continuous development and change.

References

Ballou, R.H. (1992) *Business logistics management*. New York: Prentice Hall.
Beaumont H., J. Preston and J. Shires (1996) "SORT-IT: strategic organisation and regulation in transport, summary of country reports", Version 0.2, Research project for the Commission of the European Communities, DGVII. Institute for Transport Studies, University of Leeds.
Daugherty, P.J., T.P. Stank and D.S. Rogers (1996) "Third-party logistics service providers: purchasers' perceptions", *International Journal of Purchasing and Materials Management*, 23–29.
Davies, G.J. (1987) "The international logistics concept", *International Journal of Physical Distribution and Logistics Management*, 17:20–27.
Heaver, T.D. (2000) *Liner shipping and international logistics*. Antwerp: ITMMA.
Heaver, T.D., H. Meersman, F. Moglia and E. Van de Voorde (2000) "Do mergers and alliances influence european shipping and port competition?", *Maritime Policy and Management*, 27:365–373.
Herring R.J. and R.E. Litan (1995) *Financial regulation in the global economy*. Washington, DC: Brookings Institution.

Hise, R.T. (1995) "The implications of time-based competition on international logistics strategies", *Business Horizons*, 38:39–46.

International Monetary Fund (1999) *World economic outlook*. Washington, DC: IMF.

Knetter, M. and M. Slaughter (1999) "Measuring product–market integration", NBER, Cambridge, Working Paper 6969.

Miyazaki, A.D., J.K. Phillips and D.M. Phillips (1999) "Twenty years of jbl: an analysis of published research", *Journal of Business Logistics*, 20:1–19.

Wood, D.F., A. Barone, P. Murphy and D.L. Wardlow (1995) *International logistics*. New York: Chapman and Hall.

World Bank (various years) *World development indicators*. Washington, DC: World Bank.

Young, R.R. (1995) "Book review: international logistics", *Transportation Journal*, 34:72–73.

Part 2

SUPPLY CHAIN MANAGEMENT

Chapter 6

THE DEVELOPMENT OF THINKING IN SUPPLY CHAIN AND LOGISTICS MANAGEMENT

DARREN HALL and ALAN BRAITHWAITE

Logistics Consulting Partners, Berkhamsted

"History is a hard core of interpretation surrounded by a pulp of disputable facts." (Anon.)

1. The significance of supply chain management

One of the most remarkable shifts in corporate strategy and operational activity in recent decades has been the externalization of production to an extent that corporations are now utterly reliant upon external resources. The reasons for this trend are many and varied but, at a superficial level, the core–periphery model has come to dominate corporate thinking, whereby all but principal activities are pushed outwards to external parties. The mantra of core competence and the endless search for value and variety have made the integrated organization an extinct form of corporation. Suppliers have arguably been incorporated into the fabric of corporate activity to such an extent that their role is now strategically inseparable from any internal aspect, with the result that a significant proportion of competitive advantage rests with the management of this crucial external resource.

Many once-central operational corporate activities – product design and development, services and facilities management, logistics, and manufacturing – have been taken over by suppliers. Consequently, there has been an ever-increasing focus on managing the external relations of production and the control of resource flows from source to consumer. The term "supply chain management" has been coined to represent these activities. Although supply chain management has experienced enormous growth as both a business imperative and a legitimate field of academic study, it is far from being a coherent body of thought, still less a universally adopted practice. The myriad manifestations and implications of externalization have generated a hugely complex subject matter.

Handbook of Logistics and Supply Chain Management, Edited by A.M. Brewer et al.

1.1. A matter of definition

As a result of the complexity discussed above, any attempt to construct an authoritative account of the development of thinking in supply chain management is at once confronted with the problem of scope – supply chain management does not possess unitary definition or demonstrate singular application. Indeed, the converse is true: it is a curiously eclectic subject area, the rich and diverse composition of which seems to be in a state of constant flux. Charting the development of this subject area therefore necessitates the incorporation of a wide array of disciplines and activities, ranging from the mechanical to the esoteric. Perhaps uniquely, supply chain management integrates such diverse interests as inventory planning, manufacturing operations, and consumer behavior with intercorporate strategy, global information technology architectures, and stochastic optimization modeling.

Whereas many of these individual elements have long existed as subjects in isolation, it is only comparatively recently that they have been brought together to form a distinct and purposeful subject. In fact, supply chain management can be credited with forging bonds between functional areas of business operations and activities. The essence of supply chain management is the confluence of these diverse strands according to an overarching logic.

The development of the subject can be traced in outline by considering the way in which its definition has evolved. Although the term "supply chain management" is a product of recent times, the antecedence of the subject at a conceptual level reaches further into history than might be imagined. To provide an initial definition of our topic, and to create an historical perspective, we cite here a prescient text written by Arch Shaw in 1915, held in the Harvard Business School archive. In it, Shaw identifies an agenda for business that was as relevant at the end of the century as at the beginning: "the relations between the activities of demand creation and physical supply ... illustrate the existence of the two principles of interdependence and balance" (cited in Christopher, 1985).

Christopher uses this example to argue that: "logistics has always been a central and essential feature of all economic activity and yet paradoxically it is only in recent years that it has come to receive serious attention from either the business or academic worlds. One obvious reason for this neglect is that whilst the functions that comprise the logistics task are individually recognized, the concept of logistics as an integrative activity in business has really only developed within the last twenty years" (Christopher, 1985). Christopher's work, then, contrasts a perennial awareness of the importance of integration and synchronization between supply and demand with its marginal impact throughout the majority of the 20th century.

While there is little point in seeking to document a perfect definition for supply chain management, the conceptual foundation necessarily encompasses notions of operational integration and organizational inclusiveness. Moreover, any depiction of supply chain management that seeks to be more than a rehash of antiquated materials management definitions must include notions of complexity, volatility, and operational velocity. This chapter illustrates the ways in which supply chain management has become increasingly inclusive, binding operational ties and forging strategic bonds between companies in a production system.

1.2. The layout of this chapter

In order to address this complex picture, and to plot a meaningful trajectory for the future development of the subject, a partly chronological and partly thematic structure is required. There is no neatly linear format that adequately represents the interwoven intricacies of the development of supply chain management. Its influences are truly global, featuring a rich representation across continents, industrial sectors, and academic schools, almost spanning one century.

The chapter first points to the array of elements that represent the foundations of the subject, identifying the seminal moments, and highlights the major themes in the development of the subject by considering the development and limitations of functional excellence in operations and business management. The impact and influence of Japan in the development of supply chain management is woven throughout the chapter. The chapter then appraises the development of supply chain management proper, with the emergence of lean thinking. The chapter then considers the types of supply chains that exist, by examining supply chain structures and the conceptualization of them. Finally, the chapter uses the experiences of modern corporations and thinkers to project the future of the subject.

2. The precursor to supply chain management: functional excellence and its inadequacies

The roots of supply chain management may be found in many disciplines. This section documents the ways in which the most significant functional areas have evolved from isolated activities to form part of a greater whole – a supply chain. The need for functional excellence should never be disparaged; for many industries and companies, it represented a core skill and one point of competitive advantage. Moreover, attaining functional excellence often required gaining an understanding, and eventual control, of processes that would have subsequent bearing on an ability to operate in a way accordant with the principles of supply chain management.

2.1. Operations research and manufacturing systems

The prominent highlights from the 1950s and 1960s include, first and foremost, Jay Forrester's seminal work into supply chain operations. Published in 1961, *Industrial Dynamics* illustrated the ways in which errors, inaccuracies, and volatility amplified across a supply chain. These effects were caused by a complex mixture of operational and behavioral factors. Operationally, each element of the supply chain sought to operate at a locally efficient level, without any apparent regard for the endpoint in the supply chain – which was, in any case, functionally invisible. Behaviorally, the proponents across the supply chain found the addition of buffer inventory irresistible.

The combination of these factors resulted in an imbalance of inventory and time in the chain that, despite the attempts to be as locally efficient as possible, were often catastrophic when considering the chain as a whole. The "Forrester effect" represented a powerful antidote to the wealth of algorithmic material that stemmed from operations research at that time, epitomized by the concept and calculation of the "economic order quantity". The irony is that manufacturing industry seemed to absorb the restrictive and misleading algorithms and eschewed Forrester's logic, irrespective of its value.

2.2. Materials management

Materials management was changing its brief, stemming from the growing recognition of the importance of physical distribution management to corporate profitability. In their critique of the National Council of Physical Distribution Management's definition of the subject, a rather anodyne account of physical goods handling and transportation, Bowersox et al. (1968) identified two important omissions and, in so doing, set the scene for the development of supply chain management from a logistical perspective.

First, they noted that: "for a physical distribution system to function properly it must enjoy timely and accurate dissemination and feedback of distribution information. Little of any importance happens in a logistical system without continuous and distortion-free information flow" (Bowesox et al., 1968). Next, and perhaps of even greater significance, the authors placed a "proper emphasis upon the joint activities between firms linked together in the physical distribution of a single assortment of products" (Bowesox et al., 1968). The underlying message within the Bowersox critique is the integration of operational activities across an array of organizational elements.

A note of supply chain integration was also sounded in the popular press. The *Financial Times* (27 February 1975), for example, recognized that: "It is no longer possible for companies to treat their physical distribution policies in isolation. As distribution costs loom larger, they have to be taken increasingly into account in the planning of product ranges and the siting of production units. Decisions have to be taken about the level of distributive service which companies are prepared to offer."

At this point in the story, the constituent elements of supply chain management had formed, but one essential ingredient, an overarching logic, was still missing. To find that all-important catalyst, we travel to Japan, a country which has ironically received criticism for its lack of innovation and predilection for replicating the ideas of the West. In fact, while the West slept, the Japanese were developing intercorporate strategies and operations management practices that proved to be decades ahead of anything the West had to offer. Japan had an operating philosophy that was the ideal structure for the principles that were forming. The adoption of lean production principles, when applied across a network of external companies in a production system, represented the breakthrough in the development of logistics and supply chain management.

2.3. Quality

It was also in the post-War years that four men – Feigenbaum, Deming, Juran, and Ishikawa – made a decisive contribution to the advancement of supply chain management. The group collectively was responsible for developing the principles of total quality management, and with it raising awareness of the importance of control across the entirety of a production environment – that is to say, extending to suppliers. Their work forged one more operational bond between entities in a production system through, in particular, the development of causal analysis. To solve a non-conformance issue permanently, it was necessary to change the causes of problems, rather than merely address their symptoms through the application of palliatives.

Until that time, quality (i.e., product conformance) tended to be the result of inspection procedures. In 1972, Volkswagen ran a remarkable advertisement that boasted of 800 inspectors working at its Wolfsburg plant to guarantee quality. However, the pure edicts of quality held that quality cannot be inspected into production; conformance must stem from process control. Internal quality systems would be meaningless without the control of externally supplied components. There was a consequent requirement to guarantee the conformance of raw materials, components, and assemblies.

In fact, the subsequent application of quality standards and the pursuit of total quality management in the 1990s became "... the managerial imperative of the decade" (Drummond, 1992). The significance of quality at that time cannot be overestimated. In fact, the success of Japan was commonly attributed to one source: "[j]ust four decades after digging out from under the rubble of World War

II, Japan has transformed itself – into an economic powerhouse through its single-minded focus on quality" (Green et al., 1994).

The implications for supply chain management were enormous; suddenly, there was a new exigency to ensure that the manufacturing and production processes among the supply base were controlled. A raft of initiatives were launched under the quality banner, representing surrogate attempts at supply chain management. Supplier assessment, process capability studies, quality systems, procedures and standards, statistical process control, continuous improvement – the list is almost endless.

The implementation of quality standards and the attainment of total quality management were responsible for advancing the cause of supply chain management in two key areas. First, whereas the onus initially tended to be placed on suppliers to conform, the exposure of causal mechanisms highlighted the role that customer and supplier play in collectively achieving high levels of product conformance. An awareness of the elimination of variability – a concept at the heart of total quality – thus tended to forge stronger bonds between all elements of the supply chain. Second, the monitoring of performance created another basis on which suppliers could be judged, displacing the criterion of purchase price. This theme is explored in the next section, which discusses the evolution of purchasing.

2.4. Purchasing and sourcing strategy

In the 1980s and 1990s, there was a proliferation of texts on purchasing, having common themes of selecting, using, and developing suppliers in line with strategic goals and overall business objectives. Gadde and Håkansson (1993) capture the essence of these works, noting that "purchasing has become increasingly important in the framework of any company, developing from an almost clerical function to a major strategic resource. New styles of purchasing involve careful analysis and planning, requiring an understanding of supplier relationships, product-development processes, quality-driven management and industrial networks."

By the 1990s, the growing recognition of the importance of supply chain management was beginning to permeate mainstream literature. Although many texts were still firmly rooted in functional disciplines, attempts at integration were becoming increasingly evident. For example, in their book *Professional Purchasing*, Gadde and Håkansson (1993) note that "... what goes on at the interface between individual companies has gradually gained in importance in relation to the effectiveness and development of industrial systems" and that, as a result, "... the competitiveness and profit-generating capacity of the individual firm is highly dependent on its ability to handle purchasing."

This statement alludes to an important transition in sourcing, whereby the direct impact of purchasing (i.e., procurement costs) were being subsumed within an emergent strategic importance, encompassing far broader themes. A number of authors report on this notion, one of the most significant of the decade, noting incorporation of supply chains into strategic thinking in a quite fundamental way, forming a radical departure from orthodox methodologies (Lamming, 1993). Strategic alliances and joint ventures reconceptualized the relations of production, leading to a number of stratagems for sharpening the responsiveness of supply chains. Stratagems related to quality, engineering and logistics engineering, and logistical developments were implemented alongside a suite of enabling mechanisms, or sourcing strategy, concerning the conduct of relations across supply chains.

As a consequence, sourcing strategy began to evolve from its traditionally myopic form, focused on price, to a meticulous but uncompromising practice, geared towards the attainment of an array of world-class suppliers. Organizations wanted to surround themselves with companies able to offer a broad range of capabilities in particular areas of production, leading to rationalization, and giving rise to supply-base-reduction programs, single sourcing initiatives, and the like. This process empowered specific suppliers, while deepening control of the overall supply chain (Wormald, 1989).

Reflecting the shift in strategic thinking, texts explicitly focusing on supply chain management appeared (Macbeth and Ferguson, 1994). The Cardiff Business School, fuelled by visits to Japan, and eager to understand the emergent automotive supplier communities in south Wales, published a vast array of papers and books on supplier development and supply chain management. An excellent introduction to this seminal work may be found in Hines (1994). However, the most significance work from the Cardiff Business School was yet to come. In 1996, Dan Jones, together with James Womack, wrote *Lean Thinking*, possibly one of the most important books on supply chain management (Jones and Womack, 1996).

3. Leanness: the first supply chain agenda

3.1. Muda and supply chain management

The Japanese obsession with manufacturing and production excellence has been focused on one aim – the elimination of waste (or *muda*). The operations management methodologies that have become synonymous with Japan are the practical face of this philosophy. Waste was analyzed as having seven components: inventory and overproduction, transportation and motion, defects, waiting and inappropriate processing. Latterly, an eighth component has been added: unsuitable product design. The importance of waste elimination to the development of supply chain management cannot be overstated. The combination of these key concepts represents a cogent and cohesive logic that forms an architecture for operations across a production system.

At each point in the production process, the elimination of waste requires that only those products for which there is a demand are made, in exactly that quantity, only when needed, and without defects and unnecessary motion. Any operational activity that deviates from this tenet should be continually refined in the pursuit of the ideal – instantaneous single piece flow from source to customer.

Waste-minimizing practices and the development of lean manufacturing were largely centered around the efforts of Taichi Ohno and Shigeo Shingo at Toyota, hence the particular, and advanced, form of lean manufacturing known as the "Toyota production system". Ohno and Shingo painstakingly assessed each aspect of production operations within Toyota and its suppliers in order to find and to eliminate waste. The concepts and principles that were applied in the identification of waste and the determination of value essentially exposed and redefined the relationships between cost, quality, and flexibility. The principles, techniques, and practices that Toyota developed included:

(1) process analysis,
(2) quality control,
(3) the elimination of delays through shortened production,
(4) autonomation,
(5) just-in-time,
(6) cellularization,
(7) schedule control,
(8) production balance and leveling,
(9) standardization, and
(10) kanban.

An array of work study techniques were developed to explore the applications of a new logic in production. Ohno and Shingo fundamentally rejected the established tenets of manufacturing based on economies of scale – locally efficient utilization, batch production, inventory buffers, inspection, and so on. Rather, their doctrine called for changes to be brought to all aspects of the production system; although radical in principle, they were usually small and incremental in practice.

Since long machine set-up times dictated large batch sizes for the purpose of amortization, Shingo developed a new operational philosophy – the single minute exchange of dies (SMED) system. The newly found ability to change over machines for the manufacture of different products allowed smaller quantities of a grater variety of items to be made, thus fundamentally altering the basis of production and deriving economics of scope. More responsive production systems targeted overall systemic effectiveness rather than abstract and isolated efficiency. The extension of this principle eventually gave rise to the term "mass customization", representing the ability of a production system to meet the demands of increasingly differentiated and demanding markets (see Pine, 1993).

Two authors, with dramatically different styles, are worthy of mention for bringing an awareness of the principles of supply chain management to manufacturing environments in the West. Richard Schonberger made a hugely important contribution to the development of operations management in two texts: the excellent *Japanese Manufacturing Techniques* (1982) and *World Class Manufacturing* (1986). His ideas, although having a strong manufacturing bias, are imbued with many core supply chain management principles, such as leanness, flexibility, and – most importantly – simplicity. Goldratt and Cox (1984) threw down the gauntlet to conventional wisdom in *The Goal*, a profoundly original text that contradicted a generation of manufacturing operations.

3.2. Time compression

"Time is the longest distance between two places."
(Tennessee Williams, *The Glass Menagerie*, 1945)

Perhaps first principle of the Toyota production system to have a fundamental impact on Western producers was just-in-time (JIT), the principle that goods are delivered at the right quantity at the right place immediately in advance of their requirement. Parenthetically, it would be incorrect to assume that JIT is a product of Toyota alone; Takeuchi (1990) refers to a hub of subcontractors surrounding Nissan in Tokyo, built from the 1960s, for JIT supply.

JIT has been interpreted in a number of ways, according to circumstance; its significance varies considerably across nations and industries. It is thus possible to identify a spectrum of meaning for JIT; for example, it may simply be used to indicate timely delivery, or as a total production philosophy. A wide literature exists on JIT, a testament to its compeling logic, inherent operational benefits, and impact on supply chains (see Chapter 13).

While early accounts of JIT usually cautioned against attempting to posit general trends based on isolated examples, no such restraint is necessary now. To gauge the rate at which JIT has grown in application, as recently as 1992, Mair (wrongly) noted that that the paradigm of JIT was actually a model of only one firm. Today, there is not a manufacturer that does not purport, to some extent, to practice JIT materials management or production. The flow of materials throughout a supply chain according to the tenets of JIT has completely changed the industrial landscape and economics of production. Spatial agglomeration has proceeded apace, with supplier parks and satellite operations becoming commonplace. Referring to the period of expansion that led the North American automobile industry to be transcontinental in outlook, Estall (1985) noted that JIT would "tend to restore the significance of geographical concentration."

The selection of suppliers for JIT status has been structured around the logistical characteristics of the products that are supplied. The decisive parameters are size, price, usage, and degree of derivatization. Any parts that are at the upper end of these four scales require JIT supply to limit the space required for storage, to reduce working capital, to increase flow, or to minimize buffer stocks. The automotive industry initially selected seats, exhausts, and the like for JIT supply, but an increasing proportion of parts are now supplied in this way. The principle has been extended and refined – Ford's operation in Valencia, for example, has its parts delivered just-in-sequence from overhead tunnels fed by conveyors from suppliers located less than 1 km away.

The operation of lean production required a number of changes to materials management regimes and, in turn, to the organization of wider operational aspects focusing on visibility and overall systemic control through the elimination of non-value-added activities. A JIT system afforded no opportunity for inspection, process variations, or time-consuming assembly activities, so the members of a supply chain in such an environment were required to improve their operational processes. The synchronization of production, focused on customer requirements, led to the development of pull supply chains, a theme explored in the next section.

3.3. Push and pull supply chains

Michael Porter, one of the finest corporate strategists of the modern era, made a significant contribution, albeit sometimes opaquely, to the development of supply chain thinking in his many books. In particular, his articulation of the concept of the value chain both succinctly expressed the practices of a generation of companies and gave rise to a debate about the implications of the approach and alternatives to it. Porter developed a theory of the firm predicated on value being constructed from the imposition of profit margins at successive points in the production chain. The model, applied particularly to the U.S.A., encapsulated many isolationist rigidities that have been deemed partially responsible for the long-term decline in the competitiveness of Western industry. Companies operating in this fashion are characterized by confrontational intercorporate politics encompassing marginal short-term commitment, competitive tendering, and multisourcing, as well as low levels of mutual investment and cross-equity alliances.

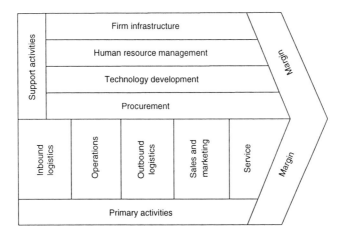

Figure 1. The push supply chain (after Porter, 1985).

In a *push supply chain* (Figure 1), costs are transmitted up the chain; input costs for the members of the chain are determined by the selling price of the preceding level. This imposition of profit margins at successive points in the production chain constructs value. While some suppliers may enjoy the fruits of this approach, it is fundamentally flawed in that there is no guarantee that the next level in the chain will be able to afford the goods, still less that the end customer will find the price attractive. There is a fundamental disconnect between realistic market selling price and the cost behavior of the supply chain. Moreover, the operation of the cost plus environment generates little incentive for suppliers in the lower echelons to improve their performance and so to reduce their costs. Customers may be inclined to seek alternative suppliers, adding to the total cost of acquisition and reducing stability. This framework therefore imperils the long-term commercial viability of the chain.

In stark contrast to the cost plus approach, a *pull supply chain* (Figure 2) operates on the principle that the supply chain must be able to deliver to market a product at an affordable level. It is the responsibility of everyone in the chain to ensure that operational costs and commercial structures support this objective. Suppliers know that, as the price is set by the customer, their profitability derives from their own input costs (i.e., to their internal efficiencies and external costs). Consequently, pull supply chains place downward pressure on suppliers to become more efficient and to operate for the common good. Often suppliers receive support and guidance in the attainment of greater levels of efficiency. Supplier development initiatives, orchestrated by concerned customers, often extend to operations specialists providing advice on efficient and effective manufacturing techniques, and suppliers also may operate in groups to develop common solutions for shared problems. The most significant texts on the development of target costing across supply chains are those by Cooper and Slagmulder (1997) and Kaplan and Cooper (1998).

4. The structure of supply chains

4.1. Western and Japanese hierarchies

Before the significance of supply chain management was fully appreciated, the structure of intercorporate organization was considered in the simplistic terms of vertical and horizontal integration or disintegration – concepts borrowed from economics. However, these descriptions were highly flawed, being polar, reductionist, and somewhat vacuous, having little inherent meaning or clear implications.

Attempts to understand the significance of supply chain structures by creating more philosophically rigorous and critical accounts did not emerge until the 1980s. The study by Hall (1983) is by far the most significant of these, recognizing that supply chain structures, and the methods of interaction between organizations, were extremely important determinants of operational performance. Hall and, to a lesser extent Sheard (1983), working in the automotive industry, highlighted the high degree of vertical integration and supply chain networking apparent in Japanese supply chains.

These studies, later supported by Dicken (1988) and Hill (1989), revealed radically different structures between East and West. For any given tier, it seemed

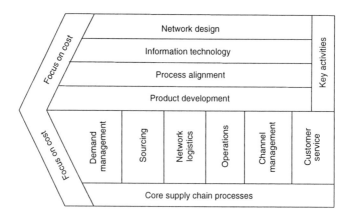

Figure 2. The pull supply chain.

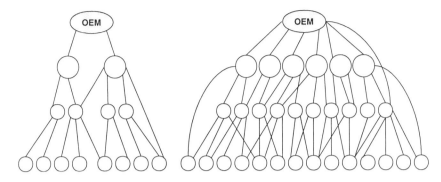

Figure 3. The *Keiretsu* and Western supply chain structures (source: Hall, 1983).

that Western organizations tended to maintain relationships with far more suppliers than did their Japanese counterparts. Moreover, Japanese companies had more levels spanning raw material to finished goods in their supply chains than did those in the West. In short, Japanese tiers were far more pyramidal in structure. These attributes can be seen in Figure 3.

Hall noted that the Japanese even had terms to describe the degree of closeness in the intercorporate relationship – from the closely tied *Shitauke*, akin to subcontractors, through the broader networked *Gaichu*, and finally to the *Kaubai*, the independent suppliers. The very fact of these distinctions committed to language speaks volumes about the disparities between the chaotic and complex Western supply chains and their Japanese cousins.

At an operational level, Western companies were required to handle far more components than were their Japanese counterparts, this carrying an onerous administrative burden. The Japanese preferred to organize inbound material flows into subassembly units and assembly systems, a method of materials management that dramatically reduced handling and assembly cycle times. Although these studies were typically based on research carried out at automotive assemblers, the principles were generically applicable, and have subsequently been applied across a wide array of sectors. Hall attributes low inventory, lean, and responsive supply chains in large part to this method of intercorporate networking (see Chapter 12). The response of Western companies was to redesign their bills of material into fewer units comprising more parts – these new systems were then procured from fewer suppliers that, in turn, dealt with component suppliers. In this way, there was a subdivision of tiers, mimicking the structures implemented in the Far East.

The origins of functionally related subcontracting hierarchies in Japan can be traced back to 1938, when legislation was introduced to reorganize the structure of the Japanese economy, such that small and medium enterprises were amalgamated and merged to be directly accountable to large firms, thus creating a coherent whole. As a result, "[t]he combination of government ordinances, pressure from giant manufacturers, and the ultimate leverage of capital control provided by the main banks was all it took to build pyramids from which subcontractors could not escape" (Miyashita and Russell, 1994). Supply chain management in the West is rather more ad hoc, and can be criticized for a return "to a mentality of dealing with problems as they come up rather than building a corporate infrastructure to avoid or resolve them" (Miyashita and Russell, 1994).

Today, Japan's industrial structure is dominated by the *keiretsu*; there are two types: *yoko* (horizontal) and *tate* (vertical). The former are large groupings of companies with common ties to a powerful bank, while the latter are large companies connected to thousands of subservient companies, linked by a production theme, and arranged in tiers. Members of such tiers are known as *kyoryoku kai*. Under the *keiretsu* structure, supply chain relations are of paramount importance. In particular, allegiance to a particular head is absolute, where traditional bonds are still very strong. The formation of supplier associations is also seen as a key aspect of creative control over suppliers.

4.2. Network supply structures

The term "supply chain" tends to suggest a linear arrangement of organizations conducting operational activities in a particular sequence. However, this view has emerged as an inadequate representation of intercorporate activity. Dicken and Thrift (1992) recognized that the production chain is a complex, dynamic system that plays host to differential political relationships. Thus, "the inter-firm structure of large corporations is increasingly better represented as a network than as a hierarchy as such firms strive to create more flexible organizational structures. Both the organizational and the spatial forms that result are infinitely more complex than the simplistic hierarchies envisaged in earlier studies ... ." This point is illustrated in Figure 4.

While the authors addressed their remarks to large corporations, recent studies of small and medium enterprises reveal a large degree of network organization to enhance flexibility, delivery, and cost factors. In these cases, the network encompasses a portfolio of virtual resources that may compensate for internal deficiencies, forming a type of mutual support grouping. So pronounced have these associations become that competitiveness must be explored in terms of the ability to manage complex networks of production.

The advent, and subsequent saturation, of information technology has intensified the penetration and significance of the network as an appropriate conceptualization of intercorporate organization. Recent literature on e-commerce points to the

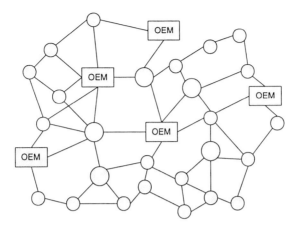

Figure 4. A network supply chain structure.

emergence of virtual communities of corporate networks, which Evans and Wurster (1999) call "hyperarchies", in which corporate structure is dissolved and the traditional distinctions between customer and supplier are blurred or even obliterated. Trading communities have formed in the automotive, retail, and aerospace sectors that comprise portfolios of customers and suppliers, often at a global scale, engaging in commercial transactions.

However, there are opposing implications from the concept of virtual supply networks. On the one hand, the nature of information is radically changing. The virtual age has facilitated the presence and accessibility of data, to be used in ways unthinkable a few years ago; and we are moving from a position of data to one of information and thence to knowledge – a process that must, over time, produce smarter individuals, companies, and actions. On the other hand, the creation, operation, and mobility of these networks defies man's understanding on all but a superficial level, and far outpaces any hitherto observed organizational structure. Companies will be tempted to "chase the dragon" of lower prices and improved profitability, switching sources through the virtual network and exposing myriad risks of unreliability and quality.

5. Enlightenment at last, but what is next?

By the latter half of the 1990s, texts explicitly focusing on supply chain management began to appear. the book by Riggs and Robbins (1998), *Supply Management Strategies – Building Supply Chain Thinking into All Business*

Processes, is typical of this new breed of literature that displays a highly developed understanding of supply chain management. The work concisely documents the failings of myopic purchasing activities, explains the importance of time and the development of networks. It also adopts a truly inclusive perspective, emphasizing the importance of identifying, and changing in line with, market requirements, design and product introduction, as well as operational activities. Moreover, the total cost of ownership is explored to debunk the myth of lower purchase prices irrevocably equating to lower overall cost. Turning attention towards implementation, the authors document the importance of streaming (also known as "logistics product groups"). The principle behind this concept – a core theme in supply chain management – is that differences in logistically significant characteristics (e.g., product size, weight, rate of sale, and supply source) require differential treatment in the supply chain. Once like groups have been identified, a logistical or business infrastructure may be constructed that is sensitive to these attributes, and supported by operational policies, such as inventory levels and flow characteristics.

Tyndall et al. (1998) expand on these themes in an outstandingly good volume that represents perhaps the first identifiable, truly focused, supply chain management text. To position the development of the subject, a logistics development path is plotted, beginning at disconnected business entities, moving through integrated operations (i.e., process oriented operations) to seamless and transparent customer-focused supply chains. Last, and most significantly, they introduce the concept of virtuality in supply chain management, an idea that encompasses a high degree of operational fluidity within a portfolio of external agents. Then, the authors meritoriously summarize the zenith of supply chain management, identifying the full suite of issues confronting the synchronization of supply and demand in complex and volatile production systems. Their twelve-point plan concisely captures the essence of supply chain management:

  (1) build in flexibility,
  (2) plan and measure accurately,
  (3) develop logistically separate operations where appropriate,
  (4) get lean by emphasizing simplicity and speed,
  (5) optimize information,
  (6) treat customers unequally (segment and stratify),
  (7) operate globally,
  (8) practice virtuality and collaborative management,
  (9) exploit electronic commerce,
 (10) leverage people,
 (11) operationalize new product introductions and phase-outs, and
 (12) mass customize and postpone.

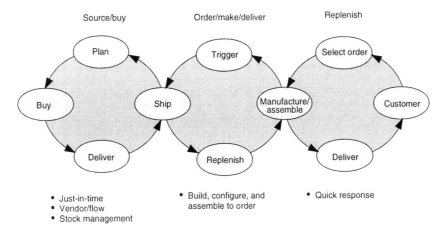

Figure 5. Schematic representation of the tight linkages that are key to effective supply chain management.

The discussion then moves on to consider sourcing strategy and something they call the "new logistics" – "moving less ... faster!" – illustrated in Figure 5.

Supply chain management has arguably reached a stage of purposeful refinement, but the array of new pressures, environmental changes, and opportunities continues to escalate. Technology has become remarkably affordable, with the price/power ratio having fallen approximately 50-fold in just 15 years. The implications for intercorporate (business-to-business) communications, stemming from the speed and volume of data transfer, over the internet has facilitated data sharing to enable a host of collaborative and synchronization processes. Nevertheless, e-commerce is still in its infancy, and the impact thus far must be understood in this context. These advancements have given rise to the development of alternative channels, but far more change is inevitable. There may be an irony that the enabling powers of the internet may unravel the tightly knit supply chains that have emerged.

References

Bowersox, D.J., E.W. Smykay and B.J. La Londe (1968) *Physical distribution management*. London: Macmillan.

Christopher, M. (1985) *The strategy of distribution management*. Aldershot: Gower.

Cooper, R. and R. Slagmulder (1997) *Target costing and value engineering*. Portland: Productivity Press.

Dicken, P. (1988) "The changing geography of Japanese FDI in manufacturing industry: a global perspective", *Environment and Planning A*, 20:633–653.

Dicken, P. and N. Thrift (1992) "The organisation of production and the production of organisation", *Transactions of the Institute of British Geogrraphy*, 17:279–291.

Drummond, H. (1992) *The quality movement, what total quality managment is really all about*. London: Kogan Page.

Estall, R. (1985) *A modern geography of the United States*. New York: Times Co.

Evans, P.B. and T.S. Wurster (1999) "Strategy and the new economics of information", in: D. Tapscott, ed., *Creating value in the network economy*. Boston: HBS Press.

Forrester, J.W. (1961) *Industrial dynamics*. Harvard: MIT Press.

Gadde, L.-E. and H. Håkansson (1993) *Professional purchasing*. New York: Routledge.

Goldratt, E.I. and J. Cox (1984) *The goal*. Aldershot: Gower.

Green, C. (ed.) (1994) *The quality imperative*. New York: McGraw-Hill.

Hall, R.W. (1983) *Zero inventories*. Chicago: Irwin.

Hill, R.C. (1989) "Comparing transnational production systems: the automobile industry in the USA and Japan", *International Journal of Urban and Regional Research*, 13:462–480.

Hines, P. (1994) *Creating world class suppliers*. London: Pitman.

Jones, D.T. and J.P. Womack (1996) *Lean thinking*. New York: Simon and Schuster.

Kaplan, R. and R. Cooper (1998) *Cost and effect*. Boston: HBS Press.

Lamming, R. (1993) *Beyond partnership: strategies for innovation and lean supply*. London: Prentice Hall.

Macbeth, D.K. and N. Ferguson (1994) *Partnership sourcing*. London: Pitman.

Mair, A. (1992) "Just-in-time manufacturing and the spatial structure of the automobile industry – lessons from Japan", *TESG*, 83(2):82–92.

Miyashita, K. and D. Russell (1994) *Kemelsn: inside the hidden Japanese agglomerates*. New York: MacGraw Hill.

Pine, J.B. (1993) *Mass customisation: the new frontier in business competition*. Boston: HBS Press.

Porter, M.E. (1985) *Competitive advantage: creating and sustaining superior performance*. New York: Free Press.

Riggs, D.A. and S.L. Robbins (1998) *Supply management strategies – building supply chain thinking into all business processes*. New York: AMACOM.

Schonberger, R. (1982) *Japanese manufacturing techniques*. New York: Free Press.

Schonberger, R. (1986) *World class manufacturing*. New York: Free Press.

Sheard, P. (1983) "Automobile production systems in Japan: organisational and locational features", *Antipode*, 21(1):49–68.

Takeuchi, A. (1990) "Nissan Motor Company: stages of international growth, locational profile and subcontracting in the Tokyo region", in: dM. e Smidt and E. Wever, eds., *The corporate firm in a changing world*. London: Routledge.

Tyndall, G., C. Gopal, W. Partsch and J. Kamauff (1998) *Supercharging supply chains*. Wiley: New York.

Wormald, J. (1989) "Manufacturer integration and supplier relationships", *European Motor Business*, May, 144–161.

Chapter 7

THE SUPPLY CHAIN MANAGEMENT AND LOGISTICS CONTROVERSY

DOUGLAS M. LAMBERT[*]
Ohio State University, Columbus

1. Introduction

One of the most significant paradigm shifts of modern business management is that individual businesses no longer compete as solely autonomous entities, but rather as supply chains. Business management has entered the era of internetwork competition. Instead of brand versus brand or store versus store, it is now suppliers–brand–store versus suppliers–brand–store, or supply chain versus supply chain. In this emerging competitive environment, the ultimate success of the single business will depend on management's ability to integrate the company's intricate network of business relationships (Christopher, 1998).

Increasingly, the management of multiple relationships across the supply chain is being referred to as "supply chain management" (SCM). Strictly speaking, the supply chain is not a chain of businesses with one-to-one, business-to-business relationships, but a network of multiple businesses and relationships. SCM offers the opportunity to capture the synergy of intra- and intercompany integration and management. In that sense, SCM deals with total business process excellence and represents a new way of managing the business and relationships with other members of the supply chain.

Thus far, there has been relatively little guidance from academia, which has in general been following rather than leading business practice (Hewitt, 1994; Cooper et al., 1997a). There is a need to build theory and develop normative tools and methods for successful SCM practice. The exploratory empirical findings reported here are part of a research effort to develop a normative model to guide future research. Executives can use the model to capture the potential of successful SCM.

The Global Supply Chain Forum, a group of non-competing firms and academic researchers with the objective to improve the theory and practice of

[*]The author would like to acknowledge the contribution of the members of The Global Supply Chain Forum in the preparation of this chapter.

Handbook of Logistics and Supply Chain Management, Edited by A.M. Brewer et al.
© 2001, Elsevier Science Ltd

SCM, define SCM as:[*] "… the integration of key business processes from end user through original suppliers that provides products, services, and information that add value for customers and other stakeholders."

This broader understanding of the SCM concept is illustrated in Figure 1, which depicts a simplified supply chain network structure, the information and product flows, and the key supply chain business processes penetrating functional silos within the company and the various corporate silos across the supply chain. Thus, business processes become supply chain business processes linked across intra- and intercompany boundaries.

2. SCM versus logistics

The term SCM was originally introduced by consultants in the early 1980s and has subsequently gained tremendous attention (La Londe, 1998). Since the early 1990s, academics have attempted to give structure to SCM (Stevens, 1989; Towill et al., 1992). Bechtel and Jayaram (1997) provided an extensive retrospective review of the literature and research on SCM. They identified generic schools of thought, and the major contributions and fundamental assumptions of SCM that must be challenged in the future.

Until recently, most practitioners (Davis, 1993; Lee et al., 1993; Arntzen et al., 1995; Lee and Billington, 1995; Camp and Colbert, 1997), consultants (Scharlacken, 1998; Tyndall et al., 1998) and academics (Lee and Billington, 1992; Bowersox and Closs, 1996; Fisher, 1997; Sheffi and Klaus, 1997; Handfield and Nichols, 1999) viewed SCM as not appreciably different from the contemporary understanding of logistics management, as defined by the Council of Logistics Management in 1986.[†] That is, SCM was viewed as logistics outside the firm to include customers and suppliers. Logistics as defined by the Council of Logistics Management always represented a supply chain orientation, "from point of origin to point of consumption." Then why the confusion? It is probably due to the fact that logistics is a functional silo within companies and is also a bigger concept that deals with the management of material and information flows across the supply chain. This is similar to the confusion over marketing as a concept and marketing

[*]Previously the Research Roundtable of The International Center for Competitive Excellence, University of North Florida. In 1996, this group moved to the Ohio State University and became The Global Supply Chain Forum.

[†]The Council of Logistics Management defined logistics management as: "The process of planning, implementing, and controlling the efficient, cost-effective flow and storage of raw materials, in-process inventory, finished goods, and related information flow from point-of-origin to point-of-consumption for the purpose of conforming to customer requirements" (Council of Logistics Management, 1986).

as a functional area. Thus, the quote from the chief executive officer: "Marketing is too important to be left to the marketing department." Everybody in the company should have a customer focus. The marketing concept does not apply just to the marketing department. It is everybody's responsibility to focus on serving the customer's needs.

The understanding of SCM has been reconceptualized from integrating logistics across the supply chain to the current understanding of integrating and managing key business processes across the supply chain (Cooper et al., 1997b). Based on this emerging distinction between SCM and logistics, in October 1998 the Council of Logistics Management announced a modified definition of logistics. The modified definition explicitly declares the Council's position that logistics management is only a part of SCM. The revised definition is: "Logistics is that part of the supply chain process that plans, implements, and controls the efficient, effective flow and storage of goods, services, and related information from the point-of-origin to the point-of-consumption in order to meet customers' requirements."*

Imagine the degree of complexity required to manage all suppliers back to the point of origin and all products/services out to the point of consumption. It is probably easier to understand why executives would want to manage their supply chains to the point of consumption because whoever has the relationship with the end user has the power in the supply chain. Intel created a relationship with the end user by having computer manufacturers place an "Intel inside" label on their computers. This affects the computer manufacturer's ability to switch microprocessor suppliers. But managing all tier 1 suppliers' networks to the point of origin is an enormous undertaking. Managing the entire supply chain is a very difficult and challenging task, as illustrated in Figure 2.

3. The marketing perspective

The early marketing channel researchers such as and Bucklin (1966) conceptualized the key factors for why and how channels are created and structured. From a supply chain standpoint these researchers were on the right track in the areas of: (1) identifying who should be a member of the marketing channel; (2) describing the need for channel co-ordination; and (3) drawing actual marketing channels. However, for the last 30 years many channels researchers have ignored two critical issues. First, they did not build on the early contributions by including suppliers to the manufacturer, and thus neglected the importance of

---

*Presented at the annual business meeting, Council of Logistics Management in October, 1998. The definition is posted at the Council of Logistics Management's homepage (http://www.CLM1.org).

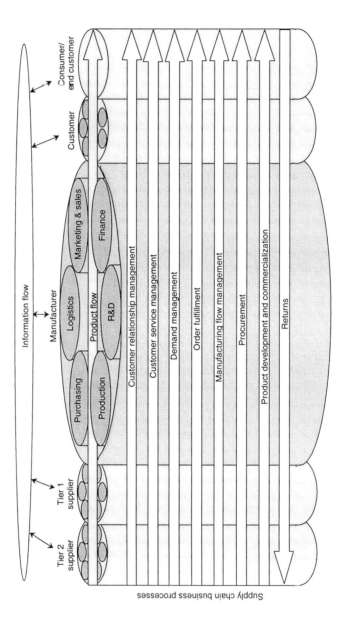

Figure 1. Interating and managing business processes across the supply chain (source: Cooper et al., 1997b).

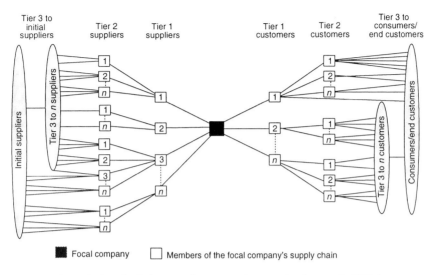

Figure 2. Supply chain network structure (source: Lambert et al., 1998a).

a total supply chain perspective. Second, they focused on marketing activities and flows across the channel, and overlooked the need to integrate and manage multiple key processes within and across companies. More recently, Webster (1992) challenged marketers and marketing researchers to consider relationships with multiple firms. He also called for cross-functional consideration in strategy formulation.

Unlike the marketing channels literature, a major weakness of much of the SCM literature is that the authors appear to assume that everyone knows who is a member of the supply chain. There has been little effort to identify specific supply chain members, key processes that require integration, or what management must do to successfully manage the supply chain.

4. A conceptual framework of supply chain management

The conceptual framework emphasizes the interrelated nature of SCM and the need to proceed through several steps to design and successfully manage a supply chain. The SCM framework consists of three closely interrelated elements: the supply chain network structure, the supply chain business processes, and the SCM components (Figure 3).

The supply chain network structure is the member firms and the links between these firms. Business processes are the activities that produce a specific output of value to the customer. The management components are the managerial variables

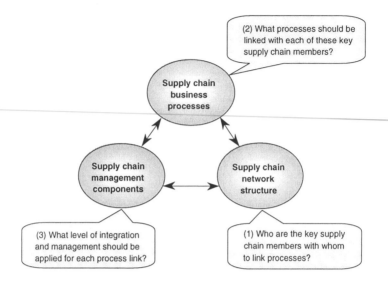

Figure 3. Elements and key decisions in the SCM framework (source: Cooper et al., 1997b).

by which the business processes are integrated and managed across the supply chain. Each of the interrelated elements that constitute the framework is described below.

4.1. Supply chain network structure

All firms participate in a supply chain from the raw materials to the ultimate consumer. How much of this supply chain needs to be managed depends on several factors, such as the complexity of the product, the number of available suppliers, and the availability of raw materials. Dimensions to consider include the length of the supply chain and the number of suppliers and customers at each level. It would be rare for a firm to participate in only one supply chain. For most manufacturers, the supply chain looks less like a pipeline or chain than an uprooted tree where the branches and roots are the extensive network of customers and suppliers. The question is how many of these branches and roots need to be managed.

The closeness of the relationship at different points in the supply chain will differ. Management will need to choose the level of partnership appropriate for particular supply chain links (Lambert et al., 1996a,b). Not all links throughout the supply chain should be closely co-ordinated and integrated. The most appropriate relationship is the one that best fits the specific set of circumstances (Cooper and

Gardner, 1993). Determining which parts of the supply chain deserve management attention must be weighed against firm capabilities and the importance to the firm.

It is important to have an explicit knowledge and understanding of how the supply chain network structure is configured. We suggest that three primary structural aspects of a company's network structure are: the members of the supply chain, the structural dimensions of the network, and the different types of process links across the supply chain. Each issue is addressed separately below.

4.2. Identifying supply chain members

When determining the network structure, it is necessary to identify who the members of the supply chain are. Including all types of members may cause the total network to become highly complex, since it may explode in the number of members added from tier level to tier level. To integrate and manage all process links with all members across the supply chain would, in most cases, be counterproductive, if not impossible. The key is to sort out some basis for determining which members are critical to the success of the company and the supply chain, and thus should be allocated managerial attention and resources.

Marketing channels researchers identified members of the channel based on who takes part in the various marketing flows, including product, title, payment, information, and promotion flows (Stern and El-Ansary, 1995). Each flow included relevant members, such as banks for the payment flow and advertising agencies for the promotion flow. The channel researchers sought to include all members taking part in the marketing flows, regardless of how much impact each member had on the value provided to the end customer or other stakeholders.

The members of a supply chain include all companies/organizations with whom the focal company interacts directly or indirectly through its suppliers or customers, from point of origin to the point of consumption. However, to make a very complex network more manageable it seems appropriate to distinguish between primary and supporting members. The definitions of primary and supporting members are based on interviews and discussions with the members of The Global Supply Chain Forum, and by applying the definition of a business process proposed by Davenport (1993). *Primary members* of a supply chain are: all those autonomous companies or strategic business units who carry out value-adding activities (operational and/or managerial) in the business processes designed to produce a specific output for a particular customer or market.

In contrast, *supporting members* are companies that simply provide resources, knowledge, utilities, or assets for the primary members of the supply chain. For example, supporting companies include those that lease trucks to the manufacturer, banks that lend money to a retailer, the owner of the building that provides warehouse space, or companies that supply production equipment, print marketing brochures, or temporary secretarial assistance. These supply chain members support the primary members now and in the future.

The same company can perform both primary and supportive activities. Likewise, the same company can perform primary activities related to one process and supportive activities related to another process. An example from one of the case studies is an original equipment manufacturer (OEM) that buys some critical and complex production equipment from a supplier. When the OEM develops new products they work very closely with the equipment supplier to assure there is the right equipment to make the new product. Thus the supplier is a primary member of the OEM's product development process. However, once the machinery is in place, the supplier is a supportive, not a primary, member for the manufacturing flow management process, since supplying the equipment does not in itself add value to the output of the process, even though the equipment itself adds value.

It should be noted that the distinction between primary and supporting supply chain members is not obvious in all cases. Nevertheless, this distinction provides a reasonable managerial simplification and yet captures the essential aspects of who should be considered as key members of the supply chain. The approach for differentiating between types of members is to some extent similar to how Porter (1984) distinguished between primary and support activities in his "value chain" framework.

The definitions of primary and supporting members make it possible to define the point of origin and the point of consumption of the supply chain. The *point of origin* of the supply chain occurs where no previous primary suppliers exist. All suppliers to the point of origin members are solely supporting members. The *point of consumption* is where no further value is added, and the product and/or service is consumed.

4.3. The structural dimensions of the network

Three structural dimensions of the network are essential when describing, analyzing, and managing the supply chain. These dimensions are the horizontal structure, the vertical structure, and the horizontal position of the focal company within the end points of the supply chain.

The *horizontal structure* refers to the number of tiers across the supply chain. The supply chain may be long, with numerous tiers, or short, with few tiers. As an example, the network structure for bulk cement is relatively short. Raw materials are taken from the ground, combined with other materials, moved a short distance, and used to construct buildings. The *vertical structure* refers to the number of suppliers/customers represented within each tier. A company can have a narrow vertical structure, with few companies at each tier level, or a wide vertical structure, with many suppliers and/or customers at each tier level. The third structural dimension is the company's *horizontal position within the supply chain*. A company can be positioned at or near the initial source of supply, be at or near to the ultimate customer, or somewhere between these endpoints of the supply chain.

In the companies studied, different combinations of these structural variables were found. In one example, a narrow and long network structure on the supplier side was combined with a wide and short structure on the customer side. Increasing or reducing the number of suppliers and/or customers will affect the structure of the supply chain. For example, as some companies move from multiple to single source suppliers, the supply chain may become narrower. Out-sourcing of logistics, manufacturing, marketing, or product development activities is another example of decision-making that is likely to change the supply chain structure. It may increase the length and width of the supply chain, and likewise influence the horizontal position of the focal company in the supply chain network.

Supply chains that burst to many tier 1 customers/suppliers will strain the resources for how many process links the focal company can integrate and closely manage beyond tier 1. In general, our research team has found that companies with immediately wide vertical structures actively managed only a few tier 2 customers or suppliers. Some of the companies studied have transferred the servicing of small customers to distributors, thus moving the small customers further down the supply chain from the focal company. This principle, known as *functional spin-off*, is described in the channel literature, and can be applied to the focal company's network of suppliers (Stern and El-Ansary, 1995).

In the companies studied, the supply chains looked different from each company's perspective, since the management of each company sees its firm as the focal company, and views the membership and network structure differently. However, because each firm is a member of the other's supply chain, it is important for the management of each firm to understand their interrelated roles and perspectives. The reason for this is that the integration and management of business processes across company boundaries will be successful only if it makes sense from each company's perspective (Cooper et al., 1997a)

5. Supply chain business processes

Successful SCM requires a change from managing individual functions to integrating activities into key supply chain processes. Traditionally, both upstream and downstream portions of the supply chain have interacted as disconnected entities receiving sporadic flows of information over time.

The purchasing department placed orders as requirements became necessary, and marketing, responding to customer demand, interfaced with various distributors and retailers and attempted to satisfy this demand. Orders were periodically given to suppliers, and the suppliers had no visibility at the point of sale or use. Satisfying the customer often translated into demands for expedited operations throughout the supply chain as member firms reacted to unexpected changes in demand.

Operating an integrated supply chain requires continuous information flows, which in turn help to create the best product flows. The customer remains the primary focus of the process. Achieving a good customer-focused system requires processing information both accurately and in a timely manner for quick response systems that require frequent changes in response to fluctuations in customer demand. Controlling uncertainty in customer demand, manufacturing processes, and supplier performance are critical to effective SCM.

In many major corporations, such as 3M, management has reached the conclusion that the optimization of product flows cannot be accomplished without implementing a process approach to the business. The key supply chain processes identified by members of The Global Supply Chain Forum are:

(1) customer relationship management,
(2) customer service management,
(3) demand management,
(4) order fulfillment,
(5) manufacturing flow management,
(6) procurement,
(7) product development and commercialization, and
(8) returns.

These processes are shown in Figure 1.

5.1. Customer relationship management process

The first step towards integrated SCM is to identify key customers or customer groups, which the organization targets as critical to its business mission. Product and/or service agreements specifying the levels of performance are established with these key customer groups. Customer service teams work with customers to further identify and eliminate sources of demand variability. Performance evaluations are undertaken to analyze the levels of service provided to customers as well as customer profitability.

5.2. Customer service management process

Customer service provides the single source of customer information. It becomes the key point of contact for administering the product/service agreement. Customer service provides the customer with real-time information on promised shipping dates and product availability through interfaces with the organization's production and distribution operations. Finally, the customer service group must be able to assist the customer with product applications.

5.3. Demand management process

Hewlett-Packard's experience with SCM indicates that inventory is either essential or variability driven (Davis, 1993). Essential inventory includes work in process in factories and products in the pipeline moving from location to location. Variability stock is present due to variance in process, supply, and demand. Customer demand is by far the largest source of variability and it stems from irregular order patterns. Given this variability in customer ordering, demand management is a key to effective SCM.

The demand management process must balance the customer's requirements with the firm's supply capabilities. Part of managing demand involves attempting to determine what and when customers will purchase. A good demand management system uses point-of-sale and "key" customer data to reduce uncertainty and provide efficient flows throughout the supply chain. Marketing requirements and production plans should be co-ordinated on an enterprise-wide basis. Thus, multiple sourcing and routing options are considered at the time of order receipt, which allows market requirements and production plans to be co-ordinated on an organization-wide basis. In very advanced applications, customer demand and production rates are synchronized to manage inventories globally.

5.4. Customer order fulfillment process

The key to effective SCM is meeting customer-need dates. It is important to achieve high order-fill rates either on a line item or an order basis. Performing the order fulfillment process effectively requires integration of the firm's manufacturing, distribution, and transportation plans. Partnerships should be developed with key supply chain members and carriers to meet customer requirements and reduce the total delivered cost to customer. The objective is to develop a seamless process from the supplier to the organization and then on to its various customer segments.

5.5. Manufacturing flow management process

The manufacturing process in make-to-stock firms traditionally produced and supplied products to the distribution channel based on historical forecasts.

Products were pushed through the plant to meet a schedule. Often the wrong mix of products was produced, resulting in unneeded inventories, excessive inventory carrying costs, mark downs, and trans-shipments of product.

With SCM, product is pulled through the plant based on customer needs. Manufacturing processes must be flexible to respond to market changes. This requires the flexibility to perform rapid changeover to accommodate mass customization. Orders are processed on a just-in-time basis in minimum lot sizes. Production priorities are driven by required delivery dates. At 3M, manufacturing planners work with customer planners to develop strategies for each customer segment. Changes in the manufacturing flow process lead to shorter cycle times, meaning improved responsiveness to customers.

5.6. Procurement process

Strategic plans are developed with suppliers to support the manufacturing flow management process and development of new products. Suppliers are categorized based on several dimensions, such as their contribution and criticality to the organization. In companies where operations extend worldwide, sourcing should be managed on a global basis.

Long-term partnerships are developed with a small core group of suppliers. The desired outcome is a win–win relationship, where both parties benefit. This is a change from the traditional bid-and-buy system to involving a key supplier early in the design cycle, which can lead to dramatic reduction in product development cycle times. Having early supplier input reduces time by getting the required co-ordination between engineering, purchasing, and the supplier prior to design finalization.

The purchasing function develops rapid communication mechanisms such as electronic data interchange and internet linkages to quickly transfer requirements. These rapid communication tools provide a means to reduce the time and cost spent on the transaction portion of the purchase. Purchasers can focus their efforts on managing suppliers, as opposed to placing orders and expediting. This also has implications for the role of the sales force when orders are not placed through the sales person.

5.7. Product development and commercialization

If new products are the lifeblood of a corporation, then product development is the lifeblood of a company's new products. Customers and suppliers must be integrated into the product development process in order to reduce time to market. As product life cycles shorten, the right products must be developed and successfully launched in ever shorter time frames in order to remain competitive.

Managers of the product development and commercialization process must:

(1) co-ordinate with customer relationship management to identify customer articulated and unarticulated needs,
(2) select materials and suppliers in conjunction with procurement, and
(3) develop production technology in manufacturing flow in order to manufacture and integrate into the best supply chain flow for the product–market combination.

5.8. Returns process

Managing the returns channel as a business process offers the same opportunity to achieve a sustainable competitive advantage as managing the supply chain from an outbound perspective (Clendein, 1997). In many countries this may be an environmental issue, but this is not always the case. Effective process management of the returns channel enables identification of productivity improvement opportunities and breakthrough projects.

At Xerox, returns are managed in four categories: equipment, parts, supplies, and competitive trade-ins. "Return to available" is a velocity measure of the cycle time required to return and asset to a useful status. This metric is particularly important for those products where customers are given an immediate replacement in the case of product failure. Also, equipment destined for scrap and waste from manufacturing plants is measured in terms of the time until cash is received.

6. Types of business process links

As noted earlier, integrating and managing all business process links throughout the entire supply chain is likely not appropriate. Since the drivers for integration are situational and different from process link to process link, the levels of integration should vary from link to link, and over time. Some links are more critical than others. As a consequence, the task of allocating scarce resources among the different business process links across the supply chain becomes crucial. The Global Supply Chain Forum research indicates that four fundamentally different types of business process links can be identified between members of a supply chain. These are: managed business process links, monitored business process links, not-managed business process links, and non-member business process links.

6.1. Managed process links

Managed process links are links that the focal company finds important to integrate and manage. In the supply chain shown in Figure 4, the managed process links are indicated by the thickest solid lines. The focal company will integrate and manage process links with tier 1 customers and suppliers. As indicated by the remaining thick solid lines in Figure 4, the focal company is actively involved in the management of a number of other process links beyond tier 1.

6.2. Monitored process links

Monitored process links are not as critical to the focal company as are managed process links. However, it is important to the focal company that these process links are integrated and managed appropriately between the other member companies. Thus, the focal company, as frequently as necessary, simply monitors or audits how the process link is integrated and managed. The thick dashed lines in Figure 4 indicate the monitored process links.

6.3. Not-managed process links

Not-managed process links are links that the focal company is not actively involved in, and that are not critical enough to justify the use of resources for monitoring. In other words, the focal company fully trusts the other members to manage the process links appropriately or, because of limited resources, leaves it up to them. The thin solid lines in Figure 4 indicate the not-managed process links. For example, a manufacturer has a number of potential suppliers for cardboard shipping cartons. Usually the manufacturer will not choose to integrate and manage the links beyond the cardboard supplier all the way back to the growing of the trees. The manufacturer wants certainty of supply, but it may not be necessary to integrate and manage the links beyond the cardboard supplier.

6.4. Non-member process links

The case studies indicated that managers are aware that their supply chains are influenced by decisions made in other connected supply chains. For example, a supplier to the focal company is also a supplier to the chief competitor, which may have implications for the supplier's allocation of manpower to the focal company's product development process, availability of products in times of shortage, and/or protection of confidentiality of information. Non-member process links are

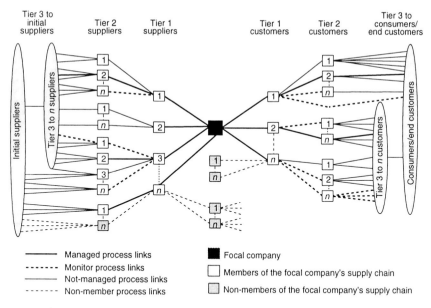

Figure 4. Types of intercompany business links (source: Lambert et al., 1998a).

process links between members of the focal company's supply chain and non-members of the supply chain. Non-member links are not considered as links of the focal company's supply chain structure, but they can and often will affect the performance of the focal company and its supply chain. The thin dashed lines in Figure 4 illustrate examples of non-member process links.

Based on the process links just described, our research reveals variation in how closely companies integrate and manage links further away from the first tier. In some cases, companies work through or around other members/links in order to achieve specific supply chain objectives, such as product availability, improved quality, or reduced overall supply chain costs. For example, a tomato ketchup manufacturer in New Zealand conducts research on tomatoes in order to develop plants that provide larger tomatoes with fewer seeds. Their contracted growers are provided with young plants in order to ensure the quality of the output. Since the growers tend to be small, the manufacturer further negotiates contracts with suppliers of equipment and supplies such as fertilizer and chemicals. The farmers are encouraged to purchase their raw materials and machinery using the contract rates. This results in higher quality raw materials and lower prices, without sacrificing the margins and financial strength of the growers.

There are several examples of companies who, in times of shortage, discovered that it was important to manage beyond tier 1 suppliers for critical times. One example involves a material used in the manufacture of semiconductors. It turned out that the six tier 1 suppliers all purchased from the same tier 2 supplier. When shortages occurred it became apparent that the critical relationship was with the tier 2 supplier. It is important to identify the critical links in the supply chain, and these may not be the immediately adjacent firms.

7. Business process chains

Davenport (1993) defined a process as "a structured and measured set of activities designed to produce a specific output for a particular customer or market". A process can be viewed as a structure of activities designed for action with a focus on end customers and on the dynamic management of flows involving products, information, cash, knowledge, and/or ideas.

Thousands of activities are performed and co-ordinated within a company, and every company is by nature in some way involved in supply chain relationships with other companies. When two companies build a relationship, certain of their internal activities will be linked and managed between the two companies. Since both companies have linked some internal activities with other members of their supply chain, a link between two companies is thus a link in what might be conceived of as a supply chain network. For example, the internal activities of a manufacturer are linked with and can affect the internal activities of a distributor, which in turn are linked with and can have an effect on the internal activities of a retailer. Ultimately, the internal activities of the retailer are linked with and can affect the activities of the end customer.

The results of empirical research by Håkansson and Snehota (1995) stress that "the structure of activities within and between companies is a critical cornerstone of creating unique and superior supply chain performance". In this research the executives believed that competitiveness and profitability could increase if internal key activities and business processes are linked and managed across multiple companies. Thus, "Successful supply chain management requires a change from managing individual functions to integrating activities into key supply chain business processes" (Lambert et al., 1997).

Our research team has found that in some companies executives emphasize a functional structure, others a process structure, and others a combined structure of processes and functions. Those companies with processes had different numbers of processes consisting of different activities and links between activities. Different names were used for similar processes, and similar names for different processes. This lack of intercompany consistency is a cause for significant friction and inefficiencies in supply chains. At least with functional silos, there is generally an understanding of what functions such as marketing, manufacturing and accounting/finance represent. If each firm identifies its own set of processes, how can these processes be linked across firms? A simplified illustration of such a disconnected supply chain is shown in Figure 5.

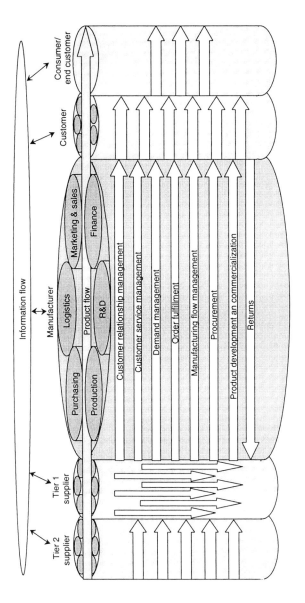

Figure 5. The disconnects in SXM (source: Lambert et al., 1998a).

The primary focus thus far has been on determining processes internal to the company. It is necessary to address which processes are critical and/or beneficial to integrate and manage across the supply chain. In the case study companies, it became clear that in some cases the internal business processes have been extended to suppliers and managed to some extent between the two companies involved. This may imply that when a leadership role is taken, a company's internal business processes can become the supply chain business processes. The obvious advantage when this is possible is that each member of the band is playing the same tune.

The number of business processes that it is critical and/or beneficial to integrate and manage between companies will likely vary. In some cases it may be appropriate to link just one key process, and in other cases it may be appropriate to link multiple or all key business processes. However, in each specific case, it is important that executives thoroughly analyze and discuss which key business processes to integrate and manage. The major components for integrating and managing a supply chain network are addressed below.

8. The management components of supply chain management

The SCM management components are the third element of the SCM framework (see Figure 3). The level of integration and management of a business process link is a function of the number and level, ranging from low to high, of componets added to the link. Consequently, adding more management components or increasing the level of each component can increase the level of integration of the business process link.

The literature on business process re-engineering (Hammer and Champy, 1993; Hewitt, 1994), buyer–supplier relationships (Stevens, 1989), and SCM (Lambert et al., 1996b) suggests numerous possible components that must receive managerial attention when managing supply relationships. Based on the management components identified in our previous work, review of the literature, and interviews with 80 managers, nine management components are identified for successful SCM: planning and control, work structure, organization structure, product flow facility structure, information flow facility structure, management methods, power and leadership structure, risk and reward structure, and culture and attitude. These components are briefly described below.

Planning and control of operations are keys to moving an organization or supply chain in a desired direction. The extent of joint planning is expected to bear heavily on the success of the supply chain. Different components may be emphasized at different times during the life of the supply chain, but planning transcends the phases. The control aspects can be operationalized as the best performance metrics for measuring supply chain success.

The *work structure* indicates how the firm performs its tasks and activities. The level of integration of processes across the supply chain is a measure of *organization structure*. All but one of the literature sources examined cited work structure as an important component. Organizational structure can refer to the individual firm and the supply chain; the use of cross-functional teams would suggest more of a process approach. When these teams cross organizational boundaries, such as in-plant supplier personnel, the supply chain should be more integrated.

Product flow facility structure refers to the network structure for sourcing, manufacturing, and distribution across the supply chain. Since inventory is necessary in the system, some supply chain members may keep a disproportionate amount of inventory. As its is less expensive to have unfinished or semi-finished goods in inventory than finished goods, upstream members may bear more of this burden. Rationalizing the supply chain network has implications for all members.

Virtually every author indicates that the *information flow facility structure* is key. The kind of information passed among channel members and the frequency of information updating has a strong influence on the efficiency of the supply chain. This may well be the first component integrated across part or all of the supply chain.

Management methods include the corporate philosophy and management techniques. It is very difficult to integrate a top-down organization structure with a bottom-up structure. The level of management involvement in day-to-day operations can differ across supply chain members.

The *power and leadership structure* across the supply chain will affect its form. One strong leader will drive the direction of the chain. In most supply chains studied to date, there are one or two strong leaders among the firms. The exercise of power, or lack of, can affect the level of commitment of other members. Forced participation will encourage exit behavior, given the opportunity (Macneil, 1980). The anticipation of sharing of *risks and rewards* across the supply chain affects long-term commitment of its members.

The importance of corporate culture and its compatibility across members of the supply chain cannot be underestimated. Meshing *cultures and individuals' attitudes* is time consuming, but is necessary at some level for the channel to perform as a chain. Aspects of culture include how employees are valued and incorporated in the management of the firm.

Figure 6 illustrates how the management components can be divided into two groups. The first group is the physical and technical group, which includes the most visible, tangible, measurable, and easy to change components. This research, and much literature on change management, shows that if this group of management components is the only focus of managerial attention, the results will be disappointing, at best (Jaffe and Scott, 1998; Hammer and Champy, 1993).

The second group is composed of the managerial and behavioral components. These components are less tangible and visible and are often difficult to assess and alter. The managerial and behavioral components define the organizational behavior and influence how the physical and technical management components can be implemented. If the managerial and behavioral components are not aligned to drive and reinforce an organizational behavior supportive to the supply chain objectives and operations, the supply chain will likely be less competitive and profitable. If one or more components in the physical and technical group are changed, management components in the managerial and behavioral group likewise may have to be readjusted. The groundwork for successful SCM is established by understanding each of these SCM components and their interdependence. Hewitt (1994) stated that true intra- and intercompany business process management, or redesign, is only likely to be successful if it is recognized as a multicomponent change process, simultaneously and explicitly addressing all SCM components.

All nine management components were found in the business process links studied, including examples of successful SCM applications. However, the number and levels of components and combinations of representations varied. A finding is that the physical and technical components were well understood and managed the farthest up and down the supply chain. For example, in one case, the focal company had integrated its demand management process across four links by applying the following components: planning and control methods, work flow/activity structure, communication and information flow facility structure, and product flow facility structure. The managerial and behavioral management

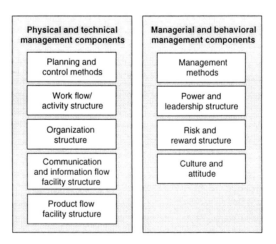

Figure 6. Fundamental management components in SCM (source: Lambert et al., 1998a).

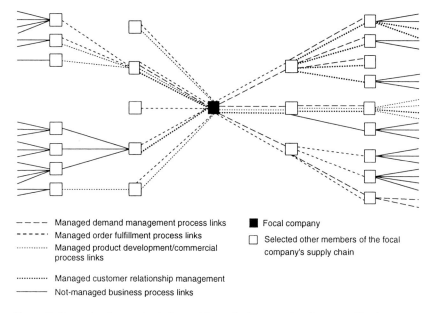

Figure 7. Example of a supply chain combining the integrated and managed business process links (source: Lambert et al., 1998a).

components were, in general, less well understood and more difficulties were encountered in their implementation. Only one example was found of managerial and behavioral management components applied to more than one link across the supply chain.

9. Mapping the supply chain

In the companies studied, the business processes were not linked across the same firms. In other words, different business processes had different looking supply chain network structures. An example is a focal company that involves supplier A, but not supplier B, in its product development process, whereas the demand management process is linked with both suppliers. Management will choose to integrate and manage different supply chain links for different business processes.

Figure 7 illustrates how the integrated and managed business process links of a focal company may differ from process to process. For simplicity, only the managed and not-managed business process links are illustrated. The monitored and non-member process links are omitted. Also, only a very few supply chain members are included. The superimposed supply chains of four individual business process chains are shown in one diagram. It is necessary first to map individual processes and then to superimpose them on one supply chain map. It is suggested that managers use this approach when mapping their supply chains.

Previous literature has indicated that some or all business processes should be linked across the supply chain, from the initial source of supply to the ultimate end customer. In this research, there were no examples of this, nor were there any in the cases described in the literature. In fact, the companies studied had only integrated some selected key process links, and were likewise only monitoring some other selected links.

10. Re-engineering improvement into the supply chain

A critical part of streamlining supply chains involves re-engineering the firm's key processes to meet customer needs. Re-engineering is a process aimed at producing dramatic changes quickly. Hammer and Champy (1993) define it as the fundamental rethinking and radical redesign of business processes to achieve dramatic improvements in critical contemporary measures of performance, such as cost, quality service, and speed. Improvement through re-engineering cannot be accomplished in a haphazard manner. These changes must be supported at the top and driven through an overall management plan.

A typical re-engineering process proceeds through three stages: fact finding, identifying areas for improvement to business process redesign, and creative improvements. The fact-finding stage is a very detailed examination of the current systems, procedures, and workflows. The key focus is placed on separating facts from opinions.

Armed with the facts collected in the first stage, re-engineering teams identify areas for improvement. They analyze where value was added for the final customer, with particular emphasis on customer contact points and product information transfers that are currently ineffective or inefficient. After identifying improvement points the creative phase of redesigning business process and information flow begins. The outcomes of the creative phase will fundamentally change both the nature of the work and how it is performed.

Figure 8 illustrates a general plan when undertaking a process re-engineering approach. Organizational energy needs to focus on the firm's mission statement. The mission statement drives the business requirements in the organization. A complete assessment is made of the firm's culture, strategies, business practices, and processes.

If this analysis proves acceptable, management implements its business solution across the supply chain. Typically, improvements are required in one of the areas to enhance supply chain performance. An example of this re-engineering is the

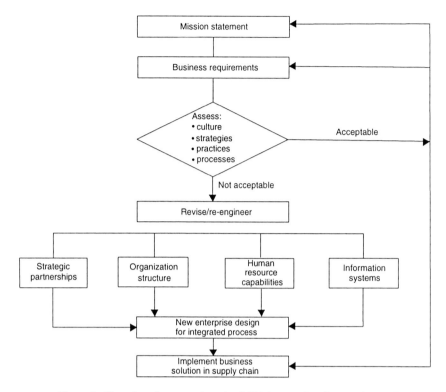

Figure 8. Flow chart for re-engineering SCM (source: Lambert et al., 1997).

new Mercedes-Benz micro car, which is based on the principle of systems supply (Coleman et al., 1995). This re-engineering of the process results in delegating more design activities to suppliers reducing the amount of engineering and labor at the primary manufacturer. The result is passing the savings of these efficiencies along to the customer in the form of increased value.

11. Implementing integrated supply chain management

Implementing SCM requires making the transition from a functional organization to a focus on process, first inside the firm and then across firms in the supply chain. Figure 9 illustrates how each function within the organization maps with the seven key processes.

In the customer relationship management process, sales and marketing provides the account management expertise, engineering provides the specifications which define the requirements, logistics provides knowledge of

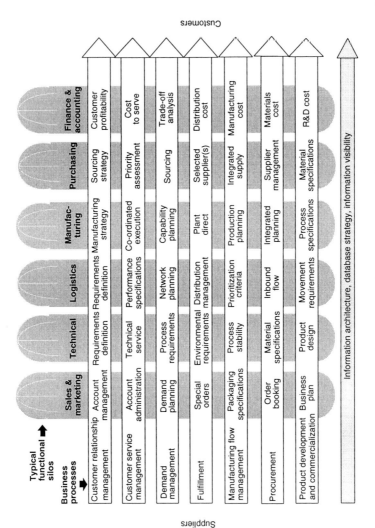

Figure 9. Implementation of SCM. Process sponsorship and ownership must be established to drive the attainment of the supply chain vision and eliminate the functional barriers that artificially separate the process flows.

customer service requirements, manufacturing provides the manufacturing strategy, purchasing provides the sourcing strategy, and finance and accounting provides customer profitability reports. The customer service requirements must be used as input to manufacturing, sourcing, and logistics strategies.

If the proper co-ordination mechanisms are not in place across the various functions, the process will be neither effective nor efficient. By taking a process focus, all functions that touch the product or provide information must work together. For example, purchasing depends on sales/marketing data fed through a production schedule to assess specific order levels and timing of requirements. These orders drive production requirements, which in turn are transmitted upstream to suppliers.

The increasing use of out-sourcing has accelerated the need to co-ordinate supply chain processes, since the organization becomes more dependent on outside contractors suppliers. Consequently, co-ordination mechanisms must be in place within the organization. Where to place this co-ordination mechanisms and which team and functions are responsible become critical decisions.

There are several process redesign and re-engineering techniques that can be applied to the seven key processes. Chrysler Corporation's development of the Neon was accomplished through the efforts of 150 internal employees. This core group leveraged their efforts to 600 engineers, 289 suppliers, and line employees. Concurrent engineering techniques required the involvement of personnel from all key functional areas working with suppliers to develop the vehicle in 42 months. The use of concurrent engineering resulted in the avoidance of later disagreements, misunderstandings, and delays.

Typically, firms within the supply chain will have their own functional silos that must be overcome and a process approach accepted in order to successfully implement SCM. The requirements for successful implementation of SCM include:

(1) executive support, leadership and commitment to change;
(2) an understanding of the degree of change that is necessary;
(3) agreement on the SCM vision and the key processes; and
(4) the necessary commitment of resources and empowerment to achieve the stated goals.

12. Conclusions

Executives are becoming aware of the emerging paradigm of internetwork competition, and that the successful integration and management of key business processes across members of the supply chain will determine the ultimate success of the single enterprise. Managing the supply chain cannot be left to chance. For this reason, executives are striving to interpret and determine how to manage the company's supply chain network and how to achieve the potential of SCM.

Research with member firms of The Global Supply Chain Forum at Ohio State University indicates that managing the supply chain involves three closely interrelated elements: (1) the supply chain network structure; (2) the supply chain business processes; and (3) the management components. The structure of activities/processes within and between companies is vital for creating superior competitiveness and profitability. Successful SCM requires integrating business processes with key members of the supply chain. Much friction, and thus waste of valuable resources, results when supply chains are not integrated and are not appropriately streamlined and managed. A prerequisite for successful SCM is to co-ordinate activities within the firm. One way to do this is to identify the key business processes and manage them using cross-functional teams. Hopefully, this chapter provides clarification on key aspects of SCM that will aid practitioners and researchers in their desire to understand and implement SCM.

It is important to distinguish between primary and supporting supply chain members, and to identify the horizontal structure, the vertical structure, and the horizontal position of the focal company in the supply chain network. There are four fundamentally different types of business process links: managed business process links, monitored business process links, not-managed business process links, and non-member business process links.

Marketing researchers were in the forefront of studying critical aspects of what we now call SCM, particularly with respect to identifying the members of a channel of distribution. The focus was, for the most part, from the manufacturer to the customer. The approach to SCM presented here ensures inclusion of suppliers and customers. There are several implications for marketing practitioners and researchers. There is a need to integrate activities across the firm and across firms in the supply chain. While marketing strategy formulation has always considered internal and external constraints, SCM makes the explicit evaluation of these factors even more critical. In addition, traditional roles of marketing and sales people are changing. Team efforts are becoming more common for developing and marketing new products, as well as managing current ones. The role of the firm's sales force is changing to one of measuring and selling the value proposition for the customer.

In combination, the SCM definition and the new framework moves SCM philosophy to its next evolutionary stage. The implementation of SCM involves identifying the supply chain members, with whom it is critical to link, what processes need to be linked with each of these key members, and what type/level of integration applies to each process link. The objective of SCM is to create the most value, not simply for the company, but for the whole supply chain network including the end customer. Consequently, supply chain process integration and re-engineering initiatives should be aimed at boosting total process efficiency and effectiveness across members of the supply chain.

References

Arntzen, B.C., G.G. Brown, T.P. Harrison and L.L. Trafton (1995) "Global supply chain management digital equipment corporation", *Interfaces*, 25:69–93.

Bechtel, C. and J. Jayaram (1997) "Supply chain management: a strategic perspective", *International Journal of Logistics Management*, 8:15–34.

Bowersox, D.J. and D.J. Closs (1996) *Logistical management – the integrated supply chain process*. New York: McGraw-Hill.

Bucklin, L.P. (1966) *A theory of distribution channel structure*. Berkeley: IBER.

Camp, R.C. and D.N. Colbert (1997) "The Xerox quest for supply chain excellence", *Supply Chain Management Review*, Spring, 82–91.

Christopher, M.G. (1998) "Relationships and alliances: embracing the era of network competition", in: J. Gattorna, ed., *Strategic supply chain management*. Ashgate: Gower Press.

Clendein, J.A. (1997) "Closing the supply chain loop: reengineering the returns channel process", *International Journal of Logistics Management*, 8:75–85.

Coleman, J.L., A.K. Bhattacharya and G. Brace (1995) "Supply chain reengineering: a supplier's perspective", *The International Journal of Logistics Management*, 6:85–92.

Cooper, M.C. and J.T. Gardner (1993) "Good business relationships: more than just partnerships or strategic alliances", *International Journal of Physical Distribution and Logistics Management*, 23:14–20.

Cooper, M.C., L.M. Ellram, J.T. Gardner and A.M. Hanks (1997a) "Meshing multiple alliances", *Journal of Business Logistics*, 18:67–89.

Cooper, M.C., D.M. Lambert and J.D. Pagh (1997b) "Supply chain management: more than a new name for logistics", *International Journal of Logistics Management*, 8:1–13.

Council of Logistics Management (1986) *What's it all about?* Oak Brook: CLM.

Davenport, T.H. (1993) *Process innovation, reengineering work through information technology*. Cambridge, MA: Harvard Business School Press.

Davis, T. (1993) "Effective supply chain management", *Sloan Management Review*, 34:35–46.

Ellram, L.M. and M.C. Cooper (1993) "The relationship between supply chain management and keiretsu", *International Journal of Logistics Management*, 4:1–12.

Fisher, M.L. (1997) "What is the right supply chain for your product?", *Harvard Business Review*, 75:105–116.

Håkansson, H. and I. Snehota (1995) *Developing Relationships in Business Networks*. London: Routledge.

Hammer, M. and J. Champy (1993) *Reengineering the corporation: a manifesto for business revolution*. New York: Harper Business.

Handfield, R.B. and E.L. Nichols, Jr. (1999) *Introduction to supply chain management*. Upper Saddle River: Prentice Hall.

Hewitt, F. (1994) "Supply chain redesign", *International Journal of Logistics Management*, 5:1–9.

Jaffe, D.T. and C.D. Scott (1998) "Reengineering in Practice: Where are the people? Where is the learning", *Journal of Applied Behavioral Science*, 34:250–267.

La Londe, B.J. (1998) "Supply chain evolution by the numbers", *Supply Chain Management Review*, 2:7–8.

Lambert, D.M., M.A. Emmelhainz and J.T Gardner (1996a) "So you think you want a partner?", *Marketing Management*, 5:25–41.

Lambert, D.M., M.A. Emmelhainz and J.T. Gardner (1996b) "Developing and implementing supply chain partnership", *International Journal of Logistics Management*, 7:1–17.

Lambert, D.M., L.C. Guinipero and G.J. Ridenhower (1997) "Supply chain management: a key to achieving business excellence in the 21st century", unpublished.

Lee, H.L. and C. Billington (1992) "Managing supply chain inventory: pitfalls and opportunities", *Sloan Management Review*, 33:65–73.

Lee, H.L. and C. Billington (1995) "The evolution of supply chain management models and practice at Hewlett-Packard", *Interfaces*, 25:42–63.

Lee, H.L., C. Billington and B. Carter (1993) "Hewlett-Packard gains control of inventory and service through design for localization", *Interfaces*, 23:1–11.

Macneil, I.R. (1980) *The new social contract, and inquiry into modern contractual relations*. New Haven: Yale University Press.

Porter, M.E. (1984) *Competitive advantage – creating and sustaining superior performance*. New York: Free Press.

Scharlacken, J.W. (1998) "The seven pillars of global supply chain planning", *Supply Chain Management Review*, 2:32–40.

Sheffi, Y. and P. Klaus (1997) "Logistics at large; jumping the barriers of the logistics functions", in: J.M. Masters, ed., *Proceedings of the Twenty-sixth Annual Transportation and Logistics Educators Conference*. Chicago: Ohio State University.

Stern, L.W. and A. El-Ansar, (1995) *Marketing channels*, 5th edn. Englewood Cliffs: Prentice Hall.

Stevens, G.C. (1989) "Integration of the supply chain", *International Journal of Physical Distribution and Logistics Management*, 19:3–8.

Towill, D.R., M.M. Naim and J. Wikner (1992) "Industrial dynamics simulation models in the design of supply chains", *International Journal of Physical Distribution and Logistics Management*, 22:3–13.

Tyndall, G., C. Gopal, W. Partsch and J. Kamauff (1998) *Supercharging supply chains*. New York: Wiley.

Webster, Jr., F.E. (1992) "The changing role of marketing in the corporation", *Journal of Marketing*, 56:1–17.

Chapter 8

THE CONCEPT OF VALUE: SYMBOLIC ARTIFACT OR USEFUL TOOL?

ANN M. BREWER
University of Sydney

1. Introduction

While interest in value has been sustained over decades, it has relatively new appeal in logistics and supply chain thinking, increasingly since the publication of *The Machine that Changed the World*, an international study of the automobile industry (Jones et al., 1990). While organizations in most industry sectors, including transport and distribution, are embracing the value concept, there is still little agreement about what "value" means and even less about how to create it. Moreover, value as a notion remains abstract and falls short on the specifics of measurement or assessment. However, it is an important concept as it goes to the core of addressing key questions such as: Why do businesses exist? Why do some organizations perform and fare better than others? However, when a concept such as value metamorphoses into a prevailing management artifact, the question is: Has it reached the point where it is starting to become less useful? Is value an artifact, a symbol of the 21st century, or a useful tool for supply chain management?

1.1. Defining value

The study of value dates back to classical economists, such as Adam Smith, who distinguished between *value in use* and *exchange value*; that is, that amount of some commodity or medium of exchange that is considered to be an equivalent for something else; a fair or adequate equivalent or return (*Oxford Dictionary*). In other words, value is something that satisfies the demand of an agent. Classical economists see value as being an inherent part of the production of commodities. Value, that is *market value*, is created from an aggregate of the costs of the principal agents of production, namely land, labor, and capital, into a given product or service. Within this view, market value is quarantined from the performance of individuals. In comparison, the Austrian economists define value as the worth that

Handbook of Logistics and Supply Chain Management, Edited by A.M. Brewer et al.
© *2001, Elsevier Science Ltd*

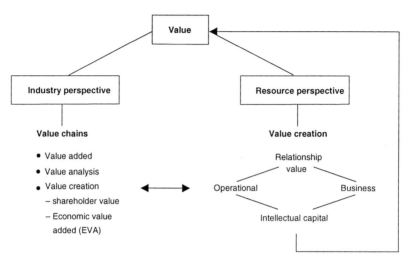

Figure 1. The concept of value.

a good or quantity of goods has for the well-being of a certain business (von Bohm-Bawerk, 1959). This view of value is similar to what is referred to as "value in use," and is the basis of market value defined as the subsequent interaction in the market of subjective estimations of commodities and the price of goods (von Mises, 1980). Subjective estimations refer to individuals deciding which products and services will benefit them and in what measure. This is referred to as *marginal utility value*; that is, the added benefit derived from each additional unit of product or service. The implications of the Austrian school on value is that it contests the classical theorists' view in that cost determines value and claims the opposite by saying costs are simply a reflection of market value.

1.2. Strategic value management

For most of the 1960s and 1970s, the concept of value was utilized in an "adding value" sense. *Added value* equals the total value created with the inclusion of a particular partner or action minus the total value created without a specific partner or action (based on Brandenburger and Nalebuff, 1996). Undoubtedly, the values of an organization are an important ingredient of its strategic position. From the 1980s onwards, attention to the link between value and competitiveness has strengthened and takes two theoretical pathways. The first pathway is the industry organization perspective (Caves et al., 1980; Porter, 1980) and the second is the resource-based view of the enterprise (Wernerfelt, 1984; Barney, 1991).

Section 2 of this chapter looks at how the concept of value has been defined, in the economic and strategic management literature, and, in turn, how this has informed the understanding of supply chain and logistics management. The concept of value opens up a new opportunity to review the time-honored concerns in the supply chain and logistics field. The chapter is divided into examining two perspectives of value: first, strategic value management, encompassing key concepts such as value added, value analysis, and creation; and, second, a resource-based view, which progresses the value debate from concept to practice. Section 3 reflects the structure depicted in Figure 1. Section 4 synthesizes all these value dimensions. The final section concentrates on the potential that the concept of value appears to embrace for supply chain management and points to the advantages that it proffers over more conventional planning and decision-making approaches.

2. The industry view

2.1. Value chain

The industry organization perspective is based on the concept of value chain, which refers to a connected series of links of core and secondary activities, comprising inbound logistics, operations, outbound logistics, marketing and sales, and service that lead to the business outcomes of each enterprise (Porter, 1980). The assumption underlying the value chain is that each activity either adds or removes value from the products or services at hand. This approach gave rise to two important notions. First, the earlier notion of *value added*; that is, the amount by which the value of an article is increased at each stage of its production by the agent or agents producing it, exclusive of the cost of materials and bought-in parts and services.

Secondly, a more recent notion of *value analysis*, the systematic and critical assessment of how to improve the relationship to the customer by focusing on total costs in relation to end-user value. A principal component of this approach today is supply chain mapping, whereby the focus is on measuring time and cost throughout the logistics pipeline (Christopher, 1998). Jones et al. (1997) offer a further critical insight into value chain strategy. They refer to an holistic lean logistics approach, which focuses on value streams across the entire supply chain rather than isolated parts of it. This approach leads to a number of value stream mapping tools, such as process activity and supply chain mapping, which facilitate the removal of waste (e.g., dissipated effort, duplication) and aids integration leading to value stream management (see Hines et al., 1998). Value is further created externally to the organization through a vertical supply chain linking suppliers of resources to customers (buyers) and through customers to consumers (end users). This approach is contingent upon each stakeholder in the supply chain adding value to the total output, that is *positive* value not *negative* value (see Davis and Kay, 1990).

2.2. Value creation

Still within the industry perspective, a third and crucial concept is *value creation*, which Porter (1980) maintains is the essential ingredient of competitive advantage based on two key strategies: low cost and differentiation. According to Porter's model, competitive advantage defined principally as higher profits, is achieved by creating higher value by driving down overall costs or providing customers with products and services that they value over the competitors' offerings, and subsequently are willing to pay a higher price for. This is termed *customer value* (Treacy and Wiersam, 1995) and is based on the subjective perceptions of the purchasers of goods and commodities. Creating value thereby rests on the understanding and interpretation of both customer perceptions and demands, as well as the capacity to build products and services with attributes that are deemed to be in the customer's interests, such as quality, efficiency, innovation, and responsiveness (Porter, 1980; Johansson et al., 1993). Values, in this sense, do not refer to reproduction costs, but to the exploration of customer values. This is referred to as a *value proposition*, and provides critical insight into a supply and demand chain strategy (Beech in Gattorna, 1998).

2.3. Shareholder value

Within the changing concept of value, the attitude towards the shareholders has also transformed substantially, as their expectations influence decisions on value creation. While the percentage of individual shareholders has increased, there has also been a rise in shareholder demands, especially among institutional shareholders (e.g., banks, fund management organizations). One of the ways that organizations have responded to boosting shareholder value is through a balanced scorecard approach that links together strategic, financial, and operational objectives (Kaplan and Norton, 1993). This approach attempts to respond to four performance arenas, including: shareholder interests, such as cash flow, sales and income growth, and return on investment; customer value; internal performance; and innovation and learning. Each of these arenas is allocated a performance scorecard so that there is a translation of business strategy into operational performance, ultimately impacting shareholder value.

According to Christopher (1998), economic value added (EVA) (i.e., operating income after tax, investment in assets, and cost of capital) is crucial to shareholder value. He argues that a negative EVA will erode shareholder value over time. One of the contributing factors to a negative EVA is non-integrated logistics pipelines, which are often accompanied by highly capital intensive facilities, leading to duplication and wastage. This realization has led to the shortening of pipelines, an integration of logistics functions, and a subsequent reduction in working capital requirements, consequently resulting in sharing, partnering, and/or out-sourcing resources (see Chapter 16). Brewer et al. (1999) argue that, although EVA alone cannot provide an organization with a sustainable source of competitive advantage, it does offer a valuable measure of wealth creation and can be used to help align managerial decision-making to strategic and operational choices. However, it is only one aspect of organizational performance measurement and needs to be used in conjunction with a balanced set of measures that provide a total framework of performance.

3. The resource-based view

The resource-based view seeks to address the issue of how to manage the supply chain to address *real* value creation, and in so doing provide a sustainable competitive advantage. Value here is seen in terms of the amount of resource as a medium of exchange, which is considered to be an equivalent for something else. Resources include not only market inputs (e.g. labor, transportation, equipment) and assets (facilities and tools owned by the organization), but also knowledge and capabilities (competencies) (Olavarrieta et al., 1997). In today's environment, where product life cycles are shortening, technologies are converging, and industry structures are becoming more diffuse, there is a growing belief that the key focus for building competitiveness no longer begins with market selection and positioning but with the development and nurturing of widely applicable distinctive internal capabilities that are relatively enduring (Rumelt, 1991). Creating value can only be attained if the stakeholders can exchange resources, share knowledge and skills, and build supply chain capability for the pursuit of goal achievement. This is at the core of competitive advantage. So why is this level of co-ordination so difficult to implement in practice?

3.1. Relationship value

To answer the above question, it is important to understand co-ordination from a relationship perspective. As previously stated, the development of relationships is central to supply chain management. Relationships can be based on a long-standing or transitory association, depending on whether it is the sharing of resources and costs and/or access to one another's markets, as well as the amount of trust that exists between the suppliers and buyers. Relationships are principally structured by how stakeholders perceive the potential benefits and outcomes as well as the risks. These perceptions do not occur by chance and are strongly influenced by the organizational values governing these relationships. It is important to consider value clusters in terms of their real impact on structural and operational characteristics, and in turn how these impact supply chain relationships and outcomes.

As Senge (1990) notes, the relationship between vision (the intended direction for a business) and its realization is frequently not strong. This difference is partially explained by the disparity between the values, norms, and artifacts that emerge among key actors in the supply chain. Effectiveness is normally the extent of the capacity to reconcile various value clusters in shaping performance. An ineffective culture is a function of poorly designed systems that are an inevitable result of erroneous perceptions of causality (Senge, 1990). This type of culture leads to a non-learning organization, which is full of design irrationalities by virtue of practitioners not being able to overcome the limitations of cognitive biases through systems thinking. These issues are crucial to value management. A number of value clusters are outlined below to provide some idea of the disparities between value clusters.

Strategic business values

Strategic business values guide the development of the supply chain and corporate strategy for navigating markets and customers. Strategic values are instrumental in the conception of the supply chain, organizing resources and people, structuring operations, and service delivery. Business strategy and actions communicate to the workforce which values are significant. Business values underpin strategic decision-making and provide the umbrella for a plethora of substructures such as planning, budgeting, distribution, and human resources. In other words, organizations represent multistructural contexts, and specific values are reflected in each of these. For example, the operational substructure is often at variance with the information substructure. Powerful belief systems are associated within these substructures (subcultures) and are mirrored in the way in which management structures the authority and work relationships (e.g., hierarchy) within the enterprise as well as external relationships (e.g., market).

Trying to manage the integration between an internal hierarchical approach and an external market one is a huge challenge in value management. The "market" culture is an example of a dynamic network that emphasizes the need to respond rapidly to customer demand. The market-based culture assumes direct relationships based on exchange of resources. Hierarchical values are centralized and partitioned by function (e.g., distribution, human resources, information). These value clusters have a significant impact on value management. In keeping with the concept of value chain, enterprise management is responding to an intensifying competitive global enterprise environment by moving away from centrally co-ordinated, multilevel hierarchies towards redesigning flexible structures that more closely represent networks (e.g., customers, distributors, suppliers, enterprises) rather than conventional enterprise pyramids. These networked structures are at the heart of relationship management. The philosophy underlying the development of global and hybrid enterprises is that they are more effective in responding to contingencies created by changing markets, technology, information, and logistics demands. For this reason network organizations, virtual corporations, hybrid organizations, and partnerships will be prominent forms of value-creating relationships in the 21st century.

A "value net" framework emerges by mapping stakeholders on their relative impact to an industry's value chain (Brandenburger and Nalebuff, 1996). Stakeholder collaboration, understood in the context of this chapter so far, is about creating value for an entire network of stakeholders by working to develop effective forms of co-operation, decentralizing power and authority, and building consensus among stakeholders (Burton and Dunn, 1996). To that end, increasing competitive pressures are fuelling concern over the extent to which different stakeholders can integrate successfully into a value partnership. One perspective is that the relationship is less likely to be fostered when exposed to competition, instead persisting in leveraging their existing market position to obtain competitive advantage (Barnett and Burgelman, 1996).

With the trend towards disaggregation and loose coupling in enterprises, both requisites of relationship management, management needs to trial new and inventive enterprise designs. Instead of using plans, schedules, and transfer prices to integrate activities, there needs to be a greater understanding of interrelated structures, processes, and technologies that contribute to new forms of relationship structures. Furthermore, these types of dynamic linkages rely on information disclosure, trust, and the freedom for partners to withdraw from one relationship and enter new ones. The key is to create an integrated structure that co-ordinates and captures relationship capacity while simultaneously maintaining flexibility and autonomy for the associated members.

Operational: professional and work values

A professional culture (Mintzberg, 1979) is based on a value cluster that influences the creation and maintenance of knowledge, information, or a skill domain (e.g., marketing versus logistics, human resource management versus finance). Professional values also symbolize the democratic processes that stakeholders within the enterprise expect. The implications of this are a greater requirement for stakeholder participation, negotiation, and mediation, instead of a reliance solely on hierarchy (i.e., chain of command).

Work design values apply to the inherent assumptions used by practitioners in the design of operational and technological processes (usually treated separately). Work values are either personally held by individual workers or collectively shared by a subgroup based on functional, gender, or other value clusters. At the personal level, work values refer to the extent to which individual workers feel that personal worth results from self-sacrificing work or occupational achievement (Blood, 1969); that is, the individual work ethic. At the collective level, the work ethic provides a framework for how members of the workforce view and approach working, and incorporate goals, standards, and beliefs that members use to preserve and justify their interests as a work and/or professional group. The work or enterprise ethic is important in understanding the manner in which the workforce responds to management and the organization overall (Aldag and Brief, 1975). When the enterprise ethic is not congruent with the personal work ethic of workers, conflict is likely to occur. The outcome of conflicting values is reduced commitment from the workforce.

Intellectual capital

The resource-based view also provides a convincing argument for management to focus their resources on organizational activities, which can attain *unique* customer value, such as intellectual capital. Intellectual capital is largely derived from knowledge, defined as "the set beliefs held by an individual about causal relationships among phenomena" (Sanchez et al., 1996). Collective knowledge can be defined as the shared set of beliefs about causal relationships held by individuals within a group (Sanchez et al., 1996). Organizational participants have to produce complex knowledge (i.e., joint intelligence) to answer complex problems or questions, and increasingly in situations in which no certainty exists beforehand (DiMaggio, 1998). This means essentially that within the production of products and values, knowledge production or organizational learning processes are becoming not only an integral part or component, but also an increasingly decisive one. The difficulty of measuring the return on intellectual capital is a result of management's inherent tendency towards overdependence on financial measures of organizational performance. Typically, however, the intellectual capital assets of the organization are less observable and not easily open to benchmarks to the same extent that financial measures are.

Core competencies, as Prahalad and Hamel (1990) emphasize, are grounded in the collective know-how that is reflected in their integration. The most proprietary elements are rooted in tacit knowledge and intuitive understanding, which cannot be simply replicated even by competitors using identical components (Venkatesan, 1992). That is why some organizations that are broad users of out-sourcing often retain some unique capability and/or internal expertise in order to protect and develop the underlying competencies that are essential to their competitive advantage. Even in the most traditional industries, organizations that seek to build their competitiveness see the necessary integration of capability and internal expertise.

As the relationship between competency and value becomes more direct, relations between the stakeholders become more integrative (and interactive) and more dependent on learning processes. This will lead to greater real-time interaction among people from diverse functions and operations that, in addition to face-to-face interaction, are increasingly conducted using information and telecommunication systems. This leads us to consider networks among stakeholders with complementary core competencies, which in turn leads to the emergence of learning and innovation processes (Dyer and Ouchi, 1993).

Although the formation of intellectual capital is not a clear-cut process, it is frequently the driver of a coherent relationship strategy. The significance of building relationships derives from the notion that it is a major hope for building a sustainable competitive advantage. All the cognitive models of participation that have been developed (e.g., Katzner, 1995) agree that intellectual capital is processed during a participative decision-making process. When viewed strategically, a collaborative learning process provides a strategic platform for change, for collective and individual exchange and renewal of ideas, as well as a vehicle for transfer of intellectual capital.

Investing in intellectual capital leads to "intelligent enterprises" (Quinn, 1992) that comprise complex, global information and decision support systems superseding many of the control and operational functions of their conventional counterparts. These issues in turn lead to a new concept of organizing in terms of recreating a "flatter" hierarchy with a membership-orientated culture concentrating on value management through the integration of assumptions, learning, and knowledge (Webster, 1992).

The issue for supply chain relationships is how to co-ordinate all parties in the relationship so as to respond to end-user customers effectively and efficiently. Management has to find a way to integrate the efforts of all functions across the supply channels, maximizing value all the way. Integration is the extent to which the stakeholders can achieve both strategic and functional value along the supply chain as well as between the supply chain and its own subunits. Value needs to be maximized in:

(1) strategic terms (vision, quality, and performance),
(2) operational terms (markets, finance, operations, transportation), and
(3) intellectual terms.

Table 1 summarizes these key components of relationship value in the supply chain.

Table 1
Dissecting relationship value

Relationship value	Structure and design	Facilitators	Performance measures
Strategic value	Planning logic Strategic resource acquisition and deployment Structuring stakeholder relationships	Design of integrated performance Vertical and horizontal supply chain management Benchmarking	*Qualitative:* Shared values concerning stakeholder value delivery *Quantitative:* Shareholder value and EVA Profitability Productivity Cash flow
Operational value	Co-ordination of resources allocation Integrated processes Cross-functional orientation Supply chain roles and tasks	*Techniques*: Category management Integrated suppliers Synchronized production Continuous replenishment Reliable operations Cross-docking EDI/EPOS e-Commerce	Market- and financial-based criteria Supply chain key performance indicators
Intellectual capital	Knowledge, intellectual property, reputation, and brand image Relationships	Information technology and telecommunications	Supply chain key performance indicators

Key: EDI, electronic data interchange; EPOS, electronic point of sale; EVA, economic value added.

4. Is value management simply symbolic artifact or useful tool?

To address this question it is important to focus on effective relationship management, since this is predicated on the process of value creation, including a redefinition of vision, network communication, building stakeholder trust, as well as collaborating and sharing knowledge and decision-making. These processes grounded in an impelling business strategy are essential from the outset of relationship formation. Ideally, it is only then that the infrastructure can be designed and implemented to support the new relationship so as to share data, information and knowledge, competencies, and all the other resources necessary to maintain the relationship.

A further consideration for relationships is to distinguish between value derived by a particular stakeholder and the overall value outcome of a relationship. The latter is even more difficult to assess due to the perceptions of the various stakeholders engaged in the relationship and their initial expectations (Anderson et al., 1993). A particular way of thinking about value that appears to be useful is that stakeholders in the relationship need to assign the effects of their decisions and actions to outcomes.

5. Conclusion

Value management rests on the premise that all stakeholders engaged in the relationship bring unique commitments to it, requiring a process of integration. To integrate member commitments, each constituency in the relationship needs to understand and share in a collective mission. Success has to be grounded in the integration of all actors and resources, which leads to a greater probability of strategic and operational attainment. However, if the interrelationship between the partners is based largely on self-interest, competition, and overt conflict, the members' attachment to the relationship is tangential. Conversely, when the relationship between the constituencies is collaborative, partners become engaged in a relationship characterized by collective interest and equality. One of the difficulties in integrating the diverse interests of various members in a relationship is the fundamental conflict over their individual control of scarce resources. Sources of conflict include information (technical expertise, quality), capital, physical resources, time (to learn), and intangible assets such as industry reputation (Barney, 1986; Hill, 1990). The relative control of these resources is reflected in each transaction within the relationship. Conflict over resources also mirrors the degree of trust between members.

Business relationships that go beyond exchanging resources and strive for an integration of interests, goals, resources, and values take on a different "rationality" from those based purely on self-interest. A strategy of integration establishes common interests among stakeholders through a process of ongoing negotiation. With the understanding that not all relationships are founded totally on conflict or calculative action, integration is the approach most likely to lead to the initiation, development, and maintenance of relationship management. An integrative strategy therefore encourages a "negotiated order" within the relationship (Strauss, 1978). Negotiation is aimed at the maximization of equitable outcomes for all stakeholders. Negotiation allows each constituency not only to preserve a cohesive social relationship but also to dissent without fear of reprisal about contribution and outcomes in the relationship. Stakeholders experience a sense of working towards a "commonality" characterized by "what is good for us is good for the relationship." Integration is associated with enhanced efficacy and, ultimately, organizational capacity of the relationship.

A strategy of integration involves a major "gelling" of distinct core values, workforces, and orientations. It also requires a collective orientation to strategic purpose, implying a mutual understanding and acceptance of the goals and strategies by various stakeholders. This approach requires a renewed "responsiveness" from relationship members who are located either at the core or the peripheries of the supply chain, and is maximized by addressing customers' requirements and demands. Exchanging and transforming resources, such as skills, capabilities, competencies, assets, infrastructure, technology, and internal relationships, are essential to value creation and management. Providing these conditions are met, then value is more than a useful tool for supply chain management, it is essential.

References

Aldag, R.J. and A.P. Brief (1975) "Some correlates of work values", *Journal of Applied Psychology*, 60:757–760.
Anderson, J.C., D.C. Jain and P.K. Chintagunta (1993) "Customer value assessment in business markets: as state-of-practice study", *Journal of Business-to-Business Marketing*, 1:3–29.
Barnett, W.P. and R.A. Burgelman (1996) "Evolutionary perspectives on strategy", *Strategic Management Journal*, 17:5–19.
Barney, J. (1986) "Strategic factors markets; expectations, luck and business strategy", *Management Science*, 32:1231–1241.
Barney, J. (1991) "Firm resources and sustained competitive advantage (The resource-based model of the firm: origins, implications, and prospects)", *Journal of Management*, 17(1):99–121.
Blood, M.R. (1969) "Work values and job satisfaction", *Journal of Applied Psychology*, 53:456–459.
Brandenburger, A.M. and B.J. Nalebuff (1996) *Co-opetition*. New York: Doubleday.
Brewer, P.C., G. Chandra and C. Hock (1999) "An economic value added (EVA): its uses and limitations", *SAM Advanced Management Journal*, 64(2):4–11.
Burton, B.K. and C.P. Dunn (1996) "Feminist ethics as moral grounding for stakeholder theory", *Business Ethics Quarterly*, 6(2):133–146.
Caves, R.E., M.E. Porter, M. Spence and J.T. Scott (1980) *Competition in the open economy: A model applied to Canada*. Cambridge, MA: Harvard University Press.
Christopher, M. (1998) *Logistics and supply chain management: Strategies for reducing cost and improving service*, 2nd edn. London: Pitman.
Davis. E. and J. Kay (1990) "Assessing corporate performance", *Business Strategy Review*, 1:1–16.
DiMaggio, P.J. and W.W. Powell (eds.) (1991) *The new institutionalism in organisational analysis*. Chicago: University of Chicago Press.
Dyer, J.H. and W.G. Ouchi (1993) "Japanese-style partnerships: Giving companies a competitive edge", *Sloan Management Review*, 35(1):51–64.
Gattorna, J. (ed.) (1998) *Strategic supply chain alignment: Best practice in supply chain management*. Aldershot: Gower.
Hill, C.W. (1990) "Co-operation, opportunism, and the invisible hand: implications for transaction cost theory", *Academy of Management Review*, 15:500–513.
Hines, P., N. Rich, J. Bichen and D. Brunt (1998) "Value stream management", *International Journal of Logistics Management*, 9(2):25–42.
Johansson, H.J., P. McHugh, A.J. Pendlebury and W.A. Wheeler (1993) *Business process engineering*. Chichester: Wiley.
Jones, D.T., D. Roos and J.P. Womack (1990) *The machine that changed the world: based on the Massachusetts Institute of Technology 5 million 5 year study on the future of the automobile*. New York: Rawson Associates.
Jones, D., P. Hines and N. Rich (1997) "Lean logistics", *International Journal of Physical Distribution and Logistics Management*, 27(3/4):153–173.
Kanter, R.M. (1989) "Becoming PALs: pooling, allying, and linking across companies", *Academy of Management Executive*, 3(3):183–193.

Kaplan, R.S. and D.P. Norton (1993) "Putting the balanced scorecard to work", *Harvard Business Review*, 71(5):134–40.

Katzner, D.W. (1995) "Participatory decision-making in the firm", *Journal of Economic Behavior and Organization*, 26(2):221–237.

Mintzberg, H. (1979) *The structure of organisations*. Englewood Cliffs: Prentice Hall.

Olavarrieta, S. and A.E. Ellinger (1997) "Resource-based theory and strategic logistics research", *International Journal of Physical Distribution and Logistics Management*, 27(9/10):559–587.

Porter, M.E. (1980) *Competitive strategy: Techniques for analysing industries and competitors*. New York: Free Press.

Prahalad, C.K. and Hamel, G. (1990) "The core competence of the corporation", *Harvard Business Review*, 68:79–91.

Quinn, J.B. (1992) *Intelligent enterprise*. New York: Free Press.

Rumelt, R.P. (1991) "How much does industry matter?", *Strategic Management Journal*, 12:167–185.

Sanchez, R., A. Heene and H. Thomas (eds.) (1996) *Dynamics of competence-based competition: Theory and practice in the new strategic management*. Oxford: Pergamon.

Senge, P.M. (1990) *The fifth discipline: The art and practice of the learning organisation*. Milsons Point, NSW: Random House.

Strauss, A. (1978) *Negotiations*. San Francisco: Jossey Bass.

Treacy, M. and F. Wiersam (1995) *Discipline of market leaders*. Reading, MA: Addison-Wesley.

Venkatesan, R. (1992) "Strategic sourcing: To make or not to make", *Harvard Business Review*, 70(6):98–108.

von Bohm-Bawerk, E. (1959) *Capital and interest. II: Positive theory of capital*. Indianapolis: Libertarian Press.

von Mises, L. (1980) *The theory of money and credit*. Indianapolis: Libertarian Press.

Webster, F.E. (1992) "The changing role of marketing in the corporation", *Journal of Marketing*, October, 1–17.

Wernerfelt, B. (1984) "A resource-based view of the firm", *Strategic Management Journal*, 5:171–180.

Chapter 9

INTERMODAL TRANSPORTATION

BRIAN SLACK
Concordia University, Montreal

1. Introduction

The concept of intermodality is at the heart of modern transportation systems. Globalization, logistics, supply chains, and hubbing, some of the main developments in contemporary freight transportation, depend at least in part on the advances that have been made over the last 40 years in bringing together separate modal systems into intermodal structures. So profound have the changes been that they constitute a revolution. In this chapter the nature and scope of this revolution is explored. The complex and diverse character of contemporary intermodal transport is examined and some of its challenges are reviewed. It can be confidently predicted that intermodality is one of the forces that will help shape the world economy of the 21st century.

2. Definitions

A great many trips involve the use of more than one mode of transport. When a plane is taken, passengers have to get to the airport by using another mode, such as a car, taxi, or bus. Similarly, when freight is transported overseas by ship, it has to be delivered to the docks by rail, barge, or truck. These types of movements are *multimodal*. Various forms of transport have been in existence for millennia. A problem with multimodal transport has always been the difficulty in transferring the freight or passengers between the modes at the point of interchange. Terminal costs are incurred that have always placed multimodal transport at a disadvantage compared with one-mode systems. This has tended to relegate multimodality to trips where a transfer is unavoidable, such as overseas shipments. Besides the physical and cost disadvantages of multimodal shipments, government regulations have imposed severe constraints on attempts to combine the modes. In the U.S.A., for example, anti-trust legislation forbids ownership between the modes, and the Interstate Commerce Commission ruled in 1931 that unitized shipments must be

Handbook of Logistics and Supply Chain Management, Edited by A.M. Brewer et al.
© *2001, Elsevier Science Ltd*

charged at carload class rates, thereby negating for some time any cost advantages of other systems.

Unlike multimodal transport, which is characterized by essentially separate movements involving different transport modes, *intermodal* transport is the integration of shipments across modes. Intermodal transport may be defined as being those integrated movements involving at least two different modes of transport under a single through rate. Its goal is to provide a seamless transport system from point of origin to the final destination under one billing and with common liability (Hayuth, 1987; DeWitt and Clinger, 1999). In such a system the relative advantages of each mode of transport are exploited and combined in order to provide the most efficient door-to-door service possible.

There are two basic components in intermodal transport. The first is the transferability of the items transported. The bottlenecks at the transfer points and the resulting high terminal costs have been long-standing obstacles to the better integration of transport modes. This has been addressed principally through technology, in which transfers have been progressively mechanized and automated. Second is the provision of door-to-door service. While there are technical problems that have had to be addressed, the major issues here have been the difficulties in establishing the organizational structures necessary to provide single liabilities and through-bills of lading. As will be demonstrated, it has been easier to overcome the technical problems than the organizational difficulties, which frequently involve regulatory restrictions.

3. Origins of intermodal transport

The long history of multimodal transport gave ample opportunity for its shortcomings to be recognized. With the growth of passenger and freight transport over the last 150 years, numerous attempts have been made to link separate mode systems more effectively. Examples include the vehicle ferries, such as those that cross the English Channel, which allow trucks to be transported between Britain and continental Europe. In most cases these developed where no alternative was available – the only way to get from England to France was by boat. Where a multimodal system was offered in competition with a modal alternative, the costs and delays incurred in intermodal transfers were usually too severe to make the multimodal service viable. While there were many early attempts at making multimodal systems (McKenzie et al., 1989), the experiments of the U.S. railways with "piggyback" were the most extensive precursors of modern intermodal systems. In the 1920s, facing growing competition from the trucking industry, several U.S. railroads began to offer services in which truck trailers were put on rail flat cars for delivery between distant cities. These services mirrored the much earlier practise of Barnum and Bailey's circus of driving circus wagons onto flat cars via ramps, and the term "circus ramp" is still used today to refer to a method of loading road trailers. The trailer on flat car (TOFC) service became the principal service by which U.S. railroads sought to confront the growing competition of the trucking industry. While TOFC traffic increased, particularly after the 1950s, it never achieved the success that had been anticipated, failing to stem the diversion of freight from rail to the roads, and never generating profits for the railroads. This was due to several factors. First, it failed to obtain the co-operation of the trucking industry, which saw TOFC as a competitor. Second, it developed a reputation for slowness. The use of circus ramps scattered in every rail center resulted in lengthy delays in assembling flat cars and hauling them to their destinations. Third, the railroads themselves never gave TOFC their full endorsement, seeing it as an expensive hybrid that diverted them from their core business, which was the shipment of bulk freight. Finally, Interstate Commerce Commission restrictions on competitive rates hamstrung the TOFC providers until the 1970s.

It is significant that the contemporary intermodal transport industry had its main roots in the U.S. shipping industry (see Chapter 27). Shipping had always experienced the highest terminal costs of any transport mode. Ships spent most of their time in port, an average of 25 days being required to turn around a ship in the early 1960s. The delays were due to the difficulties of lifting and stowing each item of cargo. Furthermore, large gangs of dockers were needed to carry out these activities, workers who were among the most militant in the labor force. As the highest cost nation, U.S. ship owners were the most receptive to innovations that could reduce costs and improve service. Interestingly, it was a former trucking executive, Malcolm McLean, who is widely credited as being the founder of intermodalism. His decision to put the freight being shipped from New York to Houston into boxes of standard dimensions and convert two World War II tankers to hold the containers has become something of a legend (*Containerisation International*, 1966; McKenzie et al., 1989; Muller, 1995).

The success of the container is due to three factors:

(1) By standardizing the dimensions of loads, machines can be employed to handle transfers. This greatly improves the efficiency of cargo handling in terms of speed and cost. Ships that once spent 25 days in port can now be turned around in less than 2 days, thereby allowing the ships to realize many more revenue-generating voyages per year.
(2) By standardizing the dimensions of the loads, ships can hold more cargo. Economies of scale in shipping are thus realized.
(3) Significant reductions in manpower are achieved. Sea-Land claim that, whereas a 40 000 ton container ship requires 750 man-hours to be unloaded, a similar amount of cargo handled by traditional methods would have required 24 000 man-hours (Slack, 1998). Ports are thus less

constrained by labor conditions, thereby achieving economies which are passed on down the transportation chain.

These conditions were greatly facilitated by the actions of the International Organization for Standardization (ISO), which in 1964, at the prompting of the U.S.A., established common dimensions for containers of 20 or 40 feet long and 8 feet wide (thereby establishing the common unit of measurement of containers as the TEU (20-foot equivalent unit). Had such standards not been established in the early years of containerization, the process could have been delayed by a proliferation of competing national standards.

4. The role of containerization in contemporary intermodal transportation

The shipping industry rapidly adopted the container for the shipment of non-bulk cargoes. The first ship employed in international commerce, a vessel owned by the company established by Malcolm McLean, Sea-Land, sailed from New York to Rotterdam in 1966. Other U.S. companies such as Matson followed quickly, forcing European and Japanese carriers to begin container services by operating joint services (*Containerisation International*, 1996). By 1970, world ocean container traffic had reached 6 million TEU. Since then growth has been explosive, and the volume of such traffic now stands at 165 million TEU. No longer dominated by the U.S. carriers (all the major U.S. lines have been purchased within the last 10 years by Asian, European, and Canadian companies), the container shipping fleet is multinational, with a very strong representation from Asian carriers (Table 1).

The container has entered virtually every ocean shipping market for a wide range of freight types. Originally seen as a solution for the shipment of break-bulk cargoes (freight that is packaged in odd shapes and dimensions), containers now hold an extremely wide range of goods, including bulk commodities such as cereals and lumber, liquids such as chemicals, and refrigerated articles such as meat and vegetables. The simplicity of the container has permitted its adaptability and flexibility in a wide range of market conditions.

Having transformed the world shipping industry, the container was slower to enter other modal systems. The trucking industry developed a wide range of units of varying dimensions, to comply with local regulatory requirements, and these serve particular market conditions. However, the unit of greatest application in the long distance market was the articulated trailer, a wheeled unit that could be easily separated from the tractor unit. It was these trailer units that the railways, particularly in North America, sought to attract for long-haul trips in their TOFC services. More generally, however, railroads continued to ship general freight using their very large fleet of covered wagons or boxcars. While wide-bodied

Table 1
The ten leading container carriers (1998)

Carrier	Country	Capacity (TEU)
Maersk	Denmark	346 123
Evergreen	Taiwan	280 237
P&O Nedlloyd	U.K./The Netherlands	250 858
MSC	Switzerland	220 745
Hanjin	Korea	213 081
Sea-Land (a)	U.S.A.	211 358
Cosco	China	202 094
APL-NOL	Singapore	201 075
NYK	Japan	163 930
MOL	Japan	133 681

Source: *Containerisation International Yearbook*, 1999.
Note: (a) Since purchased by Maersk.

aircraft can carry the ISO container, its weight is a serious problem, and thus the airline industry has had to develop other intermodal solutions (Muller, 1995).

When containers began to arrive in ports in the 1960s and 1970s, land carriers were involved inevitably with the onward shipment of the units, but this remained a very specialized and limited market. The railway companies in North America were reluctant to invest in yet another intermodal technology, when TOFC remained unprofitable. In Europe, because the state-controlled railways serve each country, there was limited interest in developing networks for containers that would be continent-wide.

In the early 1980s, however, deregulation in the U.S.A. removed the former limits on intermodal control, and several shipping lines, most notably American President Lines (APL), leased locomotives from U.S. railroads and, by providing their own cars, offered intermodal rail services from the West Coast ports to inland markets (McKenzie et al., 1989). It was at this time that the double-stack concept was introduced, whereby containers were placed one on top of the other on a rail car, potentially doubling the capacity of the train, with only a modest increase in costs. Studies indicate the savings over conventional TOFC to be 30–45% (Muller, 1995). This innovation transformed the rail intermodal industry in North America. The success of the container services offered by the shipping lines encouraged the railroads, and even the port authorities, to follow. There is now an extensive, continent-wide network of double-stack rail services, with an average of 240 eastbound trains per week.

The shipping lines were also leaders in the use of rail-hauled containers for freight shipped between markets within North America. Domestic containers are now more important than ocean-bound containers on the railways (Slack, 1998). Despite its much earlier development by the railways, container movements have since eclipsed TOFC. In addition, several trucking firms have entered into strategic alliances with the railroads to haul their shipments in containers. Over the last 15 years, therefore, the container has revitalized intermodal rail services in North America.

The incursion of containers from ocean into land markets has been more variable elsewhere in the world. In Europe, expansion of rail-based container networks has been slower than in North America. Double stacking is not possible because of clearance problems with overhead electric wires, and the existence of national railroads, each with a goal of maximizing revenues on their own network, has slowed the establishment of intermodal systems at a continental level. Nevertheless, the European Commission is promoting "combined transport" (the term it uses to refer to intermodal transport), and since the 1990s there has been a significant increase in the number of services offered, some by the railroads themselves (e.g., Intercontainer), some by rail–trucking groups (e.g., IURR), some by the shipping lines and forwarders (e.g., the consortium comprising P&O-Nedlloyd, Dan, Dubois, and Saimo Avendero), and some by the barge services on the River Rhine (see Chapter 3). In Asia, most of the remarkable economic development is taking place in coastal areas, and it is only in China where there are opportunities for intermodal systems to link interior markets with the coast. Several rail services from Hong Kong have recently been established, but the Chinese are putting a great deal of emphasis on the provision of inland waterway services, especially in the Pearl River Delta and on the Yangtze, where a number of inland container depots have been established at river ports such as Nanjing. Australia has also experienced important developments in rail-based container services since the reorganization of its national railway (Slack, 1998).

If the container dominates freight intermodal transport, it is by no means the only system available. TOFC persists on the railways, but with an increasingly smaller market share. Another system, which at one time was considered as an alternative to the container, is roll-on roll-off (RORO). In this system, road trailers are driven onto a ship and parked between decks during onward shipment. Other cargoes, such as lumber, can be handled in palletized or strapped form. This alternative still has some proponents; however, although there are certain advantages in the loading and off-loading operations, its biggest problem is the wasted space between the decks. A container ship, by comparison, is able to realize much greater stacking densities. Occupying a small, but very lucrative intermodal market niche, is the roadrailer, a truck unit that can be placed directly onto rails by a set of retractable steel wheels incorporated on the trailer. Because of its superior ride capabilities and ease of transfer between road and rail, it has found use in high-value shipments, where speed of delivery and safety are paramount. In the U.S.A., where it is used most extensively, the railroader is employed by the automobile industry to ship components and parts between plants on a just-in-time (JIT) basis (McKenzie et al., 1989).

Table 2
Container handling systems

	Dock gantry	Yard gantry	Straddle	Front end/side	Chassis
Approximate cost (U.S. $)	8 million	1 million	800 000	350 000	50 000
Land-use intensity (TEU/ha)	NA	800	400	590	170

Source: after Muller (1995).
Key: NA, not available.

5. Implications of the container

At one level, the container is a straightforward technology that has found wide use because of its simplicity. The closer it is studied, however, the more it becomes apparent that its repercussions have been extensive and profound. Some of these consequences are discussed below.

5.1. Terminal activities are highly capitalized

To facilitate the transfer of goods between the modes, a wide range of machines have been developed to lift and move containers between the modal systems. The largest, and the most expensive, are the dockside gantry cranes (Table 2). Within terminals there are: yard gantries that top lift the boxes; mobile units that lift the boxes from the top, sideways, or from beneath; straddle carriers that override the containers and lift them from above; and a host of tractor units that are used to move the boxes around the storage facility. These machines vary considerably in cost and operational efficiency (see Table 2). There is a great deal of similarity between equipment in rail, port, and road terminals.

5.2. Hubbing

Traffic concentration and the emergence of load centers is a significant feature of containerization. To justify the capital costs of the system, the shipping, air freight, and rail freight industries have developed a hub network structure (Hayuth, 1987). Freight traffic is concentrated at a relatively small number of transfer points in order to achieve scale economies. In rail transport in North America, for example, the number of rail terminals offering an intermodal service fell from 1107 in 1976 to 199 in 1993 (Slack, 1998). On the West Coast of the U.S.A., three port groups

(Los Angeles – Long Beach, Oakland, and Seattle – Tacoma) account for over 95% of traffic. Air transport, including passenger traffic, has perhaps gone furthest in developing a hub and spoke network structure (Hanlon, 1996). The reductions in terminal costs are a very important advantage for the intermodal transport industry as a whole.

5.3. Massive increase in site requirements for intermodal terminals

Intermodal terminals require ever-larger sites. The operations of the mechanized transfer systems impose considerable space demands for maneuvering and turning (see Table 2). Load centering also imposes enormous demands for storage space, since the capacities of the ships and the double-stack trains far exceeds the capacity of trucks to make the local pick-up and delivery. Compared with an old seaport berth of 50 m length and 1 ha of storage space, container terminals are now 50–100 ha in area and require berths of over 300 m length (Slack, 1998). Intermodal rail yards are 2 km long and are over 100 ha in area. Airports require 10 000 ha or more. Such sites are not easy to come by (Dempsey et al., 1997).

5.4. Scale economies in ships, railways, and planes

The container permits much higher stacking densities, particularly for ships, and thus there has been an explosion of vessel capacities. The first generation of container ships had limited carrying capacities of approximately 1000 TEU. Successive refinements have greatly increased the size of vessels. Today, ships with capacities in excess of 6000 TEU are proliferating, and there is a lively debate in the industry as to the size limits. Some see vessels of up to 15 000 TEU as feasible, while others argue that diseconomies of scale will limit capacities to 8000 TEU (Gilman, 1999). Even if the smaller figure proved to be true, the vessels would be enormous (i.e., 350 m long and drawing 17 m of water). Most existing container ports will not be capable of receiving such large vessels. On the railways too there are scale economies, with double stacking permitting unit trains to haul 400 TEU. In the airline industry, wide-bodied aircraft have significant cargo capacity advantages. For example, a Boeing 727 in all-freight mode has a capacity of 115 m^3, while a Boeing 747F has a capacity of 748 m^3.

5.5. Restructuring of the container industry

Containerization has required the ocean carriers to make significant capital outlays. Yet they face an uncertain technological future in an intensely competitive

Table 3
Alliances and mergers in the ocean container industry (2000)

Alliances	Mergers
Hapag-Lloyd – MISC – NYK – OOCL – P&O	Maersk – Sea-Land
APL/NOL – HMM – MOL	NOL – APL
CY – DSR/Senator – Hanjin – UASC	CP Ships – CAST – Ivaran – Lykes – Contship – ANZDL
Cosco – K-Line – Yangming	P&O – Nedlloyd

environment. As a result, important structural changes have occurred. Many long-established firms have simply dropped out of the market. Survivors and newer entrants from Asia have established a wide range of inter-firm linkages, ranging from slot charter arrangements, whereby a line reserves space on another's vessel, to the formation of consortia that operate on specific routes. More recently, most of the major carriers have either come together to form strategic alliances for the operation of global networks, or have been involved in equity mergers (Table 3) (Midiro and Pitto, 2000). Similar developments are evident in the rail and air industries. In the North American rail industry, the nine major class I railroads of 1990 reduced to five by 2000, and five global alliances now dominate international air transport. Although the alliances and takeovers in these industries cannot be ascribed directly to containerization, the parallels with ocean shipping are striking.

6. Organizational challenges to intermodal transportation

The technological solutions that have been developed to transfer cargoes between the modes have reduced costs and improved efficiencies. Transferability represents only one aspect of intermodalism, however. The difficulties in organizing the exchanges have been more challenging. Several of these difficulties relate to documentation and liability issues, and others are derived from problems of co-ordination between the modes. Issues relating to government regulation and control touch on nearly all these organizational obstacles. Resolving these difficulties has been slower, and the results geographically uneven.

At the heart of these difficulties is deciding who is responsible for the shipments that may be handled by different carriers. Who does the customer deal with to get a rate for delivery across the modes? How are liabilities shared? How can tracing be achieved when many carriers are involved? These are some of the basic questions that the intermodal industry has tried to solve.

6.1. Liability

When a customer wishes to ship an item he is presented with a bill of lading. This is the contract between the shipper and carrier, and details the charges for transport, the value of the goods, and the conditions of carriage and liability. Such documentation developed in the modal era, with a different regime of legal liabilities being established by different modes. In the U.S.A., rail and truck liability is the full value of the good reported on the invoice. Air shipments, however, are assessed by weight, and ocean transport uses a flat value per package. With each mode being regulated by separate commissions or boards, moves to establish through-liability provisions were difficult. Deregulation of the transport industry in the U.S.A. in the 1980s helped free some of the restrictions, but the establishment of international accords on liabilities and through documents have been slow.

6.2. Documentation

Documentation goes beyond the matter of liability. It involves allowing shippers to find where there shipments are at any time, and also the clearance of customs at international borders. Customs clearance has always been a problem for international commerce. In the days of break-bulk ocean trade, the matter was not as acute as in the post-container age, since cargo handling was so slow in the first place. Air freight, trucking, and intermodal shipments made it necessary for radical improvements. In the industry this is referred to as "facilitation." It involves a reduction in the quantity of and speeding up of paperwork. The major force in facilitation has been information technology and electronic data interchange (EDI). EDI began as a means of rapidly transferring documents, such as bills of lading, delivery notices, and invoices, with one of the most promising benefits being the swift clearance of customs. It has gone on to involve the facilitation of planning and stowing of containers in ships and on trains, and of managing the fleet of containers. It has also developed an important position in the tracing of units and maintaining control over inventories, thereby forging essential links with supply chain management. A difficulty has been that there has been a proliferation of propriety systems, in which many of the major carriers have developed a unique approach, and despite the work of the United Nations to establish a common international platform (EDIFACT), the U.S.A. continues to use its own system (ANSI X12).

6.3. Intermodal intermediaries

Who does the customer deal with to arrange intermodal shipments? Who will co-ordinate the transfer across the modes? In many jurisdictions the carriers were excluded from intermodal ownership and operations, either because of regulatory restrictions or because some carriers were state-owned monopolies (e.g., the railroads in Europe). Deregulation and privatization over the last two decades have removed several of these obstacles, and some of the leaders in intermodal operations have been companies that have extended ownership across several modes (e.g., the purchase of Sea-Land by the railroad CSX, a partnership that ended with the sale of Sea-Land to Maersk in 1999). The ocean carriers, as major parties in containerization, are playing an increasingly important role. They have moved to provide door-to-door services for customers. This vertical integration is seen as a way to improve the quality of service across the transport chain, creating seamless transport systems, and thereby enhancing customer satisfaction, as well as adding value to the service and controlling costs (*Containerisation International* 1996).

The position of customers too has evolved. In the deregulating environment in North America, shippers' associations (groups of shippers in one industry sector, such as toys, or in one geographic area, such as Seattle) are playing a growing role in transportation chains. By negotiating volume rates with the carriers they are able to offer cheaper rates to their members. Under the 1998 Ocean Shipping Reform Act (OSRA), shipper councils were given greater freedom to establish confidential contracts with carriers.

Finally, there is the proliferation of intermodal third parties. Some are the traditional freight forwarders. These have been particularly important in air freight intermodal transportation, and express companies such as UPS and Federal Express have become major providers. In the ocean transport sector there have emerged non-vessel owner common carriers (NVOCCs), which are intermediaries that offer rates and space based on their purchase of blocks of space on vessels owned by the carriers. Customs brokers, consolidators, and even insurance agencies are other types of businesses now offering customers an intermodal service.

Today, therefore, in deregulated environments there is a proliferation of intermodal providers. The growth reflects the fact that putting together intermodal services rather than merely transporting goods is a means of adding value. The ocean and rail carriers have moved into these areas because they see it as an opportunity of controlling costs and service reliability. Many, such as APL, Sea-Land, and MOL, have extended their offices inland to market their services directly; while others, such as BN-ATSF Railroad, are involved in contractual relationships with a small number of third parties to retail their services.

6.4. Regulatory issues

It is not a coincidence that the growth of intermodalism has come about in parallel with the worldwide trend towards deregulation and privatization of the transport industry over the last 20 years (Button, 1991). The relaxation of regulatory controls over rates, entry, and ownership has greatly facilitated the organization of the intermodal industry (see Chapter 23). However, the weakening or removal of these constraints has varied considerably from one part of the world to another, and the modes have been differentially impacted. For example, air transport is still constrained by bilateral treaties between nations, and ports and railroads are still nationalized in many jurisdictions (Nijkamp, 1995). While great strides have been made towards liberalization, the intermodal industry is still impacted by many transport-specific regulatory issues at the national and international level. A very partial list includes European Union restrictions on through rates from inland markets being offered by ocean shipping consortia, OSRA favouring shipping associations over NVOCCs, and restrictions on cabotage in the airline and ocean shipping industries.

There are several important regulatory issues before the intermodal industry that arise from non-transport-specific statutes. The move towards mergers throughout the transport industry are increasingly drawing the interest and intervention of monopoly commissions and anti-trust bodies. It appears that in the early years after deregulation these concerns were taken seriously, to the extent that several proposed mergers, such as that between SP and ATSF railroads, were denied. Subsequently, there appears to have been a far more tolerant regulatory regime, with many carrier mergers being approved (e.g., P&O with Nedlloyd, KLM with Northwest, BN with ATSF). The concentration, however, continues, and there appears to be increasing regulatory oversight into recent proposed takeovers, such as a proposed buy-out and merger between Canadian Airlines and Air Canada (which was rejected), the linkage of American Airlines and British Airways (conditions were set that were impossible to meet), the merger between CN and BN-ATSF (which was delayed), and the combining of the freight operations of German and Dutch railways.

6.5. Intermodal futures

Diversity appears to be the hallmark of the intermodal industry. Although the container is dominant, other systems occupy niches in specific markets. Roadrailer continues to serve markets where reliability and dependability are paramount. Pallets are important for smaller lot sizes, particularly in the air freight sector (see Chapter 28). In Europe, the swapbody, an intermodal unit with non-rigid sides, maintains an important intermodal market share. For the container itself, there appears to be a growing diversity of dimensions, providing units of varying lengths and heights for particular market sectors. There are now 28, 45, 48, and 53 feet long boxes, and "high cube" units which are 8 feet 6 inches long. This product diversity creates problems for the ocean shipping industry, whose cellular ships are designed to hold ISO standard 20 and 40 feet boxes.

Diversity is a feature of handling gear. Apart from the ship-to-shore gantry, no single system prevails. While there are commonalities between rail, port, and road intermodal terminal equipment, there are striking differences between comparable terminals, depending on local conditions. Innovation in the terminals appears to be in the direction of automation. While some time ago automated transfers were seen as a logical progression from the present mechanized methods, with the dock dedicated to Sea-Land and managed by ECT in Rotterdam as a prototype, different operating conditions and the need for flexibility has tended to work against the implementation of full automation. Instead, it is with the adoption of computer-based systems, allied with global postioning systems (GPSs), that terminal management is being automated, facilitating the storage of containers, their positioning on the ships and trains, and their tracing.

Diversity appears to be the hallmark of the carriers. In one respect this claim would appear to be misplaced. The trend towards integration between the main carriers seems to be inexorable, and thus at first glance it would appear that oligopolistic tendencies would be dominant. Certainly the trend towards alliances and mergers among the major carriers continues, but this is leading towards the establishment of a two- or three-tier level of service providers, as smaller, more regionally based carriers are appearing in all markets. In some cases, such as the regional air carriers, these smaller carriers may be allied with a major company, but there are many independents, sometimes providing feeder services, but in other cases serving smaller more specialized markets. In Asia, Robinson (1998) sees smaller carriers competing in the same market as the major carriers, by providing a cheaper, less reliable or less frequent service. The emerging associations between different players in the intermodal chain also represent this diversity of carriers. As the major actors seek to achieve vertical integration, novel and significant groupings are being produced, such as: stevedoring companies purchasing ports; shipping lines buying barge companies, trucking firms, and ports; ports buying other ports and operating inland rail services; and alliances between truckers, airlines, and railroads. The latter, in particular, appears to be one of the new phases of intermodalism. Attempts to integrate air with other modal services, from Japan to Europe via Dubai, for example, have carved out specific market niches. An important development is the emergence of freight airports (usually smaller hubs) as intermodal centers, with dedicated rail intermodal and truck terminals directly on site, as in Huntsville, Alabama, or immediately adjacent, as in Alliance, Texas.

Diversity is already the hallmark of the intermodal facilitators. Deregulation has opened up new market opportunities. There is some debate in industry circles at the moment as to whether this heterogeneity will be maintained. The growth in the size of the carrier groups, and their interest in entering into direct sales with customers or shippers' associations, is seen by some as a threat to the true third-party intermediaries. The intermediaries have shown remarkable resilience in the past and have adapted to changing circumstances. New groupings are being formed, such as the New American Consolidators Association (NACA) in the U.S.A., and Schenker-BTL in Europe. Perhaps the warnings about the imminent demise of the forwarders and brokers are premature.

The intermodal transport industry is continuing to evolve, and in so doing it is becoming increasingly integrated with supply chain management. As the chains become ever more global, through wider sourcing, intermodal solutions assume even greater importance. As DeWitt and Clinger (1999) state, customers in these global chains expect "their order to be delivered at the right place, at the right time, in the right condition, and for the right profit." These are the service challenges that intermodal transport must meet.

References

Button, K.J. (ed.) (1991) *Airline deregulation: international experiences*. London: Fulton.

Containerisation International (1996) "40 years", *Containerisation International*, April (special issue).

Dempsey, P.S., A.R. Goetz and J.S. Syliowicz (1997) *Denver International Airport: Lessons learned*. New York: McGraw-Hill.

DeWitt, W. and J. Clinger (1999) *Intermodal freight transportation*. Washington, DC: Committee on Intermodal Freight Transport, Transportation Research Board.

Gilman, S. (1999) "Size economies and network efficiency of large container ships", *International Journal of Maritime Economics*, 1:39–59.

Hanlon, J.P. (1996) *Global airlines: Competition in a transnational industry*. Oxford: Butterworth-Heinemann.

Hayuth, Y. (1987) *Intermodality: Concept and practice*. Colchester: Lloyds of London Press.

McKenzie, D.R., M. North and D. Smith (1989) *Intermodal transportation: The whole story*. Omaha: Simmons-Boardman.

Midoro, R.A. and A. Pitto (2000) "A critical evaluation of strategic alliances in liner shipping", *Maritime Policy and Management*, 27:31–40.

Muller, G. (1995) *Intermodal freight transportation*. Westport: Eno Foundation.

Nijkamp, P. (1995) "From missing networks to interoperable networks: The need for European cooperation in the railway sector", *Transport Policy*, 2:159–167.

Robinson, R. (1998) "Asian hub/feeder nets: Issues in the ownership debate", *Maritime Policy and Management*, 25:21–40.

Slack, B. (1998) "Intermodal transportation", in: B.S. Hoyle and R. Knowles (eds.), *Modern transport geography*. London: Wiley.

Part 3

LOGISTICS MANAGEMENT

Chapter 10

INTEGRATED LOGISTICS STRATEGIES

ALAN McKINNON

Heriot-Watt University, Edinburgh

1. Introduction

Integration has been one of the dominant themes in the development of logistics management. This development began around 40 years ago with the integration at a local level of transport and warehousing operations into physical distribution systems. Today, many businesses are endeavoring to integrate supply networks that traverse the globe, comprise several tiers of supplier and distributor, and use different transport modes and carriers. The process of integration has transformed the way that companies manage the movement, storage, and handling of their products. Traditionally, these activities were regarded as basic operations subservient to the needs of other functions. Their integration into a logistical system has greatly enhanced their status and given them a new strategic importance.

While logistical activities have been undergoing this fundamental restructuring, companies have been placing greater emphasis on the formulation of strategy. Business strategy has become a very fertile field of research and consultancy work, generating many new ideas, approaches, and conceptual frameworks. This has affected logistics management in two ways. First, much of the output of this research has been directly applicable to logistics, helping managers devise strategies specifically for the logistics function. Second, several of the most influential company-wide strategic models, such as that of Porter (1985), have identified a central role for logistics and confirmed that it can make a major contribution to the competitiveness and growth of a business. The real challenge is now to ensure that logistical strategies are aligned with the broader strategic goals both of individual businesses and groups of companies linked together in a supply chain.

This chapter begins by reviewing the integration of the logistics function over the past few decades, showing how it has widened its scope from the distribution of finished products to the "end-to-end" supply chain and has been elevated from an

Handbook of Logistics and Supply Chain Management, Edited by A.M. Brewer et al.
© 2001, Elsevier Science Ltd

operational to a strategic level. The following sections examine the contribution that logistics can make to wider corporate strategy, the strategic options available to logistics managers, and the impact of logistical decision-making on the freight transport system.

2. A brief history of logistical integration

The process of logistical integration can be divided into four stages:

Stage 1. The first stage in the process is generally considered to have been the "revolution in physical distribution management," which began in the early 1960s in the U.S.A. and involved the integration into a single function of activities associated with the outbound distribution of finished goods. Formerly, logistics "was a fragmented and often unco-ordinated set of activities spread throughout various organizational functions with each individual function having its budget and set of priorities and measurements" (Lambert and Stock, 1993). Separate distribution departments were created which, for the first time, were able to coordinate the management of transport, warehousing, inventory management, materials handling, and order processing. The integration of these activities within physical distribution management (PDM) had three beneficial effects:

(1) It allowed companies to exploit the close interdependence between them, establishing a "distribution mix" which could meet customer requirements at minimum cost. In designing an integrated distribution system, they aimed to achieve an optimal trade-off between the costs of the various activities. Traditional accounting structures had prevented this in the past. The development of a new "total cost approach" to distribution accounting, which became a prerequisite of PDM, permitted much more detailed analysis of distribution costs. This often revealed, for example, that a large proportion of companies' total output was being distributed in small quantities at a high delivery cost per unit. In pursuing their prime objective of maximizing revenue, sales departments were prepared to supply very small orders, in some cases at a loss. Once these inefficiencies were exposed, companies began to raise minimum order sizes, stopping deliveries to small outlets and effectively rationalizing their delivery networks (McKinnon, 1989).

(2) It gave distribution a stronger customer focus. PDM was initially motivated by a desire to cut cost, reflecting the traditional view of distribution as simply a drain on companies' resources. During the 1960s it was recognized that the quality of the distribution service could have a significant impact on sales, market share, and long-term customer loyalty (Stewart, 1965).

Distribution could therefore affect profitability on both the cost and revenue sides of the balance sheet. The new distribution departments began to develop more explicit customer service strategies based on closer co-ordination of order processing, warehousing, and delivery operations.

(3) It raised the status of distribution within the management hierarchy. When identified as a function in its own right, distribution began to take its place alongside production, marketing, and sales, with its own budget and often separate representation at company board level. A new generation of managers was appointed to oversee the full spectrum of distribution activities and devise distribution strategies for their businesses.

Stage 2. PDM was initially concerned only with the distribution of finished products. The same general principle was subsequently applied to the inbound movement of materials, components, and subassemblies, generally known as "materials management." By the late 1970s, many firms had established "logistics departments" with overall responsibility for the movement, storage, and handling of products upstream and downstream of the production operation. This enabled them to exploit higher level synergies, share the use of logistical assets between inbound and outbound flows, and apply logistical principles more consistently across the business (Bowersox, 1978). Fabbes-Costes and Colin (1999) use the term "integrated logistics" to describe the co-ordination of inbound supply, production, and distribution. They also differentiated later phases in this process, where logistics extends its influence upstream into product development and downstream into after-sales service and the recycling and disposal of waste. They called the culmination of this process "total logistics."

Stage 3. Having achieved a high level of integration within the logistics function, many firms tried to co-ordinate logistics more closely with other functions. Most businesses have a "vertical" structure built around a series of discrete functions such as production, purchasing, marketing, logistics, and sales, each with their own objectives and budgets. These functions are often represented as "silos" or "stovepipes" (Christopher, 1998). Senior managers often put the interests of their functions before the profitability of the business as a whole. Under these circumstances, logistics can play an important co-ordinating role, as it interfaces with most other functions. As Morash et al. (1996) observe, "the strong boundary spanning role found for logistics implies that logistics can be used as a vehicle for cross-functional integration, a nexus of communication and co-ordination, and for better system performance." They argue that "functional boundaries need to be made flexible and virtually transparent in the pursuit of cross-functional excellence."

With the emergence of business process re-engineering (BPR) in the early 1990s (Hammer and Champy, 1993), the relationship between logistics and

Table 1
Core business processes

Christopher (1998)	Hines (1999)	
	Electrical distributor	Chemical producer
Order fulfillment	Order fulfillment	Strategic management
New product development	Supplier integration	New business development
Marketing planning	Sales order acquisition	Customer support
Information management	New product introduction	Order fulfillment
Profitability analysis		Cost management
		Quality and environmental management
		Continuous improvement

related functions was redefined. BPR identifies a series of core processes that cut across traditional functional boundaries and are essentially customer-oriented. Effective management of these processes requires the development of new working relationships between functions and the formation of more cross-functional teams. These are acknowledged to be core processes which drive the typical business, of which order fulfillment, the raison d'être of all logistics operations, is arguably the most important (e.g., Christopher, 1998) (Table 1). As Hines (1999) points out, however, the range of key processes and their relative importance can vary between sectors and companies (see Table 1).

Bowersox and Closs (1996) have adapted the principles of BPR to logistics, emphasizing four factors "common to all logistical reengineering initiatives." The first and most important is "systems integration." The authors argue that "a logistical system with cross-functional integration should achieve greater results than one deficient in co-ordinated performance," although they concede that "effective application of systems integration in logistics is operationally difficult." The other three factors are benchmarking, "decompositional" analysis of individual logistics activities, and the quest for continuous improvement.

Stage 4. All the developments discussed so far have related to the management of an individual business. If all the businesses in a supply chain optimize their logistical activities in isolation, it is unlikely that the flow of products across the supply chain will be optimized. To achieve wider, supply chain optimization it is necessary for companies at different levels in the chain to co-ordinate their operations. This is the essence of supply chain management (SCM). The main driver of SCM over the past 20 years has unquestionably been the desire to minimize inventory. Supply chain (or "pipeline") mapping has shown that much of the inventory in a supply chain is concentrated at "organizational boundaries," where products are transferred from one company to another (Scott and Wesbrook, 1993). Uncertainty about the behavior of suppliers and customers causes firms to accumulate buffer stock. More open exchange of information and closer integration of logistical activities enables companies to cut lead times and reduce stocks, to their mutual advantage and the benefit of the supply chain as a whole. For example, Lewis et al. (1997) cite the case of a medium-sized manufacturer of mechanical/electrical equipment which doubled its stock turn (from four to eight times a year) and raised its inbound delivery service level from 60% to 98% by "re-engineering its supplier interface." While there has been some acknowledgement of the role of freight transport in the development of successful supply chain links (Gentry, 1995), this subject has attracted little research. Members of an integrated supply chain should collaborate to maximize vehicle load factors, minimize empty running, achieve an optimal allocation of freight between modes, and standardize on handling systems that make effective use of vehicle and warehouse capacity (European Logistics Association, 2000).

It is difficult to define the exact chronology of the process of integration as it has diffused at different rates across industrial sectors, countries, and company size categories. There are still many small and medium-sized business which have yet to embrace fully the principles of PDM and whose distribution management is still highly fragmented. While at an individual company level the process of logistical integration has proceeded at varying rates, the sequencing of the four stages has been more regular. There is general agreement, for example, that companies must integrate their internal logistics operations before attempting to link these operations with those of external suppliers and distributors.

Stevens (1989) has examined the nature of the transition between the various stages of integration and noted that different factors dominate at the each stage. He argues that the application of new technology has been the principal force in moving firms from stage 1 to stage 2, which he calls "functional integration." The transition to stage 3 (internal integration) involves primarily a change in organizational structure, while to attain stage 4 (external integration) management must undergo a major attitudinal change.

It is also worth noting that the multistage integration of companies' logistical operations has been reflected in the out-sourcing of logistical activities. While this process of integration has been underway, companies have been externalizing an increasing proportion of their logistics spend (McKinnon, 1999). Traditionally, they would out-source activities such as transport or warehousing on an individual basis. During the 1970s and 1980s it became common for firms to contract out their entire distribution operation, particularly in those countries with deregulated road freight markets, such as the U.K. In some cases, the processes of out-sourcing and integration were concomitant, with much of the responsibility for combining the various activities entrusted to third-party operators (Cooper and Johnstone, 1990) More recently, there have been examples of large manufacturers and retailers employing a single contractor to manage their inbound as well as outbound logistics. There is now a growing demand from multinational businesses for the services of logistics providers capable of integrating their global supply chains (Datamonitor, 1999).

3. Strategy formulation

The role of strategy is to "guide the firm in its efforts to develop and utilize key resources to achieve desired objectives within a dynamic and challenging competitive environment" (Fawcett et al., 1997). In a monograph for the Council of Logistics Management, Cooper et al. (1992) outline the process of strategic planning as it might be applied to the logistics function. They summarize this process as "identifying the long-term goals of the entity ... and the broad steps necessary to achieve these goals over a long-term horizon ... incorporating the concerns and future expectations of the major stakeholders." These goals can be defined at different levels and are usually based on a wide-ranging audit of a company's capabilities and market opportunities. At the highest level are broad corporate goals affecting the positioning of a business within its competitive environment. In companies, which have reached the third stage of integration, the logistics function will be represented in the corporate planning team, ensuring that the logistical implications of each strategic option are properly evaluated. The corporate plan drawn up to achieve these goals will define a series of logistical requirements. It will often be possible to meet these requirements in different ways, introducing a degree of flexibility into the formulation, at a lower level, of the logistics strategy. At a lower level still, separate strategies can be devised for individual logistical activities, such as transport and warehousing. The planning process must ensure that strategies developed at the different levels are closely co-ordinated.

4. Corporate goals

At the heart of all business strategy lies the desire to achieve differentiation through cost reduction and/or value enhancement (Porter, 1985). These strategic options are typically represented by a simple matrix showing four combinations of high and low ratings for cost and added value (Figure 1). By common consent, the least competitive businesses will be found in cell 1, supplying low value, undifferentiated products at relatively high cost. The most competitive companies produce high value, well-differentiated products at relatively low cost, and thus occupy cell 3. Intermediate positions are held by companies placing an emphasis

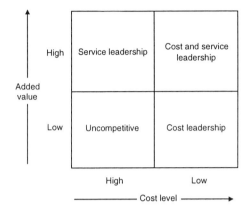

Figure 1. Strategic positioning (adapted from: Christopher, 1998).

on minimizing cost (cost leaders) or maximizing value (service leaders). Various attempts have been made to classify businesses with respect to these two criteria and to plot changes in their relative position in the matrix over time.

5. Value enhancement

Effective logistics management can help companies to gain competitive advantage through both value enhancement and cost reduction. The first of these is discussed in this section, and cost reduction is discussed in Section 6.

5.1. Product diversification

One of the most effective means of adding value to a product or service is to tailor it more closely to individual customer tastes and requirements. This involves extending the range of products or services available. The proliferation of products has major implications for logistics. There is generally a close correlation between the number of separate product lines (or stock-keeping units (SKUs)) in a company's logistical system and the amount of inventory that must be held. Highly diverse product ranges also require more complex warehousing, handling, and information systems. The process of customization can further complicate the logistics operation. It has become increasingly common for multinational manufacturers to defer the final customization of their products until they reach particular continental or national markets, in some cases adding an extra link to the supply chain (Cooper, 1993).

5.2. Development of higher value products

Higher value products are often inherently more fragile and perishable, requiring more packaging, more careful handling and, often, temperature control. More expensive products also need tighter security and are more expensive to insure while in transit. Many of the new higher value consumer products developed over the past 20 years have production and distribution systems that are intrinsically more complex and geographically extensive than those of their predecessors.

5.3. Improved service quality

Business customers and final consumers are usually prepared to pay more for faster and more reliable delivery. They also attach higher value to products supported by good after-sales service. Logistical services can therefore be used augment the basic product and help companies differentiate their offering from that of competitors.

6. Cost reduction

A recent survey of "over 200" European companies found that logistics costs represent, on average, 7.7% of sales revenue (A.T. Kearney, 2000). In some sectors, this proportion can be two or three times higher. By improving the productivity of logistics operations it is possible to cut this cost and translate some of the savings into lower prices. Over the past 20 years, the largest saving in logistics costs has accrued from a reduction in inventory levels (relative to sales). This has been achieved by the move to just-in-time/quick response replenishment, the centralization of inventory, the application of new IT systems, and the development of SCM. There have also been substantial improvements in the efficiency of freight transport operations, resulting mainly from the upgrading of transport infrastructure, liberalization of freight markets, and improved vehicle design. Warehousing costs per unit have also declined in real terms as a result of scale economies, increased mechanization, and the diffusion of new computer-based warehouse management systems (see Chapter 34). The combined effect of these trends has been to reduce the proportion of revenue spent on logistics by European firms by an average of 46% between 1987 and 1999 (A.T. Kearney, 2000).

These cost reductions were achieved during a period when product ranges were expanding and service quality steadily rising. There is little evidence of quality and value being sacrificed for cost savings, or vice versa. This is making it harder for companies to differentiate their offering in terms of both value and productivity, as benchmarks are constantly rising. Simply to remain competitive, companies are under pressure to improve both service and cost performance. As Persson (1991) explains, "logistics ... has become a win–win strategy, improving performance, quality and productivity simultaneously."

7. Logistical strategies

Several attempts have been to made classify logistical strategies. Persson (1991), for example, has identified three basic strategies and exemplified them with short case studies of Scandinavian companies. He calls them simply strategies 1, 2, and 3:

(1) *Strategy 1*. Companies use logistics to "influence competitive forces" by (i) making suppliers or customers more dependent upon them or (ii) using heavy investment in a new logistics network to discourage other firms from entering a market sector. The specialist chemical supplier Merck, for example, has developed a distribution system in the U.K. that can deliver orders varying enormously in weight, from a few grams to a tonne, in an effort to become sole supplier (or a "one-stop shop") to laboratories throughout the U.K.
(2) *Strategy 2*. Companies, using existing resources, develop innovative logistics practices to penetrate new markets or gain competitive advantage in an existing market. The abandonment of fixed depot area boundaries, for instance, and the adoption of multi-depot fleet planning can strengthen a company's competitiveness in a regional market by simultaneously cutting transport costs and delivery lead times (McKinnon, 1998).
(3) *Strategy 3*. Companies aim for across-the-board superiority in logistics by "seeking new solutions and system combinations." Such companies tend to regard logistics management as a core competence and key to future success.

An alternative typology advanced by Bowersox et al. (1989) has been much more widely quoted and subjected to greater empirical analysis. It was originally developed as part of a study of the links between logistics strategy and the organization of the logistics function. In its revised form, this classification differentiates three types of logistics strategy:

(1) *Process-based strategy*. This applies to firms at integration stage 3 and committed to the cross-functional management of business processes. The emphasis here lies in improving the efficiency of a broad range of logistical activities.
(2) *Market-based strategy*. This is concerned with a more limited group of logistical activities, often carried out by different business units, and aims to "facilitate sales and logistical co-ordination" by market sector.

Table 2
Interrelationship between strategic decisions and freight transport parameters

	Quantity of freight	Mode choice	Vehicle type	Vehicle utilization	Routing	Scheduling
Product development						
Product design	•	•	•	•		
Packaging	•	•	•	•		
Product range	•	•	•	•		
Marketing planning/sales acquisition						
Market area	•	•	•		•	
Marketing channels	•	•	•	•	•	
Sales strategy/promotion	•	•	•	•		•
Order fulfillment						
Location of production and distribution facilities	•	•			•	
Sourcing of supplies	•	•			•	
Production system	•		•	•		•
Inventory management	•	•	•	•		•
Materials handling	•	•	•	•		
After-sales service	•		•	•		•
Recycling/reverse logistics	•	•	•	•	•	

Key: •, a direct relationship exists.

(3) *Channel-based strategy.* The aim is to improve the management of logistical activities performed jointly by supply chain partners.

A sample of 375 U.S. manufacturers were asked into which of these three categories their logistic strategy fell. Approximately 54% identified with the process-based strategy, 28% with market-based strategy, and only 9% with the channel-based strategy. McGinnis and Kohn (1993) tested the validity of this typology in two surveys in which managers were asked a series of questions designed to assess the strategic orientation of their logistics function. Cluster analysis of these questionnaire data indicated that the distinction between process-based and market-based strategies was meaningful. They further refined these strategies by distinguishing, in each case, three "substrategies." A later survey by Clinton and Closs (1997) of over 1300 North American companies provided further empirical support for this typology. They used factor analysis to explore the interrelationship between 43 logistical variables to see if companies fell into reasonably coherent strategic groupings. They concluded that it was possible to detect differences in strategic emphasis and expressed "cautious optimism" that the typology proposed by Bowersox et al. was valid.

This classification is, nevertheless, highly generalized. The fact that it requires such a complex, multivariate analysis to produce strategic constructs that match the typology is in itself revealing. It highlights the multidimensional nature, and possibly uniqueness, of any strategies designed to cover the full spectrum of logistical activities. The need to tailor these activities to the particular circumstances of a business makes it very difficult to establish a series of generic strategies for the logistical function as a whole.

It is, however, possible to distinguish a range of strategic *options* relating to particular aspects of a logistical system. These are more clearly identifiable and measurable. A good example is the choice that companies must make between a postponement and a speculation strategy (Van Hoek et al., 1998). The geographical (or "place") form of postponement involves centralizing inventory and delaying its dispatch to local markets until an accurate estimate of the likely demand can be made. Speculation, on the other hand, entails dispersing inventory to local markets in the belief that you will be better able to respond to short-term increases in demand. These contrasting strategies have been examined in detail and practical tools developed to help firms determine under what circumstances they are appropriate. Both Cooper (1993) and Pagh and Cooper (1998) have constructed simple matrices to show how the preferred strategy is likely to be influenced by the nature of the product and packaging, the geography of the market, and the manufacturing strategy.

8. The role of freight transport with integrated logistics strategies

Freight transport operations are affected by a broad range of strategic decisions made at both a logistical and a corporate level within the business. These decisions impact upon different aspects of the transport operation. Table 2 is an attempt to map the interrelationships between a set of six freight transport parameters and areas of strategic decision-making grouped in relation to the three core business processes identified above, namely order fulfillment, marketing planning/sales acquisition, and product development. The presence of a dot in a cell signifies the existence of a direct relationship.

This shows that the nature of the freight transport operation is the result of a complex web of decision-making, spanning different functional areas within the business. As a result of the process of integration at functional, corporate, and supply chain levels, the strategic context within which transport decisions are made has undergone a radical change over the past 40 years. Few studies have examined the effects of this change on the physical movement of freight. Little is known, for example, about the impact of BPR or the application of the postponement principle on freight traffic levels, the modal split, and vehicle load factors?

Of the freight transport parameters listed in Table 2, only the volume of freight movement has been analyzed in detail within an integrated logistics management context. It can be argued that freight traffic levels were influenced by four levels of logistical decision-making, relating to:

(1) *Logistical structures:* numbers, locations, and capacities of factors, warehouses, terminals, and shops.
(2) *Supply chain configuration:* patterns of trading links within the logistical structures.
(3) *Scheduling of flows:* manifestation of the trading links as discrete freight movements.
(4) *Management of transport resources:* relating to the choice of vehicle, utilization of vehicle capacity, routing of delivery, etc.

The growth of freight traffic is the result of a complex interaction between decisions made at these different levels. Decisions at levels 1 and 2 determine the quantity of freight movement measured in tonne-kilometers, while decisions at levels 3 and 4 translate this movement into vehicle traffic, measured in vehicle-kilometers. This decision-making hierarchy has been adopted by several EU-funded research projects (e.g., Demkes, 1999) and been advocated by the U.K. government's Standing Advisory Committee on Trunk Road Assessment (SACTRA) (1999) as a framework for future road freight forecasting.

9. Conclusion

Freight transport is an integral part of logistical systems and supply chains. Analysis of the nature, volume, and pattern of freight movement must therefore be rooted in an understanding of the way that these systems and chains function and evolve. This chapter has outlined the development of logistics management since the early 1960s, highlighting the different stages in its integration. Over the past decade, more formal methods of strategic planning have been applied at both the corporate and the functional level. Within the strategic planning process, there is now wide recognition that logistics is a major determinant of competitiveness, profitability, and growth. Over the next decade, globalization and the growth of e-commerce will further reinforce its position within the corporate hierarchy.

References

Bowersox, D.J. (1978) *Logistical management*, 2nd edn. New York: Macmillan.
Bowersox, D.J. and D.J. Closs (1996) *Logistical management: The integrated supply chain process.* New York: Macmillan.

Bowersox, D.J., P.J. Daugherty, C.L. Droge, D.S. Rogers and D.L. Wardlow (1989) *Leading edge logistics: Competitive positioning for the 1990s.* Oak Brook, IL: Council of Logistics Management.
Christopher, M.C. (1998) *Logistics and supply chain management*, 2nd edn. London: Financial Times/ Pitman.
Clinton, S.R. and D.J. Closs (1997) "Logistics strategy: Does it exist?", *Journal of Business Logistics*, 18:19–44.
Cooper, J.C. (1993) "Logistics strategies for global businesses", *International Journal of Physical Distribution and Logistics Management*, 24:12–23.
Cooper, J.C. and M. Johnston (1990) "Dedicated contract distribution: An assessment of the UK market place", *International Journal of Physical Distribution and Logistics Management*, 20:25–31.
Cooper, M., D.E. Innis and P.R. Dickson (1992) *Strategic planning for logistics.* Oak Brook, IL: Council of Logistics Management.
Datamonitor (1999) *Global logistics.* London: Datamonitor.
Demkes, R. (ed.) (1999) "TRILOG-Europe end report", TNO, Delft, TNO-report Inro/Logistiek 1999-16.
European Logistics Association (2000) *Transport optimisation.* Brussels: European Logistics Association.
Fabbes-Costes, N. and J. Colin (1999) "Formulating logistics strategy", in: D. Waters, ed., *Global logistics and distribution planning: Strategies for management*, 3rd edn. London: Kogan Page.
Fawcett, S.E., S.R. Smith and M.B. Cooper (1997) "Strategic intent, measurement capability and operational success: making the connection", *International Journal of Physical Distribution and Logistics Management*, 27:410–421.
Gentry, J.J. (1995) "The role of carriers in buyer–supplier strategic partnerships: A supply chain management approach", *Journal of Business Logistics*, 17(2):35–55.
Hammer, M. and J. Champy (1993) *Re-engineering the corporation: A manifesto for business revolution.* London: Nicholas Brealey.
Hines, P. (1999) "Future trends in supply chain management", in: D. Waters, ed., *Global logistics and distribution planning: Strategies for management*, 3rd edn. London: Kogan Page.
A.T. Kearney (2000) *Insight to impact: Results of the Fourth Quinquennial European Logistics Study.* Brussels: European Logistics Association.
Lambert, D. and J.R. Stock (1993) *Strategic logistics management*, 3rd edn. Irwin, CA: Homewood.
Lewis, J.C., M.M. Naim and D.R. Towill (1997) "An integrated approach to re-engineering material and logistics control", *International Journal of Physical Distribution and Logistics Management*, 27:197–209.
McGinnis, M.A. and J.W. Kohn (1993) "A factor analytic study of logistics strategy", *Journal of Business Logistics*, 14:1–23.
McKinnon, A.C. (1989) *Physical distribution systems.* London: Routledge.
McKinnon, A.C. (1998) "Scottish brewers: The restructuring of a depot system", in: D. Taylor, ed., *Global cases in logistics and supply chain management.* London: International Thomson Business Press.
McKinnon, A.C. (1999) "The outsourcing of logistical activities", in: D. Waters, ed., *Global logistics and distribution planning: Strategies for management*, 3rd edn. London: Kogan Page.
McKinnon, A.C. and Woodburn (1996) "Logistical restructuring and freight traffic growth: An empirical assessment", *Transportation*, 23(2):141–161.
Morash, E.A., C. Droge and S. Vickrey (1996) "Boundary spanning interfaces between logistics, production, marketing and new product development", *International Journal of Physical Distribution and Logistics Management*, 26:43–62.
Pagh, J.D. and M.C. Cooper (1998) "Supply chain postponement and speculation strategies: How to choose the right strategy", *Journal of Business Logistics*, 19:13–33.
Persson, G. (1991) "Achieving competitiveness through logistics", *International Journal of Logistics Management*, 2:1–11.
Porter, M.E. (1985) *Competitive advantage.* New York: Free Press.
Scott, C. and R. Wesbrook, (1991) "New strategic tools for supply chain management", *International Journal of Physical Distribution and Logistics Management*, 21:23–33.
Standing Advisory Committee on Trunk Road Assessment (SACTRA) (1999) *Transport and the economy.* London: The Stationery Office.

Stevens, G.C. (1989) "Integrating the supply chain", *International Journal of Physical Distribution and Logistics Management*, 19:3–8.

Stewart, W. (1965) "Physical distribution: Key to improved volume and profits", *Journal of Marketing*, 29:65–70.

Van Hoek, R., Commandeur, H.R. and B. Vos (1998) "Reconfiguring logistics systems through postponement strategies", *Journal of Business Logistics*, 19(1):33–54.

Chapter 11

LEAN LOGISTICS

PETER HINES, DANIEL JONES and NICK RICH[*]
Cardiff Business School

1. Introduction

This chapter sets out an alternative approach to designing and managing a logistics system, which is called here "lean logistics." It draws on research that has sought to extend the production system logic pioneered by Toyota beyond the factory gate and into industries other than automobiles. The chapter starts by outlining the dilemma facing managers trying to implement new business practices and then proposes an alternative way of rethinking the logic of value creation, before illustrating how this works in a complete logistics system.

Central to this new logic is a detailed understanding of the waste or inefficiencies that lie in existing systems. Such an understanding is required so that radical or incremental improvements can then be made in the development of a lean logistics system. A framework to do just this is presented and is called "value stream mapping." The approach is illustrated with an example from the distribution industry. The chapter also raises a number of key questions for the academic community in terms of future research and applications within the broad area of lean logistics.

2. Problem definition

The last 20 years have seen a series of "new" business solutions, each of which has contributed a new perspective, but each of which may have now run its course, certainly as a stand-alone solution. These include:

[*]An earlier version of this chapter appeared as: Jones, D., P. Hines and N. Rich (1997) "Lean logistics", *International Journal of Physical Distribution and Logistics Management*, 27(3/4):153–173. The authors would like to thank the editors of the journal for allowing it to be reproduced in a modified version here.

Handbook of Logistics and Supply Chain Management, Edited by A.M. Brewer et al.
© *2001, Elsevier Science Ltd*

(1) the Tom Peters approach, which brought the customer back into prime focus and told us to "thrive on chaos" and break up traditional management structures (Peters and Waterman, 1992);

(2) the total quality management (TQM) movement, often incorrectly labeled "Japanese management," which advocated the power of variance analysis and control and the necessary involvement of shopfloor teams in eliminating root causes of variance (Deming, 1982);

(3) the production control perspective, which sought to solve the chaos and variance problems in supply chains by better forecasting and materials requirement planning (MRP) and control systems (Hill, 1995);

(4) the purchasing emphasis, which involved a transformation to partnerships with suppliers rather than the previous adversarial-style relationships (Ellram, 1991); and

(5) the business process re-engineering (BPR) viewpoint, which stressed the importance of the process and offered ways of automating these processes to cut costs (Hammer and Champy, 1993).

With hindsight, all the above tended to be partial solutions focused somewhat narrowly on particular aspects of the complex process of running a business. Popular criticisms of such approaches are that:

(1) TQM raised morale on the shopfloor, but could not cumulate the gains to real bottom-line savings, while buyer–supplier partnerships again often failed to deliver (De Meyer and Wittenberg-Cox, 1992; Sako et al., 1994);

(2) MRP and manufacturing resouces planning (MRP II) created monolithic systems that cannot respond to rapidly changing demand or unpredictable interruptions, and add cost while failing to improve utilization (Schonberger, 1990); and

(3) the credibility of BPR was undermined because it was used as a crude headcount-reduction device, particularly in the U.S.A.

One likely reason why these waves of change have had such a limited impact is that companies are still in a world of batch and queue processing, with armies of expediters and progress chasers working to beat the system and cope with the chaos. The second reason is that these remedies do not dig deep enough really to transform the ways companies operate – they have too often been seen as bolt-on extras. The third reason is that consultants have only really been able to carry out in-house change programs or to sort out the logistics between pairs of companies, when in fact the changes need to be replicated by all firms involved up and down the supply chain.

However, there are examples of organizations that stand out from the general status quo and these offer promising avenues that may help the observer to make sense of what went right and what went wrong with the solutions mentioned above.

To get beyond these partial solutions more fundamental questions about the organization of value creation have to be asked. In addition it is necessary to look behind the tools and techniques being used by these exemplar firms to discover the underlying logic behind them.

3. A new logic

A natural starting point is with *value* creation – from the customer's perspective the only reason for a firm to exist (Womack and Jones, 1996a). If subjected to a careful review, many of the steps required in the office to translate an order into a schedule and many of the steps required in the factory physically to create the product, add little or no value for the customer.

Taiichi Ohno defined seven common forms of waste (i.e., activities that add cost but no value): production of goods not yet ordered, waiting, rectification of mistakes, excess processing, excess movement, excess transport, and excess stock (Monden, 1993). It is common to find that in a factory less that 5% of activities actually add value, 35% are necessary non-value-adding activities, and 60% add no value at all (Hines and Taylor, 2000). It is easy to see the steps that add value, but it is much more difficult to see all the waste that surrounds them. The thesis is that beginning to eliminate this 60% of activities and costs offers the biggest opportunity for performance improvement today.

The provision of most goods and services stretches across several departments and functions inside a firm and across many different firms. Making an aluminum drinks can, for instance, involves six manufacturing firms from the mine to the supermarket. Each of these elements in the chain – the reduction mill, the smelter, the hot roller, the cold roller, the can maker, and the bottler – is busy trying to optimize its own performance, through typically bigger machines with faster throughput times and bigger batches. This in turn leads to a series of intermediate stores getting bigger all the time. The net result is that it takes 319 days to make the can, which itself takes only about 3 hours of processing time.

Optimizing each piece of the supply chain in isolation does not lead to the lowest cost solution. In fact it is necessary to look at the whole sequence of events, from the customer order right back to the order given to the raw materials producer, and forward through all successive firms making and delivering the product to the customer. In trying to identify possibilities for eliminating waste this makes most sense if it is done for one particular product or product family and for all the tributaries that flow into this stream of value creation.

Focusing on the whole chain is the first step, focusing on the product is the second, and focusing on the flow of value creation (not on the more traditional performance measurement of departments and firms) is the third step. This we term a *value stream* – a new and more useful unit of analysis than the supply chain or the individual firm. Focusing on the flow of value creation immediately challenges the notion that batches are necessary and better. A simplified version of this is illustrated in Figure 1. The point that this value stream concept extends both upstream from the product assembler into the "supply chain" and downstream into the "distribution chain" is reinforced in Figure 2.

Taiichi Ohno demonstrated that if one looks from the perspective of the whole value stream it is possible, and indeed vastly superior, to organize activities so that the work moves from step to step in an uninterrupted *flow* at a rate that exactly matches the *pull* of the customer. This was always thought of as a special case that only applied to very high-volume production of a standardized product, such as a car. Ohno proved the general case and developed the tools and techniques necessary to achieve it. The toolkit for the manufacturing context has come to be called the "Toyota production system." Womack and Jones (1996b) describe its generic form as "lean thinking" and have documented the impressive gains achieved by those firms who have followed his example.

The key elements of Toyota's toolkit are:

(1) level the flow of orders and work by eliminating all causes of demand distortion or amplification;

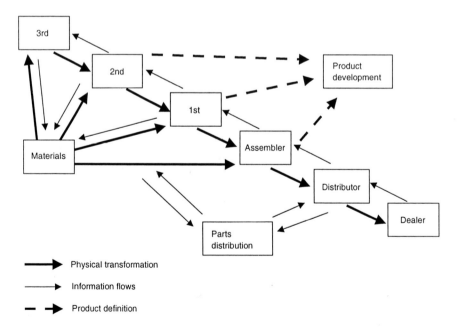

Figure 1. The automotive value stream.

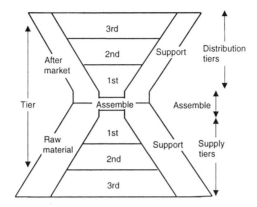

Figure 2. Total domain of the value stream.

(2) organize the work so the product flows directly from operation to operation with no interruptions – shortening set-up or response times to make or deliver every product every day or week and ensuring that no breakdowns occur though preventive maintenance;

(3) only make or deliver what is pulled by the upstream step – no more and no less – sell one, order one;

(4) work throughout the system to the same rhythm as customer demand;

(5) standardize the best work cycle for each task to ensure consistent performance;

(6) standardize and minimize the necessary safety stock between operations;

(7) make every operation detect and stop when an error occurs so it cannot be passed on, making it possible for one employee to supervise several machines or making it possible to detect rogue orders against an historic ordering profile;

(8) manage progress and irregularities using simple visual control devices; and

(9) log irregularities and prioritize in order to conduct root-cause elimination to prevent recurrences and to remove waste from the flow.

The interesting thing that happens when the lean principles are applied, using Ohno's toolkit, is that one begins to rethink not only the organization of the work but also the appropriateness of the size of machines, warehouses, and systems to fit the flow. As people, machines, warehouses, and systems are rethought and combined in different ways, layers of previously hidden waste tend to be uncovered and removed – *perfection* becomes the appropriate goal, not what your competitor is doing today. Perfection is defined as the complete removal of waste until every action and every asset adds real value for the ultimate customer. In theory, waste removal is a continuous process, operating cyclically and without end.

4. A distribution example: Toyota parts supply system

Toyota developed its Toyota Production System (TPS) in the late 1940s in its engine shop. It was fully developed and applied across Toyota's manufacturing operations by the late 1960s. Then, for the first time, it was written down and a group was formed to teach it to Toyota's first-tier suppliers, who were all members of the Toyota Group. By the mid-1970s, the TPS perspective had spread to its other first- and second-tier suppliers in Japan, primarily through their *kyoryoku kai*, or supplier associations (Hines, 1994). After the merger of Toyota's manufacturing and sales companies in 1982, they began to apply the same logic to the aftermarket parts operation in Japan. This took from 1984 to 1990, after which the logic began to spread to overseas parts suppliers and to the aftermarket distribution systems abroad.

Womack and Jones (1996b) studied the transformation of Toyota's U.S. parts distribution system (Figure 3), right back to the second-tier parts manufacturer of a replacement bumper, and compared it with its system in Japan. The picture, shown in Table 1, is an interesting one. The following sections summarize the main actions taken to make the dramatic improvements to the value streams described above.

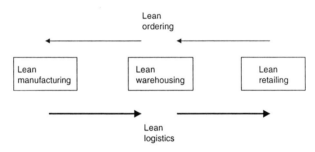

Figure 3. The downstream lean value stream.

Table 1
Toyota parts distribution system efficiency

	U.S.A. 1994	U.S.A. 1996	Japan 1990
Service rate	98% in 7 days	98% in 1 day	98% in 2 hours
System stock index	100	33	19
Throughput time	48 weeks	8 weeks	4 weeks

Source: Womack and Jones (1996b).

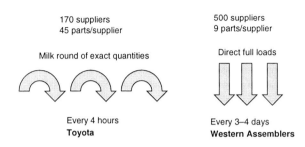

Figure 4. Toyota just-in-time logistics.

4.1. *Manufacturing: Stamping and chroming*

A six-step approach to improvement was enacted on the factory shopfloor, involving:

(1) establishing target set-up times to make every product every day,
(2) reducing set-up times,
(3) changing the layout of machines to create single-piece flow of parts through the process,
(4) moving from monthly to weekly ,and then to daily orders,
(5) stopping the use of MRP systems for production control, and
(6) teaching the chroming subcontractor to process parts in a single-piece flow.

The result of these actions was that throughput time was reduced from 6 weeks to 48 hours, with zero defects and big cost reductions.

4.2. *Delivery*

Toyota picks up both its original equipment and aftermarket parts from suppliers in the local area on a "milk round" at regular and relatively short intervals. Figure 4 compares this with a traditional auto assembly plant receiving full loads of a limited number of part numbers every 3 or 4 days directly from over 500 suppliers. As Toyota have reduced the number of first-tier suppliers it deals with by tiering, modularization, and sourcing many more part types from each supplier, it can pick up small lots of each part number from fewer locations on a milk round every 4 hours. It then never has more than a half a shift's worth of incoming parts at any one time.

The supplier has a relatively stable volume and range of parts, although the precise mix of each part required can be determined a day or two ahead. The overall volume and sequence are relatively fixed, but the mix and volumes per part are flexible. This distribution system also results in much higher transport utilization, as the capacity required can be fine tuned to relatively steady levels and anticipated increases planned for, while at the same time realizing the benefits of small lot just-in-time delivery. The extra cost of emergency shipments, which are prevalent in a batch and queue system, is removed.

4.3. Ordering

In the ordering area it is necessary to move to a daily ordering system from suppliers. In the aftermarket parts area, for instance, this is done on a "sell one, order one" basis rather than by means of traditional standard reorder quantities with long lead times. These orders are then delivered to Toyota at predictable arrival times in the warehouse, thus avoiding delays to lorries and warehouse staff alike.

At the same time a move from monthly to daily orders from the dealers is required so that a steady and relatively even flow of demand is inputted into Toyota's system. The standard frequent deliveries of goods to dealers eliminates the need for emergency vehicle off-road orders, order peaks and troughs, as well as the requirement to offload surplus stock at discount prices common in Europe and the U.S.A.

4.4. Warehouse management

A similar type of logic to that applied in the factory is also used in the warehouse involving:

(1) reduced bin sizes;
(2) storage by part type, with frequently used parts near the front or aisle end;
(3) standard binning and picking routes for each part type;
(4) a division of the working day and tasks into standard work cycles;
(5) synchronized order–pick–pack–despatch and delivery steps for each delivery route (again a milk round) out to a group of local dealers;
(6) staggered outward delivery routes;
(7) controlled progress and irregularities through binning or picking ticket bundles for each cycle (preventing working ahead), and visual control boards;
(8) the logging of irregularities and prioritization in order to conduct root-cause elimination of the most frequent problems to prevent recurrences and hence improve the process.

The result of the warehouse management and ordering improvements is that stock in Toyota's regional distribution center has been reduced from 24 to 4 weeks, while the service rate and productivity have improved to three times that of a similar, traditionally organized facility – and this with no automation. This gives a stock turn of 13, which is equivalent to the best U.S. supermarket chain.

4.5. Dealers

Daily delivery allows the dealer to reduce overall stock levels by well over half, while carrying a wider range of parts. It also improves service rates to waiting customers and eliminates delays in binning parts. It also cuts wasted walking and waiting time for the mechanics picking up parts. Freed up space can be put to revenue-earning use. One of the key roles for the dealer service manager is to level the workflow and plan regular service work in the same manner as in the warehouse.

4.6. Network structure

Toyota uses a network of regional distribution centers in Japan and the U.S.A., handling 50 000 to 60 000 part types. In Japan, the process has been extended to the elimination of almost all parts stock from the dealer. Instead this stock is held at a local distribution center, which carries about 15 000 parts and can deliver parts to dealers 4 times a day.

Dealers place timed orders on the system for regular servicing, which can flow at the right time from the regional distribution center. They can get additional parts, required after the car has been stripped down, in time to complete the job within the day. This eliminates a further 4 weeks of stock from the system. However, the ability to do this depends on the density of the vehicles and dealers within a geographical region. The system works at its best within medium to high urban population densities.

The entire Toyota parts warehousing network in Japan (Figure 5) is part owned by Toyota and part by its large dealer groups, each of which is responsible for many dealer outlets within a prefecture. Just as Toyota is dealing directly with a limited number of parts suppliers, it is also dealing with a limited number of dealer groups in Japan, and not with thousands of separately owned dealers as in the West. This makes shared ownership of the distribution chain and milk-round delivery systems easier to organize.

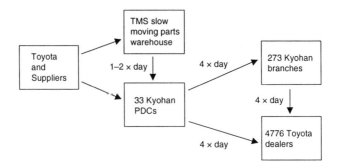

Figure 5. Toyota's parts distribution system in Japan.

5. Targeting improvements within the lean enterprise

The basis of adding value rather than just cost is the elimination of the seven wastes as discussed above. However, as illustrated in the above example this waste elimination needs to be on a wider scale than simply removing waste inside organizations, as advocated by traditional texts on the subject (Miles, 1961). It is necessary to address both the intra- and intercompany value (or waste) adding processes.

At present there is an ill-defined and ill-categorized toolkit to understand the value stream, although a few researchers have developed useful individual tools (e.g., Forza et al., 1993; New, 1993; Beesley, 1994; Jessop and Jones, 1995; Rother and Shook, 1998). In general, these authors have tended to propose their creations as *the* answer, rather than a part of the jigsaw. In addition, the existing tools derive from functional ghettos and so, on their own, do not fit well with the more cross-functional toolkit required by today's best companies.

In order to do this, a seven-element "toolkit" of methods derived from a variety of functional or academic disciplines has been assembled and applied under the term "value stream mapping" (Hines and Rich, 1997). The selection of which of the seven tools to use is made on the basis of the seven wastes prevalent in the value stream concerned and the usefulness of each tool in helping to understand such wastes. A further more detailed discussion of the practical use of these tools can be found in Hines and Taylor (2000).

At this point it is useful to review briefly the tools, before providing a case example of their application in the U.K. distribution industry.

5.1. Process activity mapping

The first of the tools is the traditional industrial engineering tool, process activity mapping (Table 2). There are five stages to this general approach:

(1) the study of the flow of processes,

(2) the identification of waste,

(3) a consideration of whether the process can be rearranged into a more efficient sequence,

(4) a consideration of a better flow pattern involving different flow layout or transportation routing, and

(5) a consideration of whether everything that is being done is really necessary.

In order to do this, several simple stages must be followed. First, a preliminary analysis of the process is undertaken, followed by the detailed recording of all the required items in each process. The result of this is a map of the process under consideration (see Table 2). As can be seen from this example, each step in the flow has been categorized within a variety of types of activity (operation, transport, inspection, and storage). The area used for each of these activities is recorded, together with the distance moved, the time taken, the number of people involved, and any additional comments.

After this, the total distance moved, the time taken, and the number of people involved can be calculated. This calculation is then recorded. The final table (see Table 2) can be used as the basis for further analysis and subsequent improvement. This is often achieved through the use of techniques such as 5W1H (Why does an activity occur? Who does it? On what machine? Where? When? How?). The basis of this approach is, therefore, to try to eliminate activities that are unnecessary, to simplify others, to combine others, and to seek sequence changes that will reduce waste.

5.2. Supply chain response matrix

The origin of the second tool lies in the time compression and logistics movement and goes under a variety of names. It was used by New (1993) and Forza et al. (1993) within a textile supply chain setting. In a more wide ranging work, Beesley (1994) applied what he termed "time-based process mapping" to a range of industrial sectors, including automotive, aerospace, and construction. A similar approach was adopted by Jessop and Jones (1995) in the electronics, food, clothing, and automotive industries. A more encompassing, macro-level approach developed by Toyota was also codified by Rother and Shook (1998).

The fundamentals of this mapping approach (Figure 6) are that it seeks to portray in a simple diagram the critical lead-time constraints for a particular process. In this case it is the cumulative lead time within a distribution company, its suppliers, and its downstream retailer. In Figure 6 the horizontal axis shows the lead time for the product both internally and externally. The vertical axis shows the average amount of standing inventory at specific points in the supply chain.

Table 2
Process activity mapping – a process industry example

Sep	Flow	Machine	Distance (m)	Time (min)	People	Operation	Transport	Inspect	Store	Delay	Comments
1. Raw material	S	Reservoir				O	T	I	S	D	Reservoir/additives
2. Kitting	O	Warehouse	10	5	1	O	T	I	S	D	
3. Delivery to lift	T		120		1	O	T	I	S	D	
4. Offload from lift	T			0.5	1/2	O	T	I	S	D	
5. Wait for mix	D	Mix area	20	20		O	T	I	S	D	
6. Put in cradle	T			2	1/2	O	T	I	S	D	
7. Pierce/pour	O	Mix area	12	0.5	1	O	T	I	S	D	
8. Mix (blowers)	O			20	1/2	O	T	I	S	D	Base material, blow, and additives
9. Test No. 1	I			30	1 + 1	O	T	I	S	D	Sample/test
10. Pump to storage tank	T	Store tank	100		1	O	T	I	S	D	Dedicated reservoir
11. Mix in storage tank	O	Store tank		10	1	O	T	I	S	D	
12. IR rest	I			10	1 + 1	O	T	I	S	D	Stamp and approve

Step	Flow	Dist	Time	Qty	O	T	I	S	D	Notes
13. Await filling	D		15		O	T	I	S	***D***	Longer if screen late
14. To filler head	T	20	0.1	1	***O***	***T***	I	S	D	Filler head
15. Fill/top/tighten	O		1	1 + 1	***O***	***T***	I	S	D	One unit
16. Stack	T	3	0.1	1	O	***T***	I	S	D	One unit
17. Delay to fill one pallet	D		30		O	T	I	S	***D***	
18. Strap pallet	O		2	1	***O***	T	I	S	D	
19. Transfer to store	T	80	2	1	O	***T***	I	S	D	
20. Await truck	D		540		O	T	I	S	***D***	Batch 360/ queue 180
21. Pick/move by forklift	T	90	3	1	O	***T***	I	S	D	Forklift
22. Wait to fill full load	D		30	1 + 1	O	T	I	S	***D***	One operator, one haulier
23. Await shipment	D		60	1	O	T	I	S	***D***	One haulier
Total	23 steps	443	781.2	25	6	8	2	1	6	
Operators			38.5	8						
% Value adding			4.93%	32%						

Key: D, delay; I, inspect; O, operation; S, store; T, transport. Bold italic letters denote a concurrence of flow and activity.

In this case the cumulative lead time is 42 working days and 99 working days of material is held within the system. Thus a total response time of 141 working days can be seen to be typical in this system. Once this is understood, each of the individual lead times and inventory amounts can be targeted for improvement activity, as was shown with the process activity mapping approach.

5.3. Production variety funnel

The production variety funnel is shown in Figure 7. This approach originates in the operations management area and has been applied by New (1993) in the textile industry. A similar method is IVAT analysis, which views internal operations in companies as consisting of activities that conform to an I, V, A, or T shape (Macbeth and Ferguson, 1994). I-plants consist of unidirectional, unvarying production of multiple identical items (e.g., chemical plant). V-plants consist of a limited number of raw materials processed into a wide variety of finished products in a generally diverging pattern. V-plants are typical in the textile and metal fabrication industries. A-plants, in contrast, have many raw materials and a

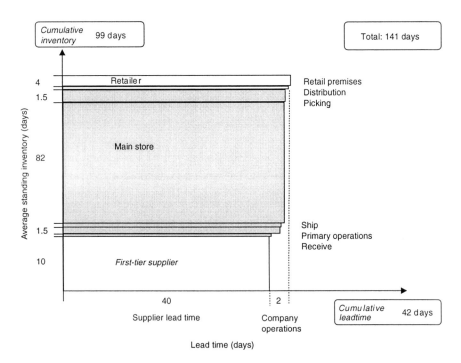

Figure 6. Supply chain response matrix – a distribution example.

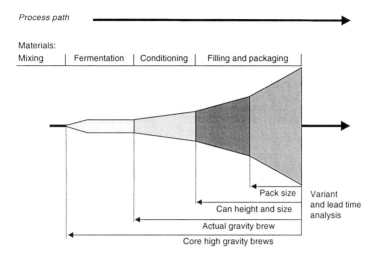

Figure 7. Production variety funnel – a brewing industry example.

limited range of finished products, with different streams of raw materials using different facilities. Such plants are typical in the aerospace or other major assembly industries. Lastly, T-plants have a wide combination of products from a restricted number of components made into semiprocessed parts held ready for a wide range of customer demanded final versions. This type of plant is typical in the electronics and household appliance industries.

Such a delineation using the production variety funnel (see Figure 7) allows the mapper to understand how the firm or the value stream operates and the accompanying complexity that has to be managed. In addition, such a mapping process helps potential research clients to understand the similarities and differences of their industry compared to one that may have been more widely researched. The approach can be very useful in helping to decide where to target inventory reduction and in making changes to the processing of products. It is also useful in gaining an overview of the company or supply chain being studied.

5.4. Quality filter mapping

The fourth value stream mapping tool is the quality filter mapping approach. This is a new tool that is designed to identify where quality problems exist in the value stream. The resulting map itself shows where three different types of quality defects occur in the value stream (Figure 8). The first of these is product defects. Product defects are defined here as defects in goods produced that are not caught by in-line or end-of-line inspection and are therefore passed on to customers.

The second type of quality defect is what may be termed service defects. Service defects are problems given to a customer that are not directly related to the goods themselves, but are due to the accompanying level of service. The most important of these service defects are inappropriate delivery (late or early) and incorrect paperwork or documentation. In other words, such defects include any problems that customers experience that are not concerned with faulty physical goods.

The third type of defect is internal scrap. Internal scrap refers to defects made within a company that have been caught by in-line or end-of-line inspection. The in-line inspection methods will vary, and can consist of traditional product inspection, statistical process control, or through "poke yoke" (or foolproofing) devices. Such methods may, respectively, be used to inspect in quality, to control quality within given tolerances, or to prevent defects from occurring in the first place.

Each of these three types of defect is mapped along the value stream. In the automotive example given in Figure 8 this value stream consists of the distributor, assembler, first-tier supplier, second-tier supplier, third-tier supplier, and raw material source. This approach has clear advantages in identifying where defects are occurring, and hence in identifying problems, inefficiencies, and wasted effort. This information can then be used for subsequent improvement activity.

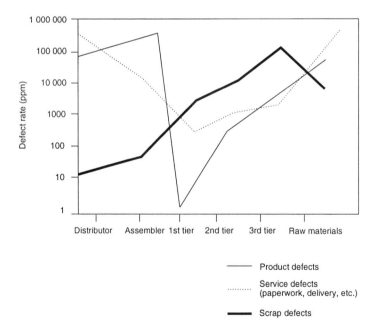

Figure 8. The quality filter mapping approach – an automotive industry example.

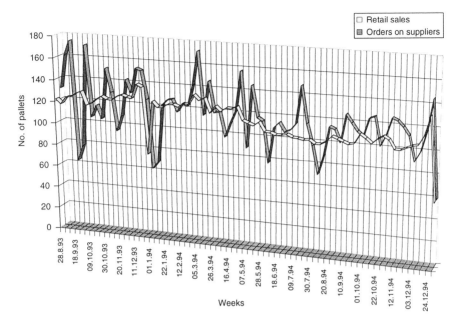

Figure 9. Demand amplification mapping – an FMCG food product example (source: Jessop and Jones, 1995).

5.5. Demand amplification mapping

Demand amplification mapping has its roots in the systems dynamics work of Forrester and Burbidge. What has now become known as the Forrester effect was first described in a *Harvard Business Review* article in 1958 by Jay Forrester. This effect is primarily linked to delays and poor decision-making concerning information and material flow. The Burbidge effect is linked to the law of industrial dynamics which states: "if demand is transmitted along a series of inventories using stock control ordering, then the amplification of demand variation will increase with each transfer" (Burbidge, 1984). Thus in unmodified supply chains, excess inventory, production, labor, and capacity are generally found as a result. It is therefore quite likely that on many day-to-day occasions manufacturers will be unable to satisfy retail demand even though they are, on average, able to produce more goods than are being sold. As a result, within a supply chain setting manufacturers have sought to hold, in some cases sizeable, stocks to avoid such problems. Forrester (1958) likens this to driving an automobile blindfolded, with instructions being given by a passenger.

The basis of the mapping tool in the value stream setting is illustrated in Figure 9. In this instance a fast moving consumer goods (FMCG) food product is being mapping through its distribution through a leading U.K. supermarket retailer. In this simple example two curves are plotted. The first represents the actual consumer sales as recorded by electronic point of sale (EPOS) data. The second curve represents the orders placed to the supplier to fulfil this demand. As can be seen, the variability of consumer sales is far lower than the supplier orders. It is also possible subsequently to map this product further upstream. An example may be the manufacturing plant of a cleaning products company or even the demand they place on their raw material suppliers.

This simple analysis tool can be used to show how demand changes along the value stream within varying time buckets. This information can then be used as the basis for decision-making and further analysis to try to redesign the value stream configuration, to manage the fluctuations, to reduce the fluctuation, or to set up dual-mode solutions where regular demand can be managed in one way and exceptional or promotions demand can be managed in a different way.

5.6. Decision point analysis

Decision point analysis is of particular use for T-plants or value streams that exhibit similar features, although it may be used in other types of industries. The decision point is the point in the value stream where actual demand pull gives way

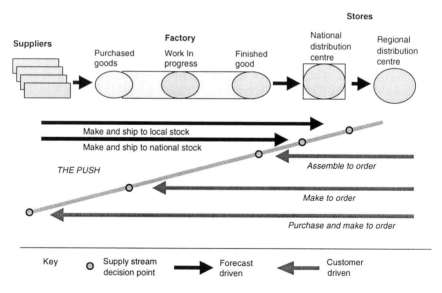

Figure 10. Decision point analysis – an FMCG example (source: Hines and Rich, 1997).

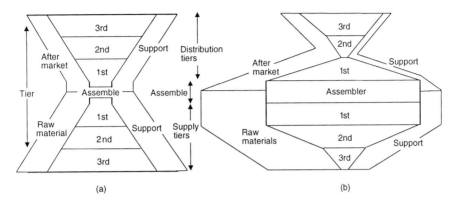

Figure 11. Physical structure mapping – an automotive industry example: (a) by number of firms involved; (b) by cost.

to a forecast-driven push. In other words, it is the point at which products stop being made due to actual demand and start being made against forecasts alone (Hoekstra and Romme, 1992). In Figure 10, which gives an example from the FMCG industry, the decision point can be at any point from regional distribution centers to national distribution centers through to any point inside the manufacturer or indeed at any tier within the supply chain.

Gaining a basic understanding of where this point lies is useful for two reasons. First, at the immediate level, it becomes possible to assess the processes that operate both downstream and upstream from this point. The purpose of this is to ensure that the processes are aligned with the relevant pull or push philosophy. Second, at a more fundamental and longer term level, it is possible to design various "What if?" scenarios to view the operation of the value stream if the decision point is moved. This may allow for a better design of the value stream.

5.7. Physical structure

Physical structure mapping is a new tool that has been found to be useful in understanding what a particular value stream looks like at an overview or industry level. This knowledge is helpful in understanding what the industry looks like, how it operates and, in particular, in directing attention to areas that may not be receiving sufficient developmental attention.

The tool is illustrated in Figure 11 and can be seen to be split into two parts: volume structure and cost structure. Figure 11a shows the structure of the industry according to the various tiers that exist in both the supplier area and the distribution area, with the assembler located in the middle point. In this simple example there are three supplier tiers and three distribution tiers. In addition, the supplier area is seen to include raw material sources and other support suppliers (e.g., tooling, capital equipment, office supplies firms). These two sets of firms are not given a tier level as they can be seen to interact with the assembler as well as the other supplier tiers.

The distribution area in Figure 11a includes three tiers as well as a section representing the aftermarket (in this case for spare parts), as well as various other support organizations providing consumables and service items. This complete industry map therefore captures all the firms involved, the area of each part of the diagram being proportional to the number of firms in each set.

Figure 11b maps the industry in a similar way, with the same sets of organizations. However, instead of linking the area of the diagram to the number of firms involved, it is directly linked to the value adding process (or, more strictly speaking, to the cost adding process). It can be seen that in this automotive case the major cost adding occurs within the raw material firms, the first-tier suppliers, and the assembler themselves. In this case the distribution costs are not significant.

However, the basis of the use of this second type of map is that it is then possible to analyze the value adding that is required in the final product as it is sold to the consumer. Thus value analysis tools employed by industrial engineers can be focused at the complete industry or supply chain structure level (Miles, 1961). Such an approach may result in a redesign of how the industry itself functions, either if the sector is dominated by one firm such as Benetton or if all the firms in the industry can be brought together to achieve such an aim. Thus, similarly to the process activity mapping tool discussed above, attempts can be made to try to eliminate activities that are unnecessary, to simplify others, to combine others, and to seek sequence changes that will reduce waste.

5.8. Summary

The use of the seven-element toolkit is not confined to any particular theoretical approach to ultimate implementation. Thus the option of whether to adopt a *kaizen*, *kaikaku* (major step change improvement), or business process re-engineering approach once the tools have been used can be left open at this stage.

6. A distribution case

In order for the reader to gain a better understanding of the approach, a brief industry case will be reviewed. The company involved is a highly profitable leading industrial distributor with over 100 000 products and an enviable record for customer service.

After undertaking preliminary discussions it was decided to focus on the · upstream value stream to the point where goods were available for distribution by the firm. Nine products were chosen based on a Pareto analysis from one particular value stream, namely the lighting product range. Preliminary interviews with key cross-functional staff showed that unnecessary inventory, defects, inappropriate processing, and transportation were the most serious wastes in the system. Based on the knowledge of the supporting research team it was decided to adopt five of the tools: process activity mapping, supply chain response matrix, quality filter mapping, demand amplification mapping, and decision point analysis.

The choice of these particular tools was based on two factors. First, of the available tools, it was felt that the five chosen would be most appropriate in identifying and understanding the particular types of waste that had been suggested by the firm as being present. Second, the use of further tools would provide only small additional benefit and so they were not used in this case.

The on-site mapping work was carried out over a 3-day period and proved that each of the tools was of value in analyzing the selected value streams. An example of how the interplay of the tools was effective was that the supply chain response matrix suggested that the key priority for the firm was supplier lead-time reduction. However, when the data from the quality filter mapping were added it was found that the real issue was on-time delivery rather than lead-time reduction. Thus, if the supply chain response matrix had been used on its own it might have resulted in shorter lead time but exacerbated the true problem of on-time delivery.

The work assisted the firm to conclude that, although there was no internal crisis warranting radical change, there was considerable room for improvement, particularly regarding the relatively unresponsive suppliers. As a result, attention has been paid to the setting up of a cross-functionally driven supplier association (Hines, 1994), with six key suppliers in one product group area for the purpose of supplier co-ordination and development. Within this supplier association is an awareness raising program of why change is required, involving ongoing benchmarking. In addition, education and implementation are being carried out using devices such as co-managed inventory, due-date performance, milk rounds, self-certification, stabilized schedules, and electronic data interchange.

The company found the ongoing mapping work very useful, and one senior executive noted that "the combination of mapping tools has provided an effective means of mapping the [company's] supply chain concentrating discussion/action on key issues." Another described the work as "not rocket science but down to earth common sense which has resulted in us setting up a follow up project which will be the most important thing we do." A conservative estimate of the savings that could be reaped as a result of the follow-up work was in excess of £10 million per year, equivalent to a 20% improvement in profitability. Such work will help position the company within a truly lean logistics value stream.

7. Where do we go from here?

This chapter has sought to provide a framework for a new way of thinking about the supply chain termed lean logistics. Lean logistics takes its fundamental philosophy from the Toyota production system and is based around extending this system right along supply chains, from customers right back to raw material extraction. Such an approach has been developed in order to overcome some of the fragmentation problems of traditional functional and business thinking. Within lean logistics the key concepts of value, value streams, flow, pull, and perfection have been discussed. Each of these key concepts has been illustrated using a case study of Toyota's parts supply system.

The starting point for Toyota in developing such a system was their work in understanding waste and inefficiency in their existing value streams. In order for the reader to see how this type of work may be done, a framework has been presented here called value stream mapping. This approach can been used to diagnose waste and help organizations and value streams to make subsequent radical or incremental improvements. A brief distribution case has been provided to illustrate the method as a basis for an extended improvement program that the company is now undertaking.

As this chapter was designed to be largely conceptual in nature, it may be useful to suggest a number of missing pieces or unresolved issues in our knowledge and application of lean logistics. This offers a basis for future work. The questions and issues are:

(1) How can we define the value and logistical implications of the new product and service bundles required by consumers?
(2) How do we define the value stream when in reality value streams overlap, split, and coalesce?
(3) How do we create an effective "wake-up call" for value streams to make the necessary improvements?
(4) What are the right warehouse types, locations, and operating systems for lean logistics?
(5) How do we design value streams that dampen chaos rather than amplify it?
(6) How do we understand the change management steps required to move from batch-and-queue to flow-and-pull logic?

While this list is not exhaustive it is designed to capture some of the important issues as supply chain management moves towards value stream management, industry moves from a mass production paradigm to a lean production approach, and the unit of competitive advantage moves from the company to the complete lean enterprise (Hines et al., 1998). The solutions to the problems, while not readily apparent, are now the subject of ongoing research at a number of institutes.

References

Beesley, A. (1994) "A need for time based process mapping and its application in procurement", in: Proceedings of the 3rd IPSERA Conference, University of Glamorgan, pp. 41–56.

Burbidge, J. (1984) "Automated production control with a simulation capability", in: Proceedings of the IFIP Conference, Copenhagen.

De Meyer, A. and A. Wittenberg-Cox (1992) *Creating product value: Putting manufacturing on the strategic agenda*. London: Pitman.

Deming, E. (1982) *Out of crisis*. Cambridge, MA: Massachusetts Institute of Technology Center for Advanced Engineering Study.

Ellram, L. (1991) "A managerial guideline for the development and implementation of purchasing partnerships", *International Journal of Purchasing and Materials Management*, 27(2):2–8.

Forrester, J. (1958) "Industrial dynamics: a major breakthrough for decision makers", *Harvard Business Review*, July–August, 37–66.

Forza, C., A. Vinelli and R. Filippini (1993) "Telecommunication services for quick response in the textile–apparel industry", in: Proceedings of the 1st International Symposium on Logistics, University of Nottingham, pp. 119–126.

Hammer, M. and C. Champy (1993) *Reengineering the corporation*. New York: Harper Business.

Hill, T. (1995) *Manufacturing strategy: Text and cases*. Basingstoke: MacMillan.

Hines, P. (1994) *Creating world class suppliers: Unlocking mutual competitive advantage*. London: Pitman.

Hines, P. and N. Rich (1997) "The seven value stream mapping tools", *International Journal of Production and Operations Management*, 17(1):44–62.

Hines, P. and D. Taylor (2000) *Going lean: A guide to implementation*. Cardiff: Lean Enterprise Research Centre, Cardiff Business School.

Hines, P., N. Rich, J. Bicheno, D. Brunt, D. Taylor, C. Butterworth and J. Sullivan (1998) "Value stream management", *International Journal of Logistics Management*, 9(2):25–42.

Hoekstra, S. and S. Romme (1992) *Towards integrated logistics structure – developing customer-oriented goods flows*. New York: McGraw-Hill.

Jessop, D. and O. Jones (1995) "Value stream process modelling: A methodology for creating competitive advantage", in: Proceedings of the 4th IPSERA Conference, University of Birmingham.

Jones, D., P. Hines and N. Rich (1997) "Lean logistics", *International Journal of Physical Distance and Logistics Management*, 27(3/4):153–173.

Macbeth, D. and N. Ferguson (1994) *Partnership sourcing: An integrated supply chain approach*. London: Pitman.

Miles, L. (1961) *Techniques of value analysis and engineering*. New York: McGraw-Hill.

Monden, Y. (1993) *Toyota production system: An integrated approach to just-in-time*, 2nd edn. Norcross, GA: Industrial Engineering and Management Press.

New, C. (1993) "The use of throughput efficiency as a key performance measure for the new manufacturing era", *International Journal of Logistics Management*, 4(2):95–104.

Peters, T. and R. Waterman (1992) *In search of excellence*. New York: Harper Collins.

Rother, M. and J. Shook (1998) *Learning to see*. Brookline, MA: Lean Enterprise Institute.

Sako, M., R. Lamming and S. Helper (1994) "Supplier relations in the UK car industry: good news–bad news", *European Journal of Purchasing and Supply Management*, 1(4):237–248.

Schonberger, R. (1990) *Building a chain of customers linking business functions to create the world class company*. New York: Free Press.

Womack, J. and D. Jones (1996a) "Beyond Toyota: how to root out waste and pursue perfection", *Harvard. Business Review*, September–October, pp. 140–154.
Womack, J. and D. Jones (1996b) *Lean thinking: Banish waste and create wealth in your corporation.* New York: Simon & Schuster.

Chapter 12

INVENTORY MANAGEMENT

DONALD WATERS
Penzance

1. Introduction

All organizations hold stocks. These are the goods they acquire and put into storage until needed. The amount of stock varies enormously, but a typical organization holds about 20% of its annual turnover. This is a major investment, which managers want to organize as efficiently as possible. Unfortunately, this is more difficult than it seems. A huge amount of work has been done to develop the principles of effective inventory control, but it is still difficult to identify the best policies. There is no ideal way of organizing stocks, and the best options depend on the type of operations, constraints, objectives, and a whole range of subjective factors. In spite of this, there is a clear trend towards lower stocks. New methods allow organizations to move goods quickly through the supply chain, working with far less stock to achieve the same levels of customer service. This chapter describes the most important work in the area.

2. Stocks and inventories

Managers often use different terms to describe the same idea or, more confusingly, use the same term to mean different things. *Stocks*, for example, are goods that are stored until they are needed, and they have no connection with "stocks and shares." An *inventory* is actually a list of the items held in stock, but many people use it to mean both the list of items and the stocks themselves.

There are several different types of stock, which we can classify as raw materials, work in progress, finished products, spare parts, and consumables. These are, of course, only convenient labels, and one organization's finished products become another organization's raw materials. Whatever names we use, the basic operations of inventory management are fairly straightforward. They follow a simple cycle from procurement through to use:

Handbook of Logistics and Supply Chain Management, Edited by A.M. Brewer et al.

(1) at some appropriate time the purchasing or procurement function places an order for an item with a supplier;
(2) after a lead time, the delivery arrives;
(3) this delivery is checked, sorted, and put into storage;
(4) over time, users within the organization need units of the item to support their operations;
(5) units are removed from stock to meet this internal demand; and
(6) stock falls, and at some point another order is placed and the cycle starts again.

You can see from this simple outline that organizations have to answer three important questions of inventory management:

(1) What items should we stock?
(2) When should we place an order?
(3) How much should we order?

The first of these is perhaps the most difficult. It is largely a matter for management judgement, and the only advice we can give is that the benefits of holding any item in stock must be greater than the penalties of not holding it. Few organizations are good at making this decision. As a result, stockholdings tend to drift upwards, as new items are added without enough thought, and old, unused items are not removed. This problem of overstocking can become severe, and managers often react with drastic stock-reduction programs that remove useful items along with the junk.

The last two questions are easier to answer, and the rest of this chapter describes some useful analyses. You might think that by focusing on these two questions we are avoiding some other important issues. How much should we invest in stock? What are the average stock levels? What service do we give to customers? How often are there shortages? What are the costs? The answers to these, and related, questions actually emerge from the ordering policy defined by the two fundamental questions above. Setting the size and timing of orders effectively fixes the average stock level, costs, service level, likelihood of shortage, and so on. We could, of course, tackle the questions the other way around and look for, say, an optimal investment in stock, which would then set the size and timing of orders. This circular argument only reinforces the view that we can fix the overall features of an inventory system by setting values to a few of its variables. How we set about this, depends on our objectives.

3. Objectives of inventory management

Problems of inventory management have been around for a very long time, but people are still developing new ideas. Despite a huge amount of previous work, the most dramatic changes have occurred within the past few years. A combination of technology and improved operations has allowed the development of new methods of inventory management, and these have been encouraged by a fundamental change of attitude towards stock.

The need to collect food when it is readily available and then store it for times of shortage is, perhaps, the oldest inventory problem, and one that confronts almost every living creature. The implicit objective is to store as much food as possible, as those with the largest stocks have the highest chance of survival. This view, that "high stock are good," has been dominant for most of our history. The accumulation of possessions is a measure of personal and collective wealth, as suggested by Pappilon in 1696 who said, "The stock of riches of a kingdom doth not only consist of our money but also in our commodities and ships for trade and magazines furnished with all necessary material" (quoted in Silver and Peterson, 1985).

The main purpose of stock, however, is not to demonstrate wealth but to give organizations a buffer between the supply of an item and demand. Stocks mean that we do not have to match the supply of an item exactly to every demand, but we have a cushion that separates the two. You can see this in a supermarket, where the stock forms a cushion between the small, frequent demands of many customers, and the large, less frequent deliveries by truck.

Having stock as a buffer between supply and demand allows an organization to continue its normal operations when there is variability or uncertainty in either. A restaurant that has to meet an unknown demand for meals keeps stocks of food and ingredients; if a delivery of materials is delayed, a factory can use stocks of raw materials to continue working normally; if customer demand for a product is unexpectedly high, a supplier can meet it from stocks of finished goods. As uncertainty and variability are inherent in most operations, high stocks seem inevitable, and even a strength. They certainly bring clear benefits, allowing for:

(1) mismatches between rates of supply and demand;
(2) demands that are larger than expected, or at unexpected times;
(3) deliveries that are delayed or too small;
(4) price discounts on large orders;
(5) items the price of which is expected to rise;
(6) items that are going out of production or are difficult to find;
(7) full loads that reduce transport costs; and
(8) cover for emergencies.

However, at the beginning of the 20th century, organizations began to question this view that high stocks were inevitably good. Industrialization made the manufacture and distribution of goods more reliable, and reduced uncertainty on the supply side. At the same time, organizations realized that holding stock typically costs 25–30% of its value a year. In other words, their policies of high stocks came with very high costs. At this point they began to look at the balance between the benefits of stock and the costs. Often it seemed that they were paying too much for the benefits, and by lowering stocks they might trade-off a reduced customer service against reduced costs. Although they would occasionally run out of stock and not meet a customer's demand, or have to reschedule operations after a late delivery of materials, they would still get a net benefit.

This has become the dominant view of stocks – they are expensive, but essential. Organizations now look at the trade-offs between holding stock and the penalties of not holding it, and they want policies that minimize the overall cost. This is where *scientific inventory control* has proved most useful.

4. Scientific inventory control

"Scientific inventory control" was a key development in stock management. It builds a quantitative model of an inventory system and calculates the order size – the "economic order quantity" – that minimizes total costs. This order size then sets the associated stock level, investment, service level, and other features. Harris (1915) is often credited with the derivation of the economic order quantity (EOQ), but many people worked in the area, including Wilson (1934) who made notable contributions.

We can describe an idealized inventory system with a single item, steady demand, fixed costs, reliable deliveries, and no shortages. It is then easy to calculate the EOQ, which is defined as the fixed order size that gives lowest overall cost. Effectively, the EOQ balances four costs associated with inventory management:

(1) the *unit cost*, which is the cost of acquiring one unit of the item – usually the purchase or transfer price;
(2) the *reorder cost*, which is the total cost of placing an order for the item and might include order preparation, follow-up, quality checks, transport, receiving, and sorting;
(3) the *holding cost*, which is the cost of holding one unit of an item in stock for one period of time, such as $20 a unit a year; and
(4) the *shortage cost*, which is the cost of not having an item in stock when it is needed.

These costs can be difficult to find, and are often little more than values agreed by managers. The shortage cost is particularly complicated. In the simplest case a retailer might lose direct profit from a lost sale, but the effects of shortages are usually wider than this, and include loss of goodwill, disruption to operations, positive action to counteract the shortage, and so on. One guideline says that the shortage cost is much higher than the others, and is the one that can, and should be, avoided. Holding stock is expensive, but it avoids the higher costs of shortages.

The EOQ assumes that shortage costs are so high that they are avoided, and it balances the other three costs. In particular, it looks for a compromise between small, frequent orders and large, infrequent ones. If we place small orders, the average stock is low, but the frequent orders become too expensive – if we place large orders, the cost of placing them is low, but the high stock level raises investment. A fairly simple derivation finds the EOQ as:

$$\text{economic order quantity} = \sqrt{\frac{2 \times \text{reorder cost} \times \text{forecast demand}}{\text{holding cost}}}.$$

Having found the best order size, we can ask the related question about when to place orders. It takes some time to process orders and for the items to arrive in stock. If both this *lead time* and the demand are relatively steady, the best time to place an order is the lead time before existing stocks will run out. In other words, you place an order when the amount of remaining stock is equal to the forecast demand in the lead time. This order will arrive just as the existing stock is running out. In practice, there is likely to be some uncertainty in both the forecast demand and the lead time, so organizations protect themselves by keeping some extra *safety stock* in reserve. Then:

reorder level = forecast demand in lead-time + safety stock.

These two equations define the main policies for inventory management. They can be used in a wide variety of circumstances, and until quite recently were used for controlling every major store.

Figure 1 shows a typical stock cycle, which we can use to calculate other values, such as the time between orders, the average stock level, investment, customer service, and so on. Although this basic analysis uses a series of assumptions and simplifications, it has several strengths. Most importantly, it is easy to use and gives good guidelines for order size and timing in a wide range of circumstance. If actual circumstances are too different from the idealized model, it is easy to modify the equations and get equivalent results. Organizations can, for example, easily allow for variable demand, price discounts, joint orders for related items, finite delivery rates, and so on. Over the years, more sophisticated models have been developed to give good results for many related problems.

5. Periodic review systems

Variations on the EOQ are widely used, but they do not give perfect answers. One common complaint from managers is the varying time between orders. The EOQ is a fixed order size that is always used and any variation in demand is dealt with by varying the time between orders. This *fixed order quantity* approach does not suit some organizations, which prefer to use a *fixed order period* or *periodic review*. They find the best time between regular orders, and then allow for any variation in demand by placing orders of varying size. You can see the difference between these two approaches in small shops. Shop A uses a fixed order quantity approach and places an order for five cases of instant coffee whenever stock on the shelves falls to ten jars; Shop B uses a periodic review approach, checking the shelves every Saturday and placing an order for the number of jars sold during the week.

The easiest way of organizing a periodic review is to look at stock levels every week, say, and then place an order that brings them up to a target stock level.

order size = target stock level − current stock.

As can be seen from Figure 2, when you place an order it must give enough stock to last until the next order arrives, which is the lead time plus the review period away. In other words, the target stock level must be high enough to meet expected demand over this time. Then if we allow for uncertainty by holding a safety stock, we get:

target stock level = forecast demand next period +

forecast demand in lead-time + safety stock.

Both the EOQ and the target stock level are calculated from forecast demand. A problem, of course, is that forecasts are never entirely accurate. The safety stock allows for this, with larger safety stocks giving more cushioning. Unfortunately, it would need very large safety stocks to cover every type of unforeseen circumstance. Some organizations saw a way of avoiding this reliance on forecasts,

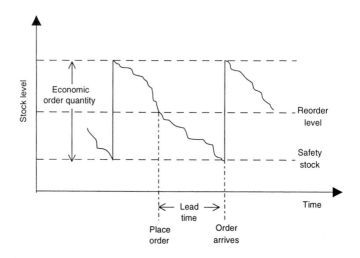

Figure 1. Stock level with the economic order quantity.

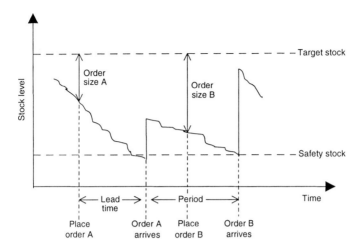

Figure 2. Stock level with a periodic review system.

but it was not until the 1960s when more powerful computer systems allowed them to introduce *dependent demand systems*.

6. Dependent demand inventory systems

The traditional way of forecasting demand assumes that the total demand for an item is made up of a large number of separate, independent demands. The total demand for bread in a supermarket, for example, is made up of independent demands from a large number of separate customers. This has the advantage of smoothing total demand, and allowing a supermarket to forecast the future demand from its historical demands. In other words, it can use the actual demands for bread over the past few Mondays to forecast the likely demand next Monday.

For many organizations, this approach to forecasting is not very reliable, particularly when the demand varies widely and is not independent. Imagine a manufacturer, which changes its production every week. It will need a different mix of materials every week, and demand for a particular material in previous weeks does not necessarily give good a forecast of demand for the next week. In addition, the demands for different materials are not independent of each other, but are related through the production plans. In these circumstances, an organization can use a completely different type of inventory management, based on *dependent demand*.

Organizations often use a master schedule to show their plans for some time in the future. For example, manufacturer will use a master production schedule to

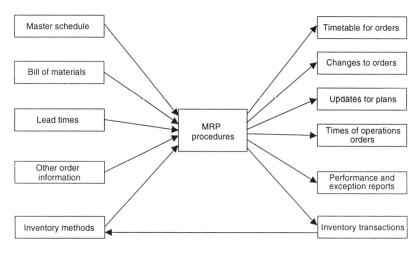

Figure 3. Outline of MRP systems.

show how many units of each product it will make over the next few weeks. If an organization knows what it is going to produce, it can easily find the materials it will need. A *parts list* or *bill of materials* gives the materials needed for each unit, and multiplying these by the number of units specified in the master schedule gives the total materials needed. Then the organization can co-ordinate its orders, so that materials arrive shortly before they are needed. This is the approach of *material requirements planning* (MRP).

You can imagine the difference between the two approaches with chefs planning the ingredients for next week's meals. With the traditional independent demand approach, the chefs might use a periodic review to see what ingredients they used in the last week and then order replacements. With MRP they look at the planned menus, see what ingredients each meal needs, and order them to arrive just before they start cooking. This approach is summarized in Figure 3.

The main benefit of MRP is that stocks of materials are exactly matched to production needs. No unnecessary materials are stored, and overall stockholdings can be reduced to very low levels. There are other benefits, particularly for aspects of planning. Suppose, for example, that the MRP calculations show that the demand for some materials is greater than the availability. When managers notice this problem, they can either change the master schedule to reduce demand for the material, or they can ask the supplier to increase the availability, perhaps by rescheduling their own operations. Now MRP becomes more than a method of controlling stocks, and plays a role in managing operations. This is an important step for inventory management. The traditional approach accepts existing conditions and organizes the stocks as efficiently as possible within these constraints. MRP suggests that there is some flexibility, and operations can be adjusted to improve the management of goods and materials (see Chapter 26). This is a change which set the scene for developments in just-in-time systems.

Despite its benefits, MRP is not without problems. It can only work in particular circumstances, notably when there is a reliable and unchanging master schedule, complete bills of materials, and accurate inventory records. There can also be problems with the suggested pattern of orders. This might give unreasonably small orders, or too much variation, or place unnecessary strains on suppliers in some other way. A common complaint is that it does not take into account the production of materials – the best schedule for using materials may be very different from the best schedule for producing them. Similarly, the bill of materials shows the order in which materials are needed, but it does not show the order in which they are made. Operations can also become less flexible – as materials are only available to make one particular schedule, this cannot be changed at short notice.

However, the most consistent problem with MRP is the size and complexity of its computer systems. The basic calculations are simple enough, but they come in huge numbers and rely on a mountain of accurate information. It sounds easy to multiply the number of units planned by the materials needed for each, but in practice many organizations have found that this is very complicated. Even small companies with a few hundred products – each requiring a few hundred parts – can find the problems of co-ordinating the whole system rather daunting. For larger companies with thousands of products, the installation and maintenance of working systems becomes both complicated and expensive.

Although it was originally developed for manufacturing industries, MRP is now used in many other types of organization. The surgical department in a hospital, for example, can find the materials it needs from its schedule of operations, in the same way that a university can plan its resources using next year's calendar of courses. MRP has proved successful in many types of organization, so it is not surprising that the basic procedures have been extended in many ways.

The first extensions were fairly small and gave better methods for dealing with supplier reliability, wastage, defective quality, variable demand, variable lead times, and so on. Later, more significant changes came with *manufacturing resources planning* (MRP II). This recognizes that the MRP approach of exploding a master schedule to plan the supply of materials can easily be extended to other resources. Basic MRP is limited to ordering and purchasing, but why not extend the analyses to other operations? To start with, the master schedule can be used to show the amount of machinery and equipment needed. This in turn sets the staffing levels and use of other resources (this is sometimes called *resource requirements planning*). The distribution of finished goods can be added (sometimes called *distribution resource planning*), and so on. Eventually, the master schedule can be used for planning most aspects of the operations. This is the aim of MRP II, which gives an integrated system to co-ordinate all aspects of production.

There is no reason to stop with MRP II, and the approach can be expanded even further to include the planning of marketing, finance, and other business functions. This becomes *enterprise resource planning*, where production plans drive integrated systems for the whole business.

7. Just-in-time operations

The traditional view of inventory management accepts that there are always mismatches between supply and demand, and stocks are essential to overcome these. MRP looks beyond this and says that supply and demand can be co-ordinated, and that some stocks can be reduced or even eliminated. Some Japanese manufacturers took this idea further by suggesting that properly designed operations can make any stock unnecessary. Their approach is known by various names including "lean operations," "stockless production," and "world class manufacturing," but it is usually called *just-in-time* (JIT) (see Chapter 13).

With JIT, all operations occur just at the time they are needed. This means, for example, that the managers do not buy materials and keep them in stock, but when production needs certain materials they arrange for the materials to be delivered to the process just as they are needed. The result is that stocks of materials are virtually eliminated. You can imagine this in the way that people buy fuel for their lawnmowers. With a petrol engine there is a mismatch between supply (bought from a garage) and demand (when the lawn is being mowed), so stocks of fuel have to be kept in the petrol tank and spare can. With an electric motor the supply of electricity is exactly matched to demand and no stocks are needed. The petrol engine uses an independent demand control system, while the electric motor uses a JIT system.

This simplified view shows JIT as a way of reducing stocks. It is, however, much more than this, and its supporters have described it as "a way of eliminating waste" or "a way of enforced problem solving." It is based on a desire to eliminate all waste from an organization. In particular, it sees all stock as a waste of resources that serves no useful purpose. It goes further than this, and says that stocks are positively damaging to an organization – they may cushion differences between supply and demand, but they also hide poor planning, co-ordination, and management. JIT says that an organization should not look for ways of controlling stock, but should eliminate them by improving management, finding the reasons why the supply does not exactly match demand, and then take whatever action is necessary to overcome the mismatch. As you can see, the key element of JIT is that it does not accept the current conditions, but looks for ways of improving them.

In principle, JIT seems fairly straightforward, but it needs a fundamental change in operations and views within the organization. Some of these changes include small, frequent orders of materials, reducing reorder costs, shortening lead times, introducing total quality management (TQM), forming alliances with suppliers, revised relations with employees, devolved decision making, and a whole range of other changes (see Chapter 10).

8. Systems with pull

It is easy to say that JIT co-ordinates the supply of materials with demand, but in practice this is rather difficult. In the 1920s many organizations experimented with minimizing stocks, but they found that the reduced levels made it impossible for them to work effectively. Fast-food restaurants introduced some aspects of JIT decades ago, but they could not work with complete systems. A key achievement of JIT is its practical way of co-ordinating supply and demand. Initially, this came through *kanbans*, which is the Japanese for "visible record or card."

JIT regards any effort put into administration as an overhead that is largely wasted. It simplifies operations so that they need as little control as possible, and this means manual systems with little paperwork and decisions made by those most closely involved. This is in marked contrast to MRP, which is computerized, expensive to control, and relies on decisions made by planners who are some distance away from the operations.

The first practical JIT systems worked in assembly lines, where products were passed down the line from one workstation to another. It is always difficult to co-ordinate this type of operation, balancing the flow of products so that stocks do not build up between workstations, and units are not delayed in the line. The traditional way of achieving this is to design a schedule for each workstation, giving a timetable of jobs that each must finish in a specified time. Finished units are then "pushed" down to the line, to form a stock of work in progress in front of the next workstation. Of course, this ignores what the next workstation is actually doing – it might be working on something completely different, or be waiting for a different item to arrive. At best, the following workstation must finish its current job before it can start working on the new material just passed to it. As you can imagine, this approach of pushing units through the line often leads to delays and large stocks of work in progress.

In contrast to this "push" system, JIT "pulls" work through the line. A workstation keeps going until it is ready to start on some new work, at which point it sends a kanban to the preceding workstation asking for more materials. When it receives this request, the preceding workstation makes the requested materials and sends them forward. In practice, there must be some lead time, so requests for materials are passed backwards during the lead time before they are actually needed. This method of pulling units through the operations can eliminate stocks of work in progress. In practice, units are passed forward in very small batches, so it is fairer to say that JIT minimizes stocks rather than eliminates them.

9. Reducing stocks in the supply chain

Early JIT systems relied on cards, the kanbans, to pull work through the process. This gives a very simple and efficient control system, but it only works over short distances. Now it is usually faster and more convenient to send messages backwards by computer, possibly using barcodes or some other form of tagging. This eases the communications, and it raises an important question. If stocks can be virtually eliminated from internal operations, why not reduce them throughout the entire supply chain? In other words, why not treat the whole supply chain as a single assembly line, linked electronically with messages passed backwards when materials are needed. This approach is growing, but it is still at a fairly early stage of development.

In an ideal situation, a final customer buying a pair of jeans in a store starts a message flowing back through the supply chain for wholesalers to replace the pair of jeans, manufacturers to make and deliver another pair, cotton weavers to deliver materials to the factory, spinners to deliver thread to the weavers, farmers to deliver cotton to the spinners, and so on. Each would arrange their operations so that lead times are short, and materials would flow very quickly through the whole chain. As usual, this approach is known by several names, most commonly *efficient customer response* (ECR). Some supply chains actually work like this – or, to be more accurate, they try to. ECR is certainly used in parts of some supply chains, but there are real problems with implementation, including the following:

(1) It is difficult to get all organizations in the supply chain working together for a common purpose, sharing information, and integrating systems. Organizations do not necessarily trust each other, they have different objectives and constraints, do not want long-term alliances, and generally see no reason for such close co-operation.

(2) JIT, and more particularly ECR, needs fundamental changes to operations, so that they can react quickly and flexibly to demands from their customers. These major changes might be impossible, too difficult, or uneconomic.

(3) Any method of inventory management needs a certain level of industrial and economic stability. Relatively few countries can supply this, and many are still at the point where high stocks are a positive advantage. If parts of the supply chain work in such areas, it becomes impossible to integrate the entire supply chain.

(4) There is always a delay before even the best new ideas are adopted. ECR can certainly reduce stocks, shorten lead times, and generally improve customer service. It is, however, still being developed and many organizations prefer to wait until all the teething problems are overcome.

(5) Ideas of inventory management are continually changing, and there are other new developments on the horizon. Some organizations do not want to adopt ECR when other advances might move them in a different direction. As a simple example, e-commerce allows final customers to enter the supply chain much earlier, possibly at the manufacturer level. This removes the need for stocks of finished goods in logistics centres and retailers, meaning that the supply chain does not need reorganizing, but can largely be removed.

Despite the practical difficulties, almost all organizations are now working to reduce their stocks. Most are happy to go along with the view that improved operations can speed the flow of goods through the supply chain, allow lower stocks, and reduce overall costs. However, the amounts of stock held are largely determined by the operations, and even the best theories for reducing stock only become feasible when there are associated developments in operations. In recent years, some of the significant changes that affect inventory management have included:

(1) universal availability of integrated computer systems with appropriate software;
(2) continuing development of models and methods for managing real inventory problems;
(3) reliable and timely data, with improved information and decision support systems;
(4) improved forecasts, reducing the need for safety stock to cover for uncertainty;
(5) widespread use of TQM, which reduces the need for safety stock to cover for defects;
(6) better educated managers who are aware of stockholding costs and the need to control them;
(7) integration of stages along the supply chain, including partnerships and alliances;
(8) concurrent engineering, rapid response, ECR, and other developments that reduce lead times and move goods quickly through the supply chain;
(9) flexible manufacturing systems and other aspects of automation that react quickly to changing demands;
(10) increased use of dependent demand systems (such as MRP and its extensions), reducing the need for raw material stocks;
(11) increased use of pull systems (such as JIT and its extensions);

(12) improvements in business-to-business (B2B) and business-to-customer (B2C) communications; and

(13) increasing use of e-commerce, bringing final customers to earlier points in the supply chain.

There is a clear desire within organizations to reduce stocks and, with effective methods of control, it is reasonable to suggest that stock levels should actually be falling. We can hypothesize that for most of the 20th century, and particularly since the widespread use of computers, improved inventory management should have reduced stocks. This does, of course, assume that the ideas and available methods are actually used in practice. To check this assumption we can look at some aggregate stock levels.

10. Changes in aggregate stocks

In the past few years, many companies have reported significant reductions in their inventories. This individual evidence is supported by wider surveys with, for example, the Institute of Grocery Distribution (1998) finding that stock levels fell by as much as 8.5% in a year, while a survey by the Institute of Logistics (1998) found that U.K. companies "managed to almost halve the stockholding requirements since the 1995 survey." This is, however, largely anecdotal evidence, and we really need to take a broader view, perhaps looking at total stockholdings at a national level.

Unfortunately, even when governments produce reliable statistics, it might be difficult to identify underlying trends in stocks. Inventories respond to a wide range of factors, only some of which are under the control of the organization. Managers in a particular company might plan to reduce their stocks of finished goods, but if the economy declines they might, despite their best intentions, be left with higher stocks. The sudden quadrupling in the price of crude oil in 1973 caused major disruptions to business, and even the best inventory policies could not cope with the scale of changes. (Interestingly, in 1973 many materials became unobtainable or very expensive, and organizations returned to the historical practice of grabbing as much stock as possible when it became available.) If the price of a raw material is about to rise, it makes sense to anticipate the rise and buy more at the lower price; if demand temporarily falls it might be better to slow production and increase stocks of work in progress; if a new advertising campaign is soon to be launched, it might make sense to build up stocks of finished goods.

Inventories clearly respond to the external influences, some of which are related to business cycles. A traditional view says that cycles start with industry being optimistic about the future – they expect sales to rise and increase production to meet this higher demand. Inventories build up as sales lag behind production, and at some point industry loses confidence and cuts back on production to use up the

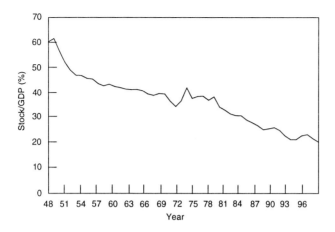

Figure 4. Aggregate stock as a percentage of GDP for the U.K. (source: Waters, 1999).

excess stocks. This causes a decline in the economy, which only picks up again when stocks are lower and production is not meeting current demand. Because it is relatively easy to change inventory levels – much easier than, say, adjusting production levels – they tend to fluctuate more than the business cycle itself.

Despite their aim of reducing stock levels, organizations might find that they are so influenced by external factors – the business cycle, inflation, changing gross domestic product (GDP), currency value, restructuring of industry, and so on – that they cannot achieve this aim. On the other hand, it might be difficult to claim that improved management alone will inevitably lead to lower stocks. Nonetheless, when we look at inventories on a national scale and over the long term, some patterns become clear. The easiest measure compares the aggregate national stock of a country with overall economic activity. To be specific, we can take the ratio of aggregate stock to GDP. This does not give an absolute measure of performance, but it allows reasonable comparisons over time. All industrial countries follow the same pattern, which we can illustrate using figures from the U.K.

Figure 4 shows the book value of aggregate stock held in the U.K. as a percentage of GDP from 1949 to 1998. These figures are collected by the U.K. Office of National Statistics and are reasonably consistent, but we should look carefully at the prevailing conventions, definitions and interpretations. Despite this warning, there is a clear underlying pattern. At the end of the 1940s and into the early 1950s there was a rapid decline in stocks as the economy returned to normal after World War II. From the early 1950s right through to the present day there has been a steady decline, which we can attribute to improving inventory management. There are, of course, some unexpected movements, such as the sudden change in the early 1970s that was caused by the rapid increase of oil prices and the economic disruption that followed. At this time the costs of raw materials rose sharply and there were frequent shortages, while declining sales left finished goods unsold. Fortunately, this can be viewed as a short–term fluctuation on the underlying downward trend.

Although the steady fall in stocks clearly suggests improving inventory management, we have to be careful that there is not some other underlying factor, such as the changing structure of industry, the move towards services, international competition, or increasing mobility. From the available evidence this seems unlikely, and we are fairly safe in saying that lower stocks come largely as a result of improved inventory management.

In the U.K. manufacturing contributes less than 20% to the GNP, but it holds 40% of the stocks. There are roughly equal amounts of materials, work in progress, and finished goods (about £20 billion of each). The amount of stock held by manufacturers has fallen much faster than other sectors of industry, suggesting that they have been at the forefront of stock reduction, and also that they are in the best position to reduce stocks. Organizations further down the supply chain have to pay more attention to their final customers and react quickly to demands – a lead time of 1 day is very good for a manufacturer, reasonable for a wholesaler, but not good enough for a retailer. You can see this effect in the type of stocks held by manufacturers, where there have been clear reductions in the amount of raw materials and work in progress, but the quantity of finished goods has remained steady or even risen slightly.

11. Review

Every organization holds stock of some kind. For most of history, these stocks were seen as beneficial and organizations kept them as high as possible. Industrialization and economic stability brought efficient supply chains that could deliver goods within a reasonable time and at an acceptable cost. This removed much of the uncertainty in supplies, and allowed organizations to rethink their inventory policies. Over the past century, organizations have seen stocks as being expensive but essential. Their main objective has been to control stocks, with independent demand models calculating "optimal order quantities" that minimize total costs. Such models are still widely used in a range of circumstances.

Since the 1960s some organizations have transferred to dependent demand systems, predominantly MRP. By relating demand for materials and other resources to planned production, these methods can virtually eliminate some stocks. They have opened the way for a new objective of stock minimization. They have also brought a change of viewpoint. Inventory managers used to assume that operations were fixed and they had to accept current practices. MRP showed that overall performance could be improved by changing operations to match the supply of materials more closely to the demand. This view was extended by JIT systems, which co-ordinate operations so that they are done at the best possible time. In particular, materials are not kept in stock, but are delivered to the operations exactly as they are needed.

Few ideas of inventory management are completely new, but they have to wait for improvements to operations before they become feasible. MRP, for example, could not be implemented until powerful computer systems were available; JIT could not be implemented until flexible operations allowed rapid response to customer demands. This is why inventory management has slowly improved over the long term, to give a continual reduction in stock levels. The U.K. is a typical industrialized country, where stock levels have fallen from 40% of GDP to 20% over a period of 40 years.

In the future, inventory management is set to continue many of the current trends. Organizations are generally agreed that they should ideally work without stocks, and they will continue to look for ways of maintaining service with significantly less stock. Leaders in the field are already moving towards zero-stock operations, but they will continue to adjust operations and squeeze out more improvements. Organizations that are currently lagging behind should improve their management, adopt better practices, and catch up with the leaders. There seems no reason to doubt that these factors will give continuing stock reductions into the future, particularly for manufacturers who are still working with relatively high stock levels.

The key issue for the future seems to be the role of technology. Flexible automation allows new operations that were considered impossible a few years ago. Improved communications and integration continue to speed the flow of goods through the supply chain. Some organizations are already reaching the limit when transport times limit their speed of response, and increasing traffic congestion might make this a more significant factor. e-Commerce is forecast to continue its dramatic growth, with B2C doubling every year, and worldwide B2B transactions approaching a value of $1 trillion. Relations between customers and their suppliers will change, as customers become involved at earlier points in the supply chain. Such changes open possibilities for fundamentally redesigned supply chains and new channels of distribution. Although it is a mature subject, inventory management is going through a period of major changes. Many of the changes are natural continuations of current trends, but the more interesting ones may still only be at the stage of theoretical concepts.

References

Harris, F. (1915) *Operations and cost*. Chicago, IL: A. Shaw & Co.
Institute of Grocery Distribution (1998) *Retail distribution 1998*. Hertford: Institute of Grocery Distribution.
Institute of Logistics (1998) *European logistics; comparative survey*. Corby: Institute of Logistics.

Papillon, F. (1696) *A treatise concerning the East India trade*. London.
Silver, E. and Peterson, R. (1985) *Decision systems for inventory management and production planning*, 2nd edn. New York: Wiley.
Waters, C.D.J. (1992) *Inventory control and management*. Chichester: Wiley.
Waters, C.D.J. (1999) *Global logistics and distribution planning*. London: Kogan Page.
Wilson, R.H. (1934) "A scientific routine for stock control", *Harvard Business Review*, 13:116–24.

Chapter 13

JUST-IN-TIME

SAMANTHA Y. TAYLOR

PPK Environment & Infrastructure Pty Ltd, Melbourne

1. The just-in-time philosophy

The logistics field is full of acronyms and jargon representing the latest production management techniques. One of the terms frequently discussed in a spectrum of disciplines is just-in-time (JIT) production. This chapter explains the concept of JIT and its development, and then explores the effect of JIT implementation on transportation.

JIT manufacturing has far-reaching effects on all areas of the production chain (e.g., accounting, purchasing, distribution) and, consequently, on goods transportation. The literal and narrower view of JIT refers to the delivery of components to the production process only when needed. This view may also be referred to as zero inventory production systems (ZIPS), point of sale driven "pull" inventory replenishment systems (IRS), or manufacture as needed (MAN). But whichever terminology is used, the effects of low inventories on all inputs to the manufacturing process are pervasive. If component parts are to be produced and delivered only when required, communication with customers and suppliers must be effective, transport must be reliable and responsive, paperwork (and hence time) must be minimized, quality must be adequate, and so on. JIT, by definition, cannot be forgiving of errors in quality or timing, and hence there is a much greater emphasis on perfect delivery every time. Inadequate quality components arriving on time for production is simply not good enough for JIT manufacturing.

As a corollary of larger inventories, inventories act as a buffer and can often hide production, distribution, or supplier problems, such as equipment breakdowns, large batch sizes, and poor co-ordination between processes. In essence, the apparently simple concept of very low (or zero) inventories leads to the philosophy of excellence in the entire manufacturing process. Indeed, many manufacturing commentators refer to the philosophy of JIT.

The fundamental objectives of JIT are indicative of its pervasiveness (Harrison, 1992):

Handbook of Logistics and Supply Chain Management, Edited by A.M. Brewer et al.
© 2001, Elsevier Science Ltd

(1) reduce or eliminate waste (e.g., unnecessary time, space, investment);
(2) empower employees, suppliers, and customers (e.g., through communication technology, electronic data interchange (EDI), and training programs); and
(3) maintain quality at all levels.

The challenges are presented in Table 1. To achieve the flexibility required of JIT, manufacture needs to be in small batches with short set-up times and rapid delivery from one activity to the next. This is further facilitated by a flexible workforce, where individuals can be moved from areas of low demand to areas of high demand.

2. Development of JIT

It was not until the economic success of the Japanese and their supremacy in production in the 1970s became evident that JIT created interest in Western countries. Depending on the disciplinary perspective in question, there are different views as to which events were most significant in the development of JIT

Table 1
Benefits, requirements and challenges of JIT

Benefits	Requirements	Challenges
Reduced inventories	Top management support	Top management support
Reduced costs	Supply-base reduction	Supplier reticence
Improved quality	Supplier development and training programs	Employee reticence
Increased productivity	Quality engineering processes	Inadequate engineering support
Improved supplier relationships	Buyer training programs	Not a panacea
Shorter lead times	Performance-based product specification	Poor product quality
Enhanced responsiveness	Blanket orders	Poor communication
Better management focus	Minimize variability of receipt quantity	Relationship building
Enhanced production scheduling	Minimize administrative paperwork	Logistics support
Greater design innovation (no need to be dependent on existing stock)	Management of inbound freight	Production schedule stability

Source: Adapted from Fawcett and Birou (1992).

Table 2
Advantages of JIT production

Area of improvement	Description
Productivity	Manufacturing productivity can improve by 20–40%
Lead times	Lead times can be reduced by more than 90%, reducing the need to carry finished goods inventory
Flexibility	Improves production flexibility, by allowing changes to the manufacturing schedule to meet changes in market demand
Market power	Market position is strengthened by the ability to supply on short notice, with greater flexibility and greater reliability
Quality	Consistency of quality improves
Inventories	Substantially reduces inventories, often by 60–8%
Floor space	Manufacturing floor space requirements are reduced by 30–4% because there is no "waiting" inventory stacked on the floor
Investment	Capital required to operate a business is significantly reduced, estimated reductions being 25–50%

Source: Adapted from Warne and Rowney (1986).

production. From a macroeconomic perspective, the Japanese government's aim of keeping the real rate of interest down after World War II led to capital being essentially rationed by lending authorities. This rationing process resulted in significant pressure on and uncertainty about the existing practice of having capital invested in stock holdings or inventories, especially since purchased material is typically the largest cost component for many products. By the early 1950s, therefore, many firms had little choice but to reduce inventories. The Japanese were simply not able to rely on lending institutions to finance their inventories, which led to the concept of JIT deliveries and JIT production. By 1986, some 30 years later, supporters of JIT had reported that by reducing inventories by 60–80%, the investment required to operate a manufacturing business could be reduced by 25–50% (Table 2).

From a manufacturing perspective there is a tendency to focus on case studies of individual companies, and specifically Japanese firms, that have successfully implemented JIT. Indeed, in the 1920s, Henry Ford used a form of JIT during production of the T-model Ford, and no doubt there are many other examples which pre-date World War II. The focus here is on modern JIT production. In particular, Toyota's success in the 1970s and 1980s has been regularly highlighted and held up as the grandfather of modern JIT production. There are a number of reasons why an automotive company is particularly suited to JIT production. In many ways, their peculiarities epitomize the complexity of manufacturing advanced products in the 21st century. Interestingly, the more complex a product, the greater the demand for excellence in the manufacturing process. The automotive industry, for example, has the following characteristics that predispose it to an integrated, flexible system such as JIT (Voss and Clutterbuck, 1989):

(1) It requires mass production assembly, where each vehicle is assembled from several thousand parts that have already undergone numerous processes. Therefore, problems with any of the processes will have a large overall effect.
(2) There are many different models, with numerous variations and large fluctuations in the demand of each variation.
(3) Every few years, the vehicles are completely remodeled and there are also often changes at the component level. In order to avoid the problems of unbalanced inventory and surplus equipment and workers, effort is focused on developing a production system, which is able to shorten the lead time from materials entry to the completion of the vehicle.

As society moves into the 21st century, there are additional pressures, such as increased globalization of sourcing, increased wealth, specificity of customer demand, shorter product life cycles, and rapid advances in technology. These pressures, and more, have led over the last 50 years, to a need for flexible approaches to manufacturing. In the early 1970s, for example, a technically advanced product in, say, consumer electronics, would last 15 years, with gradual adaptation. By 1989, product life cycles were as short as 15 months, and now they can be less than 12 months. Shorter product life cycles lead to shorter lead times, which in turn leads to economies from lower inventories.

When Toyota engineers had succeeded in streamlining their vehicle manufacturing process, and lead times between customer orders and final delivery were reduced to only a matter of days, they turned to improving their product range. In addition to supplying high-quality vehicles at low cost in a customer responsive manner, Toyota was able to satisfy customer specificity by frequently launching new products, which gave the perception of innovation. The philosophy of JIT was so far reaching that their entire business was re-engineered, and as a result Toyota is one of the 20th century's great success stories.

3. The aggregate effects of JIT and transportation

JIT implementation is typically accompanied by a rationalization of transport carriers. In a large and complex business, different units of a firm will contract transportation services as the need arises. Over time this can lead to many unco-ordinated transport services and increased costs. Through rationalization these services are streamlined and, depending on the business genre, there are a few alternatives to choose from. Either a number of transport carriers work separately to service different markets of the client's business, or a small number of companies work together taking responsibility for all distribution. In some instances a few transport companies have collaborated to form a new firm, exclusively to service a customer. This approach allows for simplification of financial investments and profits, and can usually be more easily defined and executed with a one-to-one (carrier-to-shipper) relationship. Indeed, in pan-European markets, where deregulation is still recent, shippers have complained of a lack of innovation displayed by contract carriers (Peters et al., 1998).

There are exceptions to this, for example, Ganeshan et al. (1999) suggest that by placing a fraction of an order with a discount, and less reliable, supplier, total cost savings in inventory logistics result. Consideration of this strategy, however, depends on a firm's idiosyncrasies, such as the need for propriety equipment. A discount carrier is unlikely to have specialized equipment. Furthermore, the system would not work successfully if one partnership with a reliable carrier had not been established first.

For partnerships between shippers and carriers to perform well, creative financial arrangements often need to be established, such as "put and call" options for equipment investment, and rolling contracts. Some risk sharing, through penalty payments for late deliveries, ensures the performance incentive is there for both parties. Late deliveries in JIT are costly to production. In addition, as in all relationships there needs to be a measure of the "carrot and stick" approach.

Improvements in transportation efficiency can lead to distribution with less inventory, in fewer locations and with increased confidence. Deregulation in the U.S. trucking industry increased competition and reliability, resulting in a $200 billion decrease in inventory investment over the 9-year period from 1980 to 1989 (Delaney, 1991). On the other hand, a study of urban trucking efficiency in Sydney, Australia, indicated that in 1991, in a deregulated environment, two-thirds of trucks, at best, were used on an average weekday (Taylor and Ogden, 1998). The remaining third are presumed to be idle, perhaps undergoing repair. Evidently there still appears to be scope for efficiency improvements, at least in the urban context (see Chapter 25).

In Japan, the average weight of goods carried on each trip fell by 36%, from 3.8 ton in 1980 to 2.4 ton in 1990 (Harutoshi, 1994). The greatest evidence of changes in distribution practices as a result of JIT is in statistics like this (i.e., a higher frequency of smaller quantities being delivered). The actual number of vehicles registered in each major truck class clearly shows that the numbers of small trucks and large trucks have grown much faster than the number of medium trucks. In fact the number of medium trucks on register in Australia dropped by 10% between 1975 and 1991 (Australian Bureau of Statistics, 1992).

Over the same period, large trucks increased by 30% and light vehicles increased by 80%.

With significant growth in adoption of e-commerce, already evident in the U.S.A., large volumes of goods will be delivered directly from the manufacturer or major distributor to the customer. While the worldwide trend in the last decade or so has been to increase productivity of carriers through increased mass limits and vehicle dimensions for vehicles, distribution resulting from internet shopping will benefit from innovation in smaller modular vehicles and trailers, enabling vehicles to easily distribute to local areas. Recent growth in small to medium and larger van sales in the U.S.A. has been around 20%. At the other end of the spectrum, B-double vehicles in Australia can be a maximum length of 25 m. These, eight- or nine-axle vehicles are permitted to traverse designated roads in urban areas.

Notwithstanding the possibility of costly interruptions to the manufacturing process, inefficient freight transport can increase costs, congestion, and pollution, cause customer dissatisfaction through unreliable deliveries, and reduce or close potential markets. Sustainable access to markets relies on an effective supply chain based on integrated transportation networks, seamless information exchange, and good communication and co-ordination. This requires responsive physical distribution networks and reliable information systems accessible by the whole chain. The emphasis, therefore, from a transport perspective, is on reliability, co-ordination (including multimodality), consistency (e.g., national standards), and advanced travel information. Travel times need to be predictable and regulations should not unnecessarily impede efficient freight flows.

It is relatively easy to create a list of what needs to occur for freight movement to be efficient. The challenge for infrastructure providers and traffic and transport managers is to maximize reliability and efficiency. Real-time information systems, vehicle tracking (e.g., global positioning systems (GPS)), and effective traffic engineering are all means of assisting efficient movement. Used independently, some advances can be made, but for maximum effect relatively simple traffic engineering techniques (e.g., proper signing, intersection design, adequate road maintenance, delineation) must be in place first, or at the very least alongside more sophisticated information systems and in-vehicle GPS.

4. The logistics of transportation

Transportation, in a logistics sense, involves functions such as freight bill auditing, freight payment, delivery reports, fleet maintenance, scheduling, and strategic issues in transportation (Box 1). Many companies, in an attempt to focus on core competencies, tend to out-source some, if not all, of these functions. While the concept of out-sourcing moves in and out of favor, it often presents an economically viable method of achieving productivity and/or service

Box 1
Transportation functions

Likely order of out-sourcing functions (in descending order):

- freight bill auditing
- freight payment
- transportation reports
- international shipping
- fleet maintenance
- freight consolidation
- freight tracking
- fleet management
- freight claims
- carrier routing
- carrier selection
- carrier contracting
- carrier negotiating

Source: Bardi and Tracey (1991).

enhancements. The challenge, however, is to ensure that management control, service levels, and customer/client relations are not compromised as a result of engaging a third party.

Those tasks most likely to be out-sourced are supporting administrative-type functions, because these are repetitive and clerical in nature. Core activities that require specialist knowledge and asset specificity are likely to undergo more careful analyses before being out-sourced, purely because they are integrated in the business in a strategic sense. They often include freight consolidation, freight tracking, fleet management, freight claims, carrier routing, carrier selection, carrier contracting, and carrier negotiations.

5. JIT and the supply chain

Given that the production of complex goods involves hundreds, if not thousands, of parts, total production is more than a single individual can understand in detail (Rycroft and Kash, 1999). Successful communication up and down the supply chain is essential to achieve JIT. It is, therefore, impossible to discuss the JIT philosophy without at least mentioning the role of strategic alliances in the supply chain.

According to the *Collins English Dictionary*, an "alliance is the state of being united by a common aim or common characteristics." In the context of JIT and supply chain management, an alliance can also be expressed as a shared vision of market opportunity to enhance overall business performance. Alliances can be

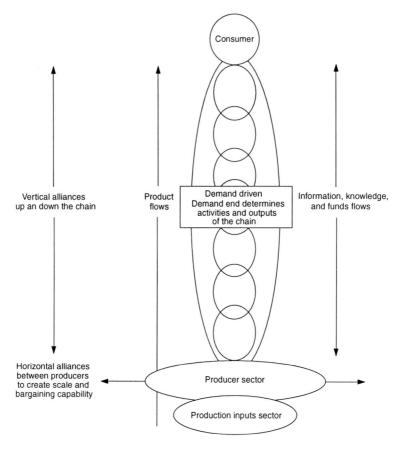

Figure 1. Product flows (source: Department of Primary Industries and Energy, 1998).

horizontal or vertical in nature. Vertical alliances are up and down the supply chain, between clients, customers, suppliers, and carriers. Horizontal alliances, on the other hand, are between individual producers or between individual carriers, to create scale and bargaining capabilities. An Australian example is the citrus industry, which comprises numerous farms owned by different firms. To deliver to the Indonesian market more successfully and with lower costs, individual firms in one region have formed a horizontal alliance. Even though each citrus firm is a competitor to the other, there are significant mutual benefits to creating economies of scale (Figure 1). Interestingly, Japanese companies, in the 1980s, did tend toward single sourcing with long-term contractual co-operative agreements (Hutchins, 1988). Their belief was that by operating in this way the supplier would develop with the customer a common interest in the achievement of market goals. It was common for customers to offer to assist with training for mutual benefit.

Not only were the Japanese very good at achieving co-ordination, their culture encourages small-scale, trial-and-error refinements to each component structure or process. Indeed they recognize that, in order to progress, small failures play a large part in building knowledge and, in particular, in reaching synergy. In fact the Japanese use a single word, *kaizen*, meaning continuous improvement from sustained and controlled experimentation. They show that significant synergistic effects are produced if many small refinements in individual processes are married with effective communication.

When the supply chain is designed efficiently and effectively, information sharing exists between supply chain members, inventory levels and product flows are transparent, and manufacturing, transportation, and distribution is co-ordinated. Thus, JIT can be achieved.

6. Trends in the industry to improve transportation for JIT

In the past, congestion was a phenomenon associated with the commute to and from work. Congestion was usually on key arterial roads and usually in one direction – inbound in the morning and outbound in the afternoon. Now, congestion is multidirectional and occurs throughout much of the day, causing major social and political concerns, and increasing business costs.

There are two categories of congestion. One kind is recurrent and predictable. The second kind occurs as a result of incidents and is stochastic. Recurrent congestion can be incorporated in travel-time estimations and in the delivery scheduling process. Incidents, such as crashes, breakdowns, and commodity spillage, by their very nature are far more difficult to accommodate, and potentially mean the difference between just-in-time and just-too-late. In the case of delivery time windows, where a specific 15-minute interval is allocated, arriving too early can ultimately be just as detrimental as arriving too late. If there is no standing room for a truck, for example, other than the delivery bay that is already occupied by the previously scheduled vehicle, the driver may be compelled to pass time by driving around the block. The driver may encounter congestion on the way and consequently return too late for the window. The alternative is to double-park, risk an infringement, restrict traffic flow, and impose costs on other drivers. The effect of unreliable travel times is, therefore, potentially disruptive and costly.

In particularly congested areas on freeways, state-of-the-art traffic management includes properly designed advanced traveler information systems (ATIS). These systems provide estimated travel times well in advance of congested areas, allowing drivers the opportunity to choose an alternative route (presuming there is one). Through smart location decisions and investments in road infrastructure, congestion costs can be reduced in the short term, but longer term benefits are difficult to guarantee. Firms that rely heavily on the road network need to make it their business to learn how the network will develop, so that they can make more informed location and investment decisions.

Alternatively, it may be prudent to schedule some deliveries for the night-time. While simple in concept, there are many considerations, including hourly rates for drivers, store hours, and objections from residents. Depending on the route and the nature of deliveries, these issues may be surmountable. One carrier, for example, developed a schedule to avoid high-volume traffic windows in a major Australian city by programming meat deliveries over, what they termed, "extended hours." The shipper required meat to be delivered from abattoirs in rural locations to butchers in the metropolitan area. Through working together, the volume delivered each night could be increased without dramatically increasing the delivery costs. By providing drivers with the keys to each shop, and making deliveries in the early hours of the morning, local truck curfews were avoided and the following benefits accrued (Taylor, 1999):

(1) a reduction in the number of vehicles on the road in peak periods,
(2) increased utilization of vehicles,
(3) reduced travel times for journeys, and
(4) fewer disruptions to customers during operating hours.

The above is an example of trust between shipper and carrier, through the provision of shop keys, and creativity in scheduling, around local truck bans. This is not to say that reaching a solution was simple, but rather that, with effort, it was possible.

Building on the complementarity of telecommunications and transportation, vehicle manufacturers are working enthusiastically to fit advanced communications technology to automobiles. Some models of prestige cars are fitted with a monitor, fax capability, e-mail, and internet links, and before too long these facilities will be available in trucks. This kind of adaptation for trucks will empower both drivers and logistics managers, potentially providing information on the state of the road network, the locations of vehicles, credential or permit updates for enforcement purposes, and the anticipated arrival time of deliveries, allowing a more dynamic system of scheduling. In addition, given the high capital cost of trucks, the marginal cost of fitting telecommunications technology is likely to be small.

The road transport industry is often slow in implementing new technology, partly because the industry operates on small margins and has a low average standard of education. For owner-drivers, the powerful image of a driver is often related by the size of his or her engine and rig. Furthermore, the industry is at the mercy of so many external influences, including macroeconomic factors, social and environmental concerns, regulatory reform, and institutional change. One area where advancement has occurred is the area of road-friendly (usually air) suspensions, which cause less damage to expensive road pavements than the oft-fitted spring suspensions. Given that a truck can do many thousands times more damage to road pavement than a car, this is no trivial concern. Trucks fitted with air suspensions can employ automatic methods of measuring axle loads, based on the changes in air pressure after loading. Methods such as these speed up enforcement, and benefit the enforcing agency, the shipper, and the carrier by reducing delays.

The opportunity for engineers and marketing personnel is to change the focus from greater power to smarter power, environmentally friendly engines, and the integration of smart technologies to assist business.

7. Final remarks

The apparently simple concept of zero or low inventories with deliveries supplied just in time for the production process has far-reaching effects internal to the firm and externally throughout the supply chain. Over the years, the increasing number of small quantities delivered has resulted in more frequent deliveries and contributed to increased congestion. The challenge is not only for logistics managers in scheduling and controlling stock flows, but also for traffic and infrastructure decision-makers to consider how they can work with logisticians, as an extension of the supply chain, in improving the reliability of the road network.

The philosophy of JIT is particularly expedient to the modern global climate, where competition means that a firm's success is inextricably linked to its responsiveness to the prevailing social, economic, and technological environment. This, in turn, means "managing" the supply chain. The challenges of JIT are, therefore, to:

(1) assure frequent, on-time deliveries,
(2) define and develop the buyer–supplier relationship (which may include government lobbying),
(3) limit loss and damage,
(4) maintain quality standards, and
(5) share relevant information.

In a strategic sense, developing a shared vision of market opportunity in concert with suppliers helps reduce waste and improve customer responsiveness. Better communication, as opposed to more stringent rules, facilitated through telecommunication technology and software development improves information and product flow. While at the same time, sharing the costs of late deliveries or, for that matter, poor quality goods, provides performance incentives. Finally, JIT is a "total" system, referring to the fact that all company members and all service providers, whether public or private, have a role to play in the supply chain.

References

Australian Bureau of Statistics (1992) "Survey of motor vehicle use, Australia", Australian Bureau of Statistics, Canberra, Catalogue 9208.

Bardi, E.J., and Tracey, M. (1991) "Transportation outsourcing: A survey of U.S. practices", *International Journal of Physical Distribution and Logistics Management*, 21(3):15–21.

Delaney, R.V. (1991) "Trends in logistics and U.S. world competitiveness", *Transportation Quarterly*, 45(1):19–41.

Department of Primary Industries and Energy (1998) "Chains of success – case studies on international and Australian food businesses cooperating to compete in the global market", Commonwealth Department of Transport and Regional Development, Canberra.

Fawcett, S.E. and Birou, L.M. (1992) "Exploring the logistics interface between global and JIT sourcing", *International Journal of Physical Distribution and Logistics Management*, 22(1):3–14.

Ganeshan, R., J.E. Tyworth and Y.M. Guo (1999) "Dual supply chains: the discount supplier option", *Transportation Research E: Logistics and Transportation Review*, 35:11–23.

Harrison, A. (1992) *Just-in-time manufacturing in perspective*. London: Prentice Hall.

Harutoshi, Y. (1994) "Demand forecast and financial feasibility of new freight transport systems", in: *Seminar TT3 on advanced road transport technologies, Omiya, Japan,* pp. 295–310. Paris: OECD.

Hutchins, D. (1988) *Just in time*. Aldershot: Gower.

Peters, M.J., Lieb, R.C. and Randall, H.L. (1998) "The use of third-party logistics services by European industry", *Transport Logistics*, 1(3):167–179.

Rycroft, R.W. and Kash, D.E. (1999) *The complexity challenge: technological innovation for the 21st century*. London: Pinter.

Taylor, S.Y. (1999) "Review of freight case studies", ARRB Transport Research, South Vermont, Austroads Report RC 7132-1.

Taylor, S.Y. and Ogden, K.W. (1998) "The utilization of commercial vehicles in urban areas", *Transport Logistics*,1(4):265–277.

Voss, C. and Clutterbuck, D. (1989) *Just in time – a global status report*. Washington, DC: ISF Publications.

Warne, J. and Rowney, M.J. (1986) *The Toyota just in time production system and early experience with ZIPS (zero inventory production system)*. New York: TTC.

Chapter 14

WAREHOUSING: A KEY LINK IN THE SUPPLY CHAIN

KENNETH B. ACKERMAN and ANN M. BREWER
University of Sydney

1. Introduction

The first role of warehousing is storage, which is defined as the assignment of goods in a selected location. For example, warehousing includes:

(1) accumulation of primary raw materials pending distribution to other locations in the supply chain,
(2) provisional storage of in-process inventory at various points in the logistics pipeline,
(3) storage of finished goods inventory near to the point of production, and
(4) storage of wholesale or retail inventory pending distribution to customers and end-users.

The second role of warehousing is the implementation of flows of goods from part of the supply chain to another, resulting today in the transformation from warehouses to distribution centers. Given these two main roles, the essential issue in warehousing is the management of space and time (Mulcahy, 1993). First, investing in space in the optimal location in the supply chain is increasingly a scarce resource, and therefore when this type of investment is made the space has to be maximized efficiently. There can be significant financial implications for warehouse strategy that does not fit in with the overall business and marketing strategies of either the owner or its customers. Second, time is also an important resource to manage given that all goods need to be delivered on time, and has implications for how warehouses are designed and their systems organized. Just-in-time is a response to this challenge. Today, a warehouse management system achieves this through the uses of electronic data interchange (EDI), real-time tracking of orders and inventory, labor/activity reports, and the simultaneous direction of multiple processes with the distribution center itself (see Chapter 34).

The earliest known use of warehousing was as a reservoir or a buffer against uncertainty of supply. Some of the oldest writing in Western civilization, the book of Genesis, describes the role of warehousing to prevent famine in ancient Egypt.

Handbook of Logistics and Supply Chain Management, Edited by A.M. Brewer et al.
© *2001, Elsevier Science Ltd*

Today, warehousing is done for many other reasons. The biggest change is that the stored inventory is most often used as a buffer against the uncertainty of demand and to improve customer service. The closer the warehouse or distribution center is to its customers, the better service it can provide. The proliferation of distribution centers in the 20th century was aimed at improving customer service. However, the substantial improvement in delivery capabilities through the use of overnight air transport has made it possible for some distributors substantially to reduce the number of distribution centers without compromising customer service.

A further change has occurred in the type of warehousing, to varying general purpose, customized, owned, leased, or operated by third parties. The latter are used when an enterprise has out-sourced this function to a third-party logistics provider. Furthermore, as transportation developed in the 19th and 20th centuries, warehousing was used as a means of achieving greater economies of scope and scale, with significant implications for transportation. For example, warehouses can hold a high number of stock-keeping units (SKU) of varying scope and scale, and thus present challenges for the way in which orders are both filled and transported accurately and quickly.

However, the transformation of warehousing occurred in the mid-1970s when the focus of logistics moved to supply chain management and the core of operations changed from an emphasis on productivity improvement to a reduction in inventory. Developments in information technology, such as manufacturing resources planning (MRP II), facilitated this change. There was also a new emphasis on quality in manufacturing. Japanese corporations enthusiastically adopted statistical quality control and the manufacturing theories of Deming (1960) and Juran (1981), with others following suit. In the 1990s, these trends strengthened to place a greater prominence on distribution, with the advent of efficient consumer response (ECR). The growing collaboration among supply chain partners (e.g., networks of suppliers and buyers) requires new initiatives in distribution. As a result, enterprise resource planning (ERP) and warehouse management systems have been developed. Today, the main processes of distribution include:

(1) *Receiving:* this includes the monitoring of goods, both in terms of quantity and quality, often with depalletizing and transferring to a storage area; and the receipt of returned goods.
(2) *Storage:* this includes organizing goods according to bulk, long and short lead time products, seasonal or promotional products, and replenishment storage.
(3) *Order picking:* this includes assigning order lists to teams and teams to zones, and the organization of pickers, whether picking is done manually or electronically (e.g., pick to light).

(4) *Despatch:* this comprises packaging and order consolidation, which may include staging or interim storage, and shipping.

In the past, these processes were labor intensive. A modern distribution center is more likely to be technologically intensive, but still, in many instances, relies on labor.

2. The challenges of warehouse management

Warehousing requires understanding new strategic challenges. The way that a warehouse is managed will depend on the prime objective, whether this is growth, superior service, cost efficiency, or continuous improvement. Management in a growth-oriented context will be judged by the ability to meet peak shipping needs, the capability of meeting seasonal or annual surges in volume, and the ability to find additional space or additional workers at short notice. In a distribution center where customer service has top priority, the warehouse management system is judged by its capacity to provide service that is superior to its competition. In some organizations, gaining cost efficiencies through controlling or reducing costs is the primary mission. In such organizations, management will be judged by its capacity to reduce freight costs, or perhaps to reduce the overall warehousing cost. In a production-oriented enterprise, a prime function of warehousing is to improve manufacturing effectiveness by storing production overflow or staging inbound materials efficiently.

The question is how can management leverage resources to improve customer service, operate more efficiently and cost-effectively, and ensure competitive success? The warehouse task involves matching its capabilities with the forces imposed by the competitive context. For example, the warehousing task is confronting a blitz of new business pressures. e-Commerce is creating a global marketplace that serves any customer anywhere. Moreover, customers today demand personalized service, both in terms of product and delivery. Warehouse management systems manage these resources, including receiving and handling inventory, through to service delivery to the end-user. It goes a long way to ensure business success through adding value in terms of time, place, and possession.

The objective of warehouse management system is twofold: efficiency (the best use of resources) and effectiveness (addresses customer demands). In order to address these goals, a number of key factors need to be taken into account, including service design (which is directly associated with location), capacity, warehouse design, systems, and management of processes, as well as quality, cost (e.g., demand), variation, and scheduling.

Warehouse management confronts a complex array of decision-making that needs to be linked to the business strategy. Strategic management decisions shape tactical and operational decisions. For example, strategic management needs to decide whether to emphasize high-quality regardless of cost, lowest conceivable cost regardless of quality, or some other mix. The decision made in this respect will determine factors such as warehouse policy, planning, design, and organization, as well as inventory management of technology, facilities, and service design, and specification, quality control, and price. Part of the strategic overview includes addressing the following questions:

(1) How well does the distribution center support the corporate mission?
(2) How well does the operation compare and compete with other warehouses elsewhere?
(3) How do the results compare with last year?

3. Warehousing quality

Planning and control of warehouse facilities and systems are made even more complex when considering the issue of quality. One of the most important measurements of warehouse quality is concerned with the perceptions of the customers who either work with the warehouse and/or receive deliveries from the warehouse. There are six quality metrics that are based on customer perceptions:

(1) customer complaints,
(2) on-time delivery,
(3) timely receiving,
(4) accurate and timely documentation,
(5) compliance with rules for loading and marking, and
(6) internal organization.

Customer complaints are usually caused by errors, and some enterprises have placed error reduction at the heart of their quality programs. As errors are identified, the ratio between mistakes and error-free activity is noted, as well as the probable cause of the error. For example, a shipping or receiving error might be caused by badly marked packages. Quality improvement then involves the supplier that shipped the product to the warehouse.

On-time delivery can be checked without regard to customer complaints. If every delivery carrier is required to indicate the time when the delivery was actually completed, a database can be established. In the case of receiving, a typical measurement is the "dock to stock" time, and this includes the process of entering the receipt in the inventory record. In today's information-based economy, speed and accuracy in updating inventory is absolutely critical. The same applies to other special requests that are made by customers and end-users.

Many customers have procedures for loading, marking, or tagging, and the warehouse should be measured by its ability to handle these customer rules.

Some quality measures are primarily internal, but if neglected they may influence customer perceptions. Internal measures include the following:

(1) work practices,
(2) accidents,
(3) employee turnover,
(4) equipment downtime,
(5) product damage, and
(6) compliance with government regulations.

3.1. Vehicle turnaround time

There are several advantages to decreasing the amount of time that vehicles spend at the warehouse dock. First, the dock doors in many operations are limited, and each door needs to be vacated as quickly as possible to allow its use for handling another shipment. Second, expediting both loading and unloading saves driver time as well as warehouse time. Third, the vehicles themselves are scarce and expensive resources, and their value is wasted when they are not moving.

Consider the practices used by the airlines. A commercial jet aircraft has a value of tens of millions of dollars, so the airline company is well motivated to keep it flying rather than on the ground. Gates at many airports are seriously overcrowded, and failure to turn each aircraft quickly will aggravate the congestion problem. Finally, many of the planes that carry cargo also carry passengers, and these customers become annoyed by ground delays. Therefore the most successful air carriers are constantly working to speed the turnaround time of every aircraft.

There has also been progress in marine cargo handling, otherwise known as "containerization." When ships were unloaded one piece at a time, it was normal for a vessel to be at a pier for a number of days. Today the turnaround time for a container ship is measured in hours rather than days. Similar technology is available to speed the turnaround time for vehicles at dock doors. Some companies produce systems that allow a vehicle to transfer its cargo from truck bed to the warehouse floor in about 5 minutes. These systems use special equipment built over the floor of the trailer and the dock. to allow the cargo to slide from the vehicle to the warehouse, or vice versa. The cost is about U.S. $75 000. Several vendors produce extendable conveyors that also reduce the time spent in loading and unloading.

Increasing the size of units handled is the easiest and least costly way to reduce vehicle turnaround time. Think about the marine containership example. What will happen to turnaround time at the truck dock if it is insisted that every vendor put everything on pallets or some other type of unitized load? If all inbound goods are in unit loads rather than floor loaded, the unloading process is much faster. The same advantage will apply if all outbound product can be shipped in unitized loads.

When the vehicle turnaround time is reduced, both transportation and ware-housing are improved. Therefore, it is this aspect of throughput that should receive first priority. If the units handled are pallets rather than cases or bags of products, the impact of future changes will be minimized. The pallet size currently used in many places is 48 inches by 40 inches (1.2 m by 1.0 m), a standard that was adopted more than three decades ago. It is a safe bet that this standard size will not be abandoned in the foreseeable future. A unitized handling procedure using the standard pallet should remain functional even if the size or weight of the product on the pallet is altered. As long as the cases can still be palletized, changes in product design should not substantially alter the materials-handling procedure.

3.2. Selecting stable stock

When a chain retailer installs a pick-to-light flow rack system, the use of the equipment is limited to the most stable items in the product line. Consider the case where the product is menswear, and merchandise chosen for the new system is limited to underwear, blue jeans, socks, and other items that are sold throughout the year with minimal changes of style. The higher fashion items cannot be effectively handled in a mechanized system because of the probability of constant change. The pick-to-light system uses computer technology to illuminate a display that shows the number of pieces to be picked from each lane of gravity flow rack. By using this system, the order selector can fill the order without using a paper-packing list. In effect, by following the blinking lights, the picking instructions are conveyed to the person doing the job.

The distribution center for a chain store uses the "A-frame" automatic order-picking machine for selection of pharmaceutical items. The A-frame functions like a giant vending machine, but it is most practical for items of the same size. Chain store staff recognize that pharmaceuticals have been and probably will continue to be adaptable to the A-frame in the foreseeable future. However, they would not consider using this equipment for order filling of general merchandise that has a wide variation of sizes and seasonal changes.

Every warehouse operator should consider whether the full line of stable stock should go into a mechanized order picking system. Extremely fast-moving items may turn so quickly that replenishment cost will be excessive. For example, if a lane of gravity flow rack must be refilled three times a day, it is probably less costly to pick this item from a warehouse pallet than from a flow rack system. At the other extreme, if there is one item that is picked less than once a week, the capabilities of the mechanized system are wasted. The very slow moving stock can be picked from shelves rather than from the mechanized system.

3.3. Identifying labor-intensive tasks

Picking orders requires the greatest amount of travel, and therefore the greatest amount of time spent per item handled. For this reason, order filling is the most labor-intensive task in most warehouse operations. Unfortunately, most of the systems that mechanise order filling also reduce flexibility in doing this job.

In today's direct-to-consumer market, there are some non-mechanical ways of controlling order-filling costs. For example, batch picking requires little or no equipment. When several orders are filled at once there can be a substantial reduction in travel time. Zone picking is another way to reduce travel without investing in mechanization.

Handling merchandise returned to the warehouse is also a highly labor-intensive task. One vendor of mechanized equipment has a system specifically designed to expedite the flow of customer returns through improved ergonomics, a superior sorting system, and conveyors to move the product through various stages of processing. Each case of returned product is opened for inspection and then separated into items that can be refurbished at the warehouse, items that need to be repackaged, and items that need to be returned to the vendor or destroyed.

Building and stripping pallets is a labor-intensive task for some warehouses. High-speed palletizing machines can expedite the task of building pallets, but most of these are designed to handle pieces of uniform size and weight. Some palletizers use robot technology to allow more flexibility in handling products of different size, but the usual trade-off is lower speed for greater flexibility. A miniature carton clamp allows the warehouse operator to select an entire tier of cartons rather than handle one case at a time. A load transfer station is designed to move an entire pallet load of merchandise from one pallet to another or from pallets to slipsheets.

4. e-Commerce and warehousing

Enabled by capabilities of the internet, e-commerce allows consumers to order products directly from manufacturers. Warehousing, in contrast, is one of the oldest activities in recorded history, predating the industrial revolution and going all the way back to the agricultural era. The new field of e-commerce seems to be a contributing factor to the growth and prosperity of the much older warehousing business.

4.1. Constant growth in warehousing occupancy

There has been a fairly steady growth in warehouse occupancy since 1992, when the low point of 91% was reached (Doerflinger et al., 1999). Warehousing occupancy was moving rapidly toward 94% as the year 2000 approached. Computer technology allows manufacturers to control inventory better today than ever before. Logically, this should show that the need for warehousing would decrease rather than increase. As producers become more skilled in controlling finished goods inventories, the glut of obsolete and slow-moving material, which clogged most warehouses a few decades ago, does not exist today.

Furthermore, improved communications and transportation allow distributors to provide good service to the entire country with far less warehouses then ever before. At the middle of the 20th century, it was common for many consumer-products companies to have over 100 warehouses to serve the U.S.A. Many of the same companies get the job done today from five or six locations, and a few get it done from only one. Since a smaller number of warehouses usually means a smaller inventory, the need for warehouse space should be going down, not up. The fact that it is increasing is generally attributed to the world wide web, or e-commerce. The phenomenon that may cause this is "disintermediation," which means cutting out the middleman. Through use of the web, customers practice two types of disintermediation:

(1) they buy directly from the manufacturer, and
(2) they buy from web retailers.

Manufacturers who are using web technology to change their distribution are looking for new warehouses around the country to replace the services formerly provided by wholesale distributors. Retailers who are newly engaged in e-commerce are looking for new resources to handle fulfillment, which is the supply of merchandise ordered directly by consumers.

4.2. e-Commerce and freight patterns

A report from Merrill Lynch (1999) lists key issues that will be addressed by the logistics industry over the next 5 years:

(1) rapid acceptance of the internet as a business medium will dictate changes in manufacturing flow and distribution patterns,
(2) e-commerce represents a shift in purchasing habits that probably will not create a material amount of new volume, and
(3) the residential delivery function will become a service necessity, not an option.

Merrill Lynch (1999) is less than optimistic about the profit potential for freight companies: "To put this in perspective for freight investors, we believe this will be similar to the excitement generated over the logistics revolution of the last decade. Almost overnight everybody was in the logistics business, citing consultant studies identifying double-digit growth curves. Yet, in the end, nobody ever came close to realizing the growth at the profits advertised. Most companies approached the market under the value added moniker, acquiring volume but falling short on profitability, eventually exiting or scaling back their investments."

At the same time, Merrill Lynch (1999) predicts that e-commerce will usher in a new era of redesign of supply chains, and this could be a growth stimulus for third-party logistics providers. This idea is supported with a prediction that third-party logistics is set to grow 20% annually over the next 5 years. Merrill Lynch (1999) also claims that third-party operators currently handle only 10% of the potential market, and that there is not a single company that offers logistics services on a global scale. Furthermore, a growing number of large customers will demand a single source out-sourcing solution for logistics needs. The role of warehousing will change as a growing number of companies abandon private warehouse investments and handle the function with services purchased from third-party operators (Merrill Lynch, 1999).

4.3. The new/old business of fulfillment

Those who proclaim e-commerce warehousing as a radically new activity forget the fact that the warehousing function has been around for many decades. A mail-order catalogue published by Sears Roebuck in 1913 contained 1495 pages and tens of thousands of items. While no one is available today who saw how orders were filled then, we can presume that fulfillment in 1913 was somewhat similar to fulfillment in 1999. Six features make product fulfillment warehousing different from other kinds of warehousing, and they have not changed much over the decades:

(1) the warehouse operator must have direct contact with consumers;
(2) information requirements are critical;
(3) order sizes are smaller than other kinds of warehouse orders;
(4) the order-taking function is much more precise, because it involves contact with consumers;
(5) customer service requirements are different, and typically more demanding; and
(6) the transportation function is more complex.

The 1913 Sears catalogue contained this promise on the cover, and it well describes the reverse logistics challenges of fulfillment warehousing: "If for any reason whatsoever you were dissatisfied with any article purchased from us, we expect you to return it to us at our expense. We will then exchange it for exactly what you want or will return your money, including any transportation charges you have paid."

Transportation and communication methods are greatly different now from the day when an American farmer sent his order to Sears by U.S. mail, and Sears delivered it by parcel post or Railway Express. However, out in the warehouse, much of the job remains quite similar. In commenting on the difference between "e-tailing" and traditional retailing, Doerflinger et al. (1999) pointed out: "the real barrier to entry is the back end – fulfilment – not the web site itself."

Paper flow for a fulfillment warehouse is more complex than for other warehouses. Nearly all orders are received electronically, and a significant amount of time is spent handling customer returns. Because a fulfillment operator deals directly with individual consumers, the customer service function is particularly critical. Some users want the fulfillment center to create invoices, or even dunning notices. Nearly every fulfillment center must handle major credit cards, and some fulfillment warehouses even handle banking for their customers. Compared with a conventional warehouse, order volume is high. The ability to automate tasks with computer technology is more critical in the fulfillment center than the conventional warehouse. The fulfillment operation is responsible for an unusually high number of stock-keeping units.

A growing number of observers recognize the critical nature of fulfillment. Writing in *Materials Handling Engineering*, editor Knill (1999) implies that nothing happens by magic in e-commerce, especially fulfillment. He points out that newcomers to fulfillment warehousing are likely to build new plants with unrealistic implementation schedules, misapplied technology, and engineering mingled with construction: "to the point that the welding torch becomes a design tool." He reminds those who feel that this could not happen to consider the baggage handling system at Denver Airport, where high-technology systems were rapidly replaced after severe damage was done to baggage on the robotic transportation system.

4.4. The residential delivery challenge

While most items sold in today's fulfillment center can be delivered by a parcel delivery service to the front door, there are others where a more complex residential delivery is necessary. One example is furniture and major appliances. The buyer of such commodities frequently expects and needs more than a box dropped at the front door. When the refrigerator or laundry appliance is bought, the buyer needs an installation service. The provision of such residential services is not always easy, and this could be one of the major growth areas for providers of third-party logistics services.

Since shipments under e-commerce are primarily to consumers who have decided to bypass traditional retail stores, residential delivery capabilities are absolutely essential for success. Yet residential delivery is more costly and less profitable than commercial delivery. The entry of third-party logistic providers into residential delivery is complicated by the fact that some countries provide subsidies for national postal services, and most also restrict the residential delivery capabilities of competitive private organizations.

The internet company that controls its own warehousing faces the task of learning to master the technicalities of the business. Early errors made by Webvan, Inc., are attributable to inexperience in warehousing. While some e-commerce companies will avoid the learning curve by using third parties, others feel that they must get into the logistics business themselves.

4.5. The nature of revolutions

According to Doerflinger et al. (1999), revolutions taking place in U.S. commerce have had some peculiar qualities. One is the fact that the revolution is always slower than everyone thinks it will be. Railroads were introduced in 1826, but their full impact was not evident until the 1880s. Electricity was first commercialized in the 1880s, but the real pay-off in industrial productivity came in the 1920s. The information revolution began with the first commercial use of computers in the 1950s, but it was not until the end of the 20th century that the productivity improvements arising from commercialization of the internet began to be realized. Just as the railroads revolutionized production and distribution in the 19th century, the internet promises to do similar things at the beginning of the 21st century. More and more products are made to order, such as Dell now does with computers. This reduces the need for the manufacturer to speculate about demand.

4.6. The social impact

Just as Henry Ford's assembly line changed the social nature of the U.S.A., the world wide web has the potential to change the lives of many workers. First, a growing number of businesses must operate on the 24/7 schedule. The web is available around the clock, and orders flow in at all hours from all over the globe. To keep pace, many fulfillment operations will run on a four-shift, 168-hour working week. A higher percentage of warehouse workers will do their work at night or on weekends.

A special report "The rise of the infomediary" published in *The Economist* (Anonymous, 1999) described infomediaries and their purpose. One example is in trucking. An organization called National Transportation Exchange uses the internet to connect shippers who have loads to fleet managers and owner operators who have excess trucking capacity. National Transportation Exchange collects a commission for each transaction. There are many such exchanges using the internet, as well as companies that operate as auctioneers or aggregators. What they have in common is the ability to consolidate buyers and sellers in fragmented markets. The infomediaries sell information about markets and thus create a platform where buyers and sellers can do business.

Today there are software programmers in India who are working for companies in the U.S.A. Tomorrow, there could be fulfillment workers in Michoacan, Mexico, shipping orders to Michigan, U.S.A. For warehousing professionals, there will be more winners than losers. However, the e-commerce world promises that life will not be as it was before the web was introduced.

5. Conclusion

Over many decades, many processes have been developed in warehousing in response to supply chain challenges. All these changes have been made in attempt to improve warehouse performance and to create strategic advantages over competitors Good performance is judged by streamlining processes in response to customer demands, while achieving cost efficiencies. Continuous improvement is an important part of accomplishing this. According to Hurdock (2000), the warehouse trends of the future will focus on the following:

(1) customers;
(2) operation and time compression – fewer distribution centers equates to larger distribution centers with more orders to process daily;
(3) continuous flow – logistics systems that prevent huge inventory accumulation;
(4) electronic transactions – the paperless warehouse;
(5) customized warehouses – compliance labeling, ticketing, bagging, dunnage, and palletization;
(6) cross-docking – fewer warehouses handling more orders will mean that more warehouses will adopt cross-docking (typically used to fill back-orders);
(7) third-party warehousing – it is likely that more companies will use third-party warehousing in order to increase service levels and leverage capital;
(8) complete automation – in order to deal effectively with increasing work volumes;

(9)   shrinking size of order of inventory; and
(10) continued upskilling of the workforce to keep in touch with technological advancements in the industry.

One of the big challenges for warehousing remains in the area of work organization, specifically with regard to staffing, flexible scheduling, and training.

References

Anonymous (1999) "The rise of the infomediary", *The Economist*, 351(8215).

Deming, W.E. (1960) *Sample design in business research*. New York: Wiley.

Doerflinger, T.M., M. Gerharty and E.M. Kerschner (1999) "The information revolution wars", Paine-Webber, New York, newsletter.

Hurdock, B. (2000) "Ten 21st century warehouse trends", *DSN Retailing Today*, 39(l5):14.

Juran, J.M. (1981) "Product quality – a prescription for the West", *Management Review,* 70:9–14.

Knill, B. (1999) "e-Commerce: Relearn the lessons", *Materials Handling Engineering*, August.

Merrill Lynch (1999) "e-Commerce: Virtually here", New York, consultant's report.

Mulcahy, D.E. (1993) *Warehouse distribution and operations handbook*. New York: McGraw-Hill.

Chapter 15

CONSOLIDATION AND TRANS-SHIPMENT

MICHEL BEUTHE
Facultés Universitaires Catholiques de Mons

EKKI KREUTZBERGER
Delft University of Technology

1. Introduction

In transport economics *consolidation* means bundling flows of passengers or goods from different origins and/or to different destinations on common parts of their routes. This operation requires a *trans-shipment* from one transport mode or vehicle to another, which takes place at a *terminal* or other type of exchange node. Trans-shipment operations are costly, but consolidation of flows gives access to important economies of density and scale in transportation, which can be best obtained by designing well-organized *networks* of various forms.

The first part of this chapter describes the different types of consolidation networks, while the second part analyses the economics of trans-shipping and network organization. For simplicity of exposition, the chapter focuses mainly on freight transportation , but most of the content can be readily extrapolated to passenger transportation.

2. Consolidation in logistical networks

The consolidation of passenger or freight flows is an economic necessity and a core business of the transport sector. The reason for this is that, given a network of roads, rail tracks, and waterways, many flows are too small to fill large transport units (trains, barges, metros, etc.) on the desired frequency level. If all origins and destinations were linked by direct transport services, either the frequency or the size of transport units would have to be restricted. This would reduce the service quality and/or increase transport costs to unacceptable levels. To avoid such an outcome, the flows can be consolidated. The principle is shown in Figure 1, which shows two direct rail services on the left-hand side where the trains (or metros) are only half

Handbook of Logistics and Supply Chain Management, Edited by A.M. Brewer et al.
© 2001, Elsevier Science Ltd

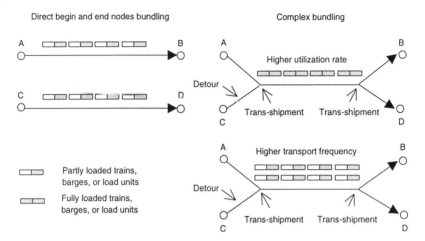

Figure 1. The impacts of consolidation (source: Kreutzberger, 1999).

loaded. The right-hand side of the figure shows two consolidation solutions: the upper model allows fully loaded trains, and the lower one higher transport frequencies. Obviously, a combination of both improvements is also possible.

The higher degree of loading allows benefit to be gained from economies of density. The costs per passenger or container are reduced, as total costs are not proportional to volume transported and can be passed on to more passengers or containers. Instead of increasing the loading degree, the transport units can also be enlarged (e.g., longer trains can be made up). An enlargement most often will reduce the costs per unit, while a higher frequency corresponds to an important improvement in the quality of transport services. These issues are discussed in more detail in Section 3.

The consolidation advantages are accompanied by a number of disadvantages, such as having detours and additional exchange nodes. The node activities require time. In many cases they also generate costs. The challenge for the transport sector is to optimize between these advantages and disadvantages. Which is the best way to consolidate the flows? The answer depends on different factors. Most important are the height of transport costs per kilometer or time unit, the size of flows, and the costs and time requirements of exchange operations. As these factors are continuously changing, transport actors have to assess the situation regularly and decide whether and how to organize consolidation by setting up and adjusting an optimal network of relations between origins and destinations. The following paragraphs of this section give a few examples of consolidation processes and networks.

At this point it is worth pointing out that the word "network" can be used with different meanings in the context of transportation. First, it may just correspond to a set of physical infrastructures, such as a road or a rail network, without any reference to its technical or commercial organization. Such networks, like the roads networks, may be fully open to users who can travel on them without any control to speak of; they may just have to pay a toll to use particular links. Some others may have a restricted access if they are operated by a single firm that schedules their use within a number of technical and financial constraints. This is most often the case for railway or pipeline networks, the use of which is usually restricted and controlled. In both cases, however, there remains a direct association with a particular set of modal infrastructures.

Besides these more or less controlled physical networks, there are also logistic networks that aim at organizing certain types of transport services, such as the transport of containers and swap bodies, or the fast dispatch of letters and small packages. These organized networks are not necessarily based on the operation of a physical network, but rather on transport services delivery using all modes and means that are efficient for the transport task. They may or may not be organized by firms owning some transport infrastructure, such as a terminal or a piece of railway network, or owning a fleet of specific vehicles, such as airlines. In the most extreme case, the logistic network is mainly an organizer of transports, which buys services from various carriers. In what follows, most of the time, it will be this second sense of the word "network" that will be used.

The consolidation of flows primarily refers to logistical networks (i.e., the networks of services). The physical networks give only a framework for the design of consolidation networks. Nevertheless, the design of a consolidation network may lead to the conclusion that some important link or node infrastructure is missing. We will come back to this question, after a discussion of some consolidation types.

In addition to *begin and end (BE) networks*, which have direct connections, there are basically four types of consolidation network, as illustrated in Figure 2. These have intermediate exchange nodes and therefore are here called *complex consolidation networks*:

(1) *Line (L) networks* comprise several segments linking two begin or end nodes. The terminals at these nodes and the intermediate ones have multimodal exchange (e.g., rail–road, barge–road).* In its simplest form a link network comprises only one segment. An example of this type of

---

*In this chapter the transport of goods in one and the same loading unit (containers, swap bodies, semi-trailers) is described by the term *intermodal transport*, in accordance with the definitions of European Conference of Ministers of Transport (1993). Railways prefer to use the term *combined transport* for the combination of road and rail transport. The term "intermodal transport" should not be confused with *multimodal transport*, which simply means that more than one modality is used in a transport chain, and may refer to any freight flow. An example is the transport of iron ore shipped from overseas to main harbors, and then transported to inland factories, first by barge and then by train.

configuration is the railway freight shuttle connection organized by several national railways between Antwerp and Italian seaports through Bettembourg, Metz, and Lyon (Jourquin et al., 2001). At each of these intermediate nodes there is a terminal where various operations can be performed (e.g., at the begin or end nodes). However, there may be some restrictions on the services provided at intermediate nodes (e.g., a

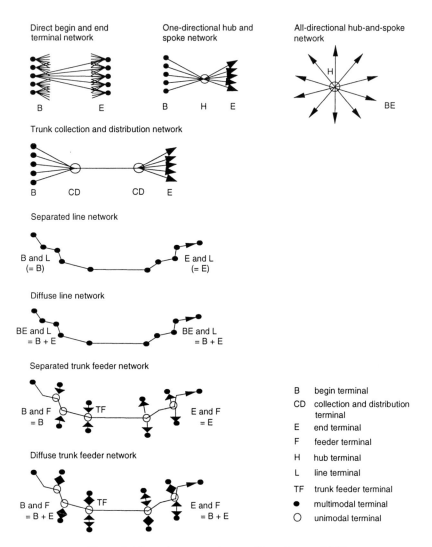

Figure 2. Network concepts (source: Kreutzberger, 2000).

distribution network of fruits which operates highly specialized trains shuttling from harbors to continental destinations for delivering loads only of, say, bananas).

(2) *Hub-and-spoke (HS) networks* are a well-known type of consolidation network. They comprise a set of links that all converge to a main hub, where transfer operations and bundling of flows can take place. The hub has unimodal exchange (e.g., between trains). Airline hubs are well known for passengers transport, but such hubs also play an important role in freight transport. An example of the latter are the railway shunting yards at Metz, France, and Muizen, Belgium, where railcars are shunted between trains to or from Zeebrugge, Gent, Antwerp, and Rotterdam harbors and Lille, Athus, Paris, Metz, Duisburg, Manchester, and Birmingham.

(3) *Trunk collection and distribution (TCD) networks* comprise a main trunk, which may be composed of several segments, and a set of links at the begin or end nodes for collecting and distributing loads to various origins and destinations in the same area. Normally, the size of transport units is smaller on the collection and distribution network than on the main trunk line. The nodes of collection and distribution normally have unimodal exchange. An example was the old Albatros network, where railcar groups from different German harbors were coupled to long trains at a north German siding functioning (the collection and distribution node). After traveling inland these long trains were split into small trains at another collection and distribution terminal. This network has now been transformed into a hub-and-spoke network. Many line networks are completed by at least a short collection and distribution link, but the begin and end nodes are not necessarily part of the same organization.

(4) *Trunk feeder (TF) networks* comprise begin or end nodes and intermediate feeder and trunk feeder nodes. The size of the transport units is generally smaller on the feeder network than on the trunk network. The trunk feeder nodes have unimodal exchange. This network type is very well known in public transport. A typical metro network contains: line nodes, where car, bicycle, or pedestrian passengers change transport modality; a trunk feeder network, linking the main metro line to smaller public transport lines (metro lines, trams, busses); and feeder nodes (the stations at junctions, where passenger transfer is unimodal between two public transportation rail lines).

Obviously, there are many possible combinations of these four basic types of network. Intermediate nodes of a line network may function as feeder nodes for the line traffic. Two hub-and-spoke networks can be connected by a main trunk line. Good examples of such a set-up are the linkage of the main hubs of two international airlines on each side of the Atlantic and the role played by main harbors on both sides of the Atlantic. Both these operate like hubs, forwarding and receiving loads to or from many places on the continent, over long distances, and by different modes. Indeed, different modes may be linked to a hub or involved in a network. Much of the distribution and collection at rail terminals is done by trucks, sometimes under the supervision of the terminal management. Networks based on important harbors often have two main inland modalities.

As already indicated, each consolidation type has typical exchange nodes. In short:

(1) All networks have begin and end nodes. These always have multimodal exchange, such as rail–road or barge–road exchange in the freight sector.
(2) The intermediate nodes are different in each type of consolidation network. Line and feeder nodes are multimodal, like begin and end nodes. Hubs, collection and distribution nodes, and trunk feeder nodes generally have unimodal exchange.

From an operational point of view at the network level, these exchanges between different links can be organized simultaneously or sequentially. In the first case, the exchanging transport units are present at the node during the same time period, and the exchange operations are concentrated in that period of time, which requires a heavy exchange infrastructure. In the case of sequential exchange, the transport units need not to be present at the same time, and thus more load units have to be stored at the exchange node.

From a technical point of view, many different solutions can be used at exchange nodes:

(1) Railcars and their loading units can be exchanged in a shunting yard, as in the case of the Belgian NEN network, which is a hub-and-spoke network for intermodal flows, or in the case of the German InterCargo network, which is a trunk collection and distribution network and a mixed network for intermodal and non-intermodal flows.
(2) Groups of railcars and their loading units can be exchanged on a siding, as in the case of Cargo Sprinters trains, which can be latched to make up longer trains, and then split again on a siding.
(3) Loading units alone can be exchanged at a terminal without exchange of the railcars. Cranes or sliding techniques are used, and the trains go through the node without any change length or composition. This solution was adopted for the new rail hub in the Antwerp harbor.
(4) When consolidating freight, loading units may not be used at all, but only the freight exchanged at specialized freight centers. Consolidation networks with freight centers are quite common in the road sector, especially for the transport of less-than-container loads (LCLs), parcels, and express deliveries. Freight centers can also play a role in intermodal

networks. For example, SNCF Fret has invested in a network of modern rail freight centers, mainly for rail–road exchange. Another example is the German Bahntrans, which has developed freight centers interconnecting rail and road networks for the distribution and consolidation of swap bodies. On the rail side, hub-and-spoke consolidation is applied, making use of shunting yards.

It is obvious that the technical and operational design of exchange nodes must match the exchange requirements of the different consolidation networks. The design may also vary with the type of freight (e.g., bulk or break-bulk, high- or low-value commodities). Small begin and end or line terminals may operate with simple equipment, such as reach stackers or mobile equipment on trains (or barges) for the trans-shipment, with paved ground serving as storage area. Large begin and end or line terminals will have heavier equipment, such as rail-mounted gantry cranes for trans-shipment, and may have sophisticated facilities for storage. The layout of line terminals must focus on the quick entry and departure of trains. At larger rail terminals with unimodal exchange, such as hub terminals, there may be a need for internal transport systems in the terminal. Such systems allow fast movement of load units between cranes and increases the capacity of the terminal.

Matching exchange requirements and node specifications is also of importance in passenger networks. Air passenger hub exchange is mostly done simultaneously, and thus the gate capacity must be large, whereas the waiting capacity may be restricted. Bus stations may have a simultaneous or sequential transfer of passengers between busses or between busses and trains. In the first case, a large number of bus parking places is required; in the second case, a small number of places may already lead to a large capacity.

As mentioned earlier, the design of consolidation networks depends to a large extent on the physical network that may need to change; the available transport technologies and the evolving requirements of industrial organization and production also influence it. The following factors are worth mentioning:

(1) The building of new bridges and tunnels, such as the Channel Tunnel and the Scandinavian links, introduces into the overall network new interfaces between modalities. Such interfaces are often "natural" points for reconsolidating freight flows, and determine a new organization of networks.

(2) The expansion of international freight traffic resulting from production specialization and economic globalization feeds the congestion of roads around major cities and in deep-sea harbors. Space becomes scarce and it is expensive to further expand the transport networks and to develop additional economic activities in the surrounding areas of harbors. A situation is reached where it is no longer possible to respond successfully to the circumstances using the existing sizes of transport units, means of transportation, and locations.

(3) The relative ease of communication, the just-in-time organization of production, and the increasing interdependence of production processes give rise to demands for higher speed in the circulation of trains, barges, trucks, metros, trams, and buses, and shorter lead times for both freight and passengers.

At present, the above-described developments are inducing a search for new network organizations and transport solutions:

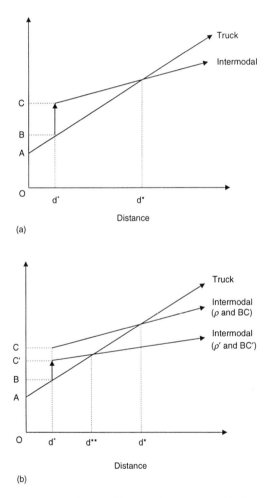

(a)

(b)

Figure 3. (a) The cost per ton of direct and intermodal transport. (b) The cost per ton of direct and intermodal transport after a decrease in the trans-shipping cost and the railway rate.

(1) New nodes on hinterland routes of large deep-sea harbors are built to function as regional inland satellite terminals with frequent connections to the main terminal by rail or waterways. These terminals organize the collection and distribution of shipments, take over storage functions and space intensive activities, and perform various services in order to alleviate the congestion of the main terminal. At the same time, they are used to enlarge the main harbor hinterland and drain additional traffic. Valburg and Duisburg along the Rhine, and Venlo and Born along the Maas are good examples. Some of these terminals may grow so important that they develop into "master hubs" with connections extending to several harbors, as is the case for some inland rail terminals (Notteboom and Winckelmans, 1999; Slack, 1999).

(2) Freight transport nodes progressively enlarge their functions beyond loading, unloading, trans-shipping, and consolidation of transport flows. If they are important enough, the nodes may develop into logistic platforms, where a large range of services is supplied by different firms. These nodes can offer temporary storage, assume the function of a distribution center for consumer goods or parts, packing and assembly of pallets and containers, manage and maintain fleets of vehicles and container stocks, and provide all kinds of logistics and administrative services. For passenger transport, they provide all the services needed by travelers, from banking to restaurants and conference facilities.

(3) There are projects to split the collection and distribution by barges within Rotterdam harbor from the barge trunk transport on waterways by the setting up of a container exchange platform where consolidation can be made.

(4) New terminals and transport units are being developed to increase the efficiency of the roll-on, roll-off transport of semi-trailers and trucks on waterways and railways.

(5) A new generation of rail terminals is being developed using advanced technologies. Examples are the French Commutor project, the Noell Megahub, the Tuchschmid Compact terminal, and the German Drehscheiben concept. A good description of various new-generation terminal concepts involving different degrees of automatic handling and robotization is given by Bontekoning and Kreutzberger (1999).[*]

---

[*]It must be recognized, however, that many advanced innovative concepts have met with some implementation difficulties. They are often designed for very important freight flows, and require a high degree of standardization of loading and transport units. On the other hand, at the present time in Europe, the financial consolidation of transport companies has a higher priority than the development of new-generation solutions in the networks.

3. The economics of trans-shipping and network organization

Terminal operations take time and are costly. Often they also correspond to a discontinuity point in logistic chains, where different agents and organizations take over the transport task. This may entail delays and disruption in information flows, because the terminal organization has its own rationality and handles many different transport loads at the same time. Moreover, handling in terminals may lead to damage.

Naturally, all transports require an initial loading operation regardless of the following transport stages. Goods are most often loaded first on a truck, except in cases where plants are located on a waterway or have a direct rail link to a railway network. Likewise, all transports require an unloading operation at destination. As different types of commodities have different transport requirements, the loading and unloading operations will be have different associated levels of cost and convenience. Nevertheless, loading and unloading are necessary for all transports, and so are not generally a source of important relative advantage for a particular transport mode.

However, compared with direct transport, transport that involves a transfer at one point from one transport mode or means to another involve the additional cost of trans-shipping. This extra cost is not just in terms of money, but also involves possible handling damage, information discontinuity, and an additional unreliability of schedule. This extra cost is a factor in all complex network organizations, and particularly limits the use of intermodal transport solutions (Janic et al., 1999), which in most cases involve some initial or final trucking of the commodities. Also to be taken into consideration is the fact that many such transport solutions involve a detour that lengthens the trip distance and the cost.

It follows that intermodal transport can be competitive only in cases where the mode or means onto which the goods are transferred has a generalized cost advantage that compensates for the additional trans-shipping cost. This point is illustrated in Figure 3a, which compares the cost of direct transport by truck with the cost of combined truck–rail transport. In this simple example, the cost of the direct transport is made up of OA (the loading cost at the origin + the unloading cost at the destination) plus a trucking rate r per ton-km transported. The cost of the intermodal transport is made up of OA plus AB (the cost of trucking over a distance $d°$), a trans-shipping cost BC, and a railway rate ρ per ton-km (ρ is smaller than r). It is seen that combined rail transport becomes cheaper for distances longer than d^*, despite the additional cost of trans-shipping, while trucking is the least costly choice for shorter distances.

Obviously, for a given trucking rate r, the threshold distance d^* is determined by the level of the railway rate ρ and the trans-shipping cost. As an example, assume that these parameters could be reduced to BC′ < BC and $\rho' < \rho$, respectively (Figure 3b). It can be seen that the intermodal transport has become more competitive over shorter distances, since the threshold distance has decreased to $d^{**} < d^*$. The fixed trans-shipping cost and the transport rate level are thus the two factors that determine the economic feasibility of intermodality and network organization involving intermodal transport (see Chapter 9). Beyond this simple example, it should be well understood that this statement must be interpreted in terms of generalized cost, meaning that indirect transport solutions require a very efficient set-up in order to achieve sufficiently high transport volumes by offering attractive tariffs, speedy service, convenient frequency, safety, reliability, and in-transit information. Clearly, the relative weights of these factors depend to a large extent on the characteristics of the commodities and the requirements of the full logistic chain, including the industrial or commercial activities of shippers and consignees. Nevertheless, the fact remains that intermodal transport solutions can be competitive only for longer transport distances. For example, road–rail–road intermodal transport in Europe is not competitive over distances shorter than 350 km, or even 500 km, due to the existing tariffs, corridor characteristics, and network organization (Demilie et al., 1998; Eberhard, 1999).

It follows that the organization of the terminals and the design and operational strategies of the networks must as a whole be well adjusted to the type and volume of traffic that the system is aimed at, as well as to the topology of the overall network and spatial localization of industry. In this respect there are no simple rules, as was demonstrated by the case studies of the TERMINET (1999) research project. In each case reviewed in that project the configuration of the network and the equipment provided at the terminals had to be designed to fit the particular circumstances of the exchange terminal and network.

The speed and reliability of operations at terminals are certainly important factors for the network users. Immobilizing people, goods, or vehicles and crew during trans-shipping operations imposes a considerable opportunity cost on the shippers and carriers. Delays impose further opportunity costs along the transport chain and industrial activities. Furthermore, time constraints resulting from business working hours and production planning may be decisive in the choice of a transport solution. In the case of railway operations in Europe, where priority is given during the day to passenger transportation, much rail freight must be carried during the night. This time constraint reduces the time span available for transporting goods, and makes it all the more important to have speedy terminal operations in order to extend as much as possible the distances covered during the night and thus the market area. In this respect, intermodal transport and trans-shipment is penalized by the rather low degree of standardization of vehicles and loading units, which hampers the use of automatic handling at terminals.

Good adjustment to local circumstances and market demands is particularly important since, for a given network administration, technical infrastructure, and types of service supplied, the sum of the capital cost and of that part of the operation cost that is fixed in the short run (as long as the service frequency and characteristics are maintained) tends to be relatively high. There are important economies of density in network utilization: the higher the load factor of the vehicles and the utilization rate of the terminals, the lower the unit cost per ton. Hence, the network infrastructure and service provided should be well adapted to market demand if the service is to be supplied at competitive tariffs.

The design of networks and terminals should also take advantage as much as possible of whatever economies of scale and economies of scope can be made. In general, there are economies of scale in the use of larger vehicles, both for engineering reasons and because crew size does not increase in the same proportion as the vehicle size. Larger boats, heavier trucks, longer trains, and bigger aircraft allow for the costs per unit to be lowered substantially if they are fully loaded (Kreutzberger, 2000). However, there are limits to size of transport units that can be used, due to, for example, the depth of harbors and waterways, the length of quays and railway sidings, the clearance of railways and waterways, and the width and load resistance of roads. Nevertheless, larger transport units offer interesting opportunities in the design of networks. Some economies of scale can also be made by using heavy equipment at terminals (e.g., cranes, trans-shipment systems).

These economies of scale are an important source of benefits for a network organization around, for example, a hub (Button, 1993). In the case of a hub-and-spoke network, freight loads being transported to several destinations can be forwarded from a begin terminal on the same large transport unit, towards the hub, where they are sorted and trans-shipped on a transport unit with other loads from different origins but for the same destination. Clearly, such consolidation of loads at the hub allows the use of larger transport units, and may contribute to lowering the operating cost per unit transported between some origins and destinations. The frequency of services may also be raised and transports organized in a more systematic way so that the quality of service is enhanced. In these cases, the alternative of direct transport would only be feasible at a higher cost with smaller vehicles.

To some extent, economies of scale can also be achieved by having more extensive or more dense networks. Again, such networks may be able to drain additional traffic, which permits the use of larger transport units and allows a higher service frequency, as well as a better organization of the transport. Hence the management of networks has some incentive to enlarge their network further (McCalla, 1999). This partly explains the development of inland regional satellite terminals, where shipments can be bundled. Provided that the network design and equipment is well adjusted to market demand, so that it operates close to its operational capacity, the benefits obtained from all these economies of scale may translate into lower tariffs. In Figure 3b, this would correspond to a lower rate ρ

for combined transport. In such a case, the network organization would be all the more attractive and drain even larger volumes.

Thus, economies of scale in transport operations are the main factor explaining the development of organized networks. However, their more active development, particularly in Europe over the last two decades, has also been the result of several general economic factors that have permitted good use to be made of them. In the first place, technological progress in the transport of goods and passengers has made communication and transport over continents and around the world easier and less costly. At the same time, the low relative cost of transport (compared with important differences in labor cost) in different countries has induced a decentralization of production. This trend towards economic globalization has been reinforced by economies of scale, which could be achieved by means of highly specialized production lines using advanced technology. Obviously, the evolution towards complex and decentralized production technology has required a satisfactory supply chain, this being all the more necessary because firms strove towards a just-in-time delivery system in order to reduce stocks of inputs (Rodrigue, 1999). All these elements have given strong incentives to develop co-ordinated transport networks, while the trend towards liberalization and deregulation of transport in many countries has progressively opened the doors to international co-operation and restructuring of networks. At the same time, they have given rise to huge traffic volumes, which have allowed gains to be made from further economies of scale in network development. These developments are particularly significant in the inter-continental transport of containers over very long distances, a sector in which very large operators have emerged, through mergers and alliances, to take full advantage of economies of scale in ship sizes and network organization from inland origin to inland destination (Marchese et al., 1999; Shashikumar, 1999).

There are also economies (and diseconomies) of scope that are worth taking into account in the design of a transport activity and network. These economies relate to the service mix that is proposed by a firm. To some extent, a diversification of services may contribute to a higher volume of activity and a fuller utilization of available capacity; it may also open the way to additional economies of scale. However, diversification may also induce additional costs if it hampers other services, complicates management, or requires substantial additional investments. As an example, it appeared that road carriers in France derived some benefit from combining the transport of general goods with either refrigerated transports or transport by tankers, but lost out when combining the last two transports (Beuthe and Sayez, 1994). Hence, benefits from economies of scope are the result of a careful combination of complementary processes and services.

Thus, in summary, economies of density and scope and, above all, economies of scale are the factors supporting the development of organized networks, despite the costly trans-shipments they involve.

References

Beuthe, M. and Sayez (1994) "The multiproduct cost function of long-distance trucking in France", Rapport GTM.

Bontekoning, Y.M. and E. Kreutzberger, eds. (1999) "Concepts of new-generation terminals and terminal-nodes", TERMINET research program of the European Commission, The Netherlands TRAIL Research School, Delft University of Technology, Delft.

Button, K.J. (1993) *Transport economics*, 2nd edn. Cambridge: Cambridge University Press.

Demilie, L., V. Dupuis, B. Jourquin and M. Beuthe (1998) "On the crossing of the Alpine chain and the Swiss regulation of trucking", in: A. Reggiani, ed., *Accessibility, trade and locational behaviour*. Aldershot: Ashgate.

Eberhard Cl.E. (1999) "Options for the integration of combined transport into European distribution system – the case of trans-European rail freight freeways (TERFF)", *Transport Modes and Systems: Proceedings of the 8th World Conference on Transport Research*, 1:435–448.

European Conference of Ministers of Transport (1993) *Terminology on combined transport*. Paris: OECD.

Janic, M., A. Reggiani and T. Spicciarelli (1999) "The European freight transport system: Theoretical background of the new generation of bundling networks", *Transport Modes and Systems: Proceedings of the 8th World Conference on Transport Research*, 1:421–434.

Jourquin, B., M. Beuthe and L. Demilie (2001) "Freight bundling networks models: Methodology and application", *Transportation Planning and Technology*, in press.

Kreutzberger, E. (2000) "Promising innovative intermodal networks with new-generation terminals", TERMINET research program of the European Commission, OTB, Delft, Deliverable 7.

Marchese, U., E. Musso and C. Ferrari (1999) "The role for ports in intermodal transports and global competition: A survey of Italian container terminals", *Transport Modes and Systems: Proceedings of the 8th World Conference on Transport Research*, 1:141–154.

McCalla, R.J. (1999) Global change, local pain: Intermodal seaport terminals and their service areas", *Journal of Transport Geography*, 7:247–254.

Notteboom, Th. and W. Winckelmans (1999) "Spatial (de)concentration of containers flows: The development of load center ports and inland hubs in Europe", *Transport Modes and Systems: Proceedings of the 8th World Conference on Transport Research*, 1:57–71.

Rodrigue, J.-P. (1999) "Globalisation and the synchronisation of transport terminals", *Journal of Transport Geography*, 7:255–261.

Shashikumar, N. (1999) "Container port dilemma on the U.S. east coast: An analysis of causes and consequences", *Transport Modes and Systems: Proceedings of the 8th World Conference on Transport Research*, 1:87–100.

Slack, B. (1999) "Satellite terminals: A local solution to hub congestion?", *Journal of Transport Geography*, 7:241–246.

TERMINET (1999) "Towards new-generation operations", 4th framework research program of the European Commission, final report.

Chapter 16

LOGISTICS OUT-SOURCING

MICHAEL BROWNE and JULIAN ALLEN
University of Westminster, London

1. Introduction

Every organization has to make the decision of whether to perform each of the activities that it requires itself or to pay another organization to carry out these activities on its behalf. "Out-sourcing" refers to the strategic decision to contract out one or more activities required by the organization to a third-party specialist.

When an activity is carried out by an organization itself it is usually referred to as being performed "in-house," while those activities that have been out-sourced and are carried out on behalf of the organization by a specialist provider are usually referred to as "third-party" services. This chapter examines logistics out-sourcing and addresses several key issues, including: the types of activities out-sourced, recent trends in the rate of out-sourcing, the potential advantages and disadvantages of out-sourcing, and how logistics decision-making and globalization are affecting out-sourcing strategies. Before focusing attention on logistics out-sourcing, the decision to contract out particular activities and the trend in this strategy is discussed in general.

2. Deciding whether to out-source activities

The decision about whether or not to out-source an activity currently performed in-house by the organization is often referred to as the "make or buy decision." Traditionally, many companies decided to carry out a very wide range of activities internally. This involved companies in directly employing staff and purchasing resources to provide for all its own needs (e.g., public relations, advertising, financial accounting, research and design, information technology (IT), transport, warehousing, market research, maintenance, repair, catering). This resulted in the development of large, vertically integrated manufacturing and retailing organizations, which had the capability to carry out all these activities with internal resources.

Handbook of Logistics and Supply Chain Management, Edited by A.M. Brewer et al.
© *2001, Elsevier Science Ltd*

More recently, many companies have been deciding to pay a third-party specialist to carry out the activity on their behalf. According to Quinn and Hilmer (1994), an organization should:

(1) identify its "core competencies" (these being "those activities in which it can achieve definable pre-eminence and provide unique value for customers") and commit the organization's resources to these activities; and
(2) out-source all the other activities required for which the organization "has neither a critical strategic need nor special capabilities."

Core competencies tend to be activities and skills in which the organization has long-term competitive advantage. These competencies are activities that the organization can perform more effectively than its competitors, and which are of importance to customers and tend to be knowledge-based rather than simply depending on owning assets. Any other non-core activities, which are not of fundamental importance to the organization's competitive edge, can be considered for out-sourcing.

A strategy of out-sourcing all non-core activities has its own risks and problems, as failure of these activities could jeopardize the organization's core business. Also, the cost of choosing a suitable company to out-source an activity to and managing this arrangement can be high, and may in fact be greater than the cost of performing the activity in-house. When the potential for vulnerability and competitive edge with respect to an activity are high, the need for tight control over sourcing is

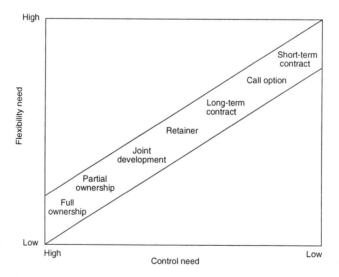

Figure 1. Potential contract relationships (source: Quinn and Hilmer, 1994).

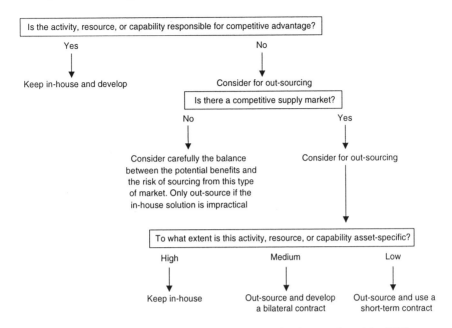

Figure 2. A risk management model for out-sourcing (source: Lonsdale, 1999).

required, which suggests either carrying out the activity in-house, through joint ownership, or through detailed long-term out-sourcing contracts. Conversely, when the potential for both vulnerability and competitive edge are low, the activity requires little sourcing control, and there are likely to be many adequate providers of the activity. In the case of activities that fall between these extreme cases, there are a number of sourcing options for the organization. The most appropriate choice will depend on the trade-off between sourcing control and the flexibility required by the organization (Quinn and Hilmer, 1994) (Figure 1).

In most out-sourcing arrangements there is at least a degree of uncertainty about whether the conditions that existed when the contract was entered into will still hold in the future. As Lonsdale (1999) notes, "out-sourcing contracts that are drawn up in an uncertain environment are necessarily incomplete – with the blanks left where the potential for uncertainty is greatest." Although this leaves scope for flexibility as circumstances change, it can also lead to a situation in which neither party is able to reach agreement when the contract is renegotiated. If it is relatively easy for the customer to switch to another third-party provider then this does not represent a significant problem. However, when the customer has had to make large investments in the out-sourcing partnership, this can prevent the customer from rational sourcing in future. The risk management model for out-sourcing developed by Lonsdale (1999) (Figure 2) is helpful in making decisions about which activities to out-source, and the most suitable type of out-sourcing arrangement.

The attractiveness of out-sourcing is enhanced when the market for an activity is mature and there are many suppliers competing with one another to provide the activity. If the market is immature and third-party suppliers of the service or product are weak, out-sourcing can prove to be very expensive in terms of the customer bringing the supplier up to the required standard. Also, in a situation in which a small number of suppliers control the entire market for an activity, they are capable of holding a customer to ransom. The organization out-sourcing an activity is vulnerable in both of the latter situations.

3. Growth in out-sourcing in recent years

Indications suggest that out-sourcing is growing rapidly across a wide range of activities. The 1996 annual assessment of out-sourcing activity in the U.S.A., for example, showed that U.S. companies with revenues (turnovers) in excess of $80 million expected to increase their expenditure on services that they out-source by 26% in 1997. These U.S. companies spent a combined total of $85 billion on out-sourced services in 1997. In the U.K., the overall out-sourcing market is estimated to be growing at an annual rate of 22% (Roberts, 1999).

Dun & Bradstreet (1998) estimated that worldwide out-sourcing expenditure would be approximately $235 billion in 1999. Another study commissioned by the Outsourcing Research Council and PricewaterhouseCoopers found that about two-thirds of the executives from multinational companies interviewed already out-sourced one or more activities, and that the typical executive spent, on average, approximately one-third of their operating budget on all forms of services provided by third-party companies (M. Corbett & Associates, 1999).

The activities that are most commonly out-sourced by organizations include IT, finance, administration, sales and marketing, human resources, logistics, security, cleaning, and maintenance.

4. Logistics out-sourcing

Turning attention to logistics out-sourcing, there are a wide range of logistics activities that can be provided by specialist third-party logistics companies. Freight transport and warehousing services have been widely available for many decades, together with documentation services to support the flow of these products (e.g., delivery and customs documentation). However, in recent years, logistics companies have begun to offer an ever-expanding range of services, such as final assembly of products, inventory management, product and package

Box 1
Logistics out-sourcing: services offered

Classical out-sourcing (a)	**Service portfolio** (b)
Warehousing	Transport
Transport	Storage
Goods despatch	Break-bulk
Delivery documentation	Load consolidation
Customs documentation	Order picking
	Order processing
Advanced services (a)	Stock control
Pick and pack	Pick and pack
Assembly/packaging	Track and trace
Returns	Vehicle maintenance
Labeling	Labeling
Stock count	Palletization
	After-sales service
Full services (a)	Consultancy advice
Order processing	Packaging/repackaging
Order planning	Return of packaging/handling equipment
Systems/IT	Quality control/product testing
Invoicing	Customization
Payments collection	
Consulting	
Shipment tracking	
Materials planning	

Sources: (a) Holland International Distribution Council (1998); (b) McKinnon (1999).

labeling, product tracking and tracing along the supply chain, order planning and processing, and reverse logistics systems (which tackle the collection and recovery of end-of-life products and used packaging in the supply chain). In addition, over the last 10–15 years a large number of logistics consultancies have been established that offer a diverse range of services, as well as logistics software and hardware companies offering logistics IT solutions. Box 1 shows the range of logistics services available from third-party specialists.

It is felt by some commentators to be more effective for companies to out-source clusters of related logistics activities, thereby contracting entire processes rather than individual activities. These clusters of activities tend to share specialized skills, physical assets, processes, and technologies, and can therefore enable economies of scale and scope when out-sourced in unison (Rabinovitch et al., 1999).

5. Trends in out-sourcing of logistics activities

In recent years there has been a marked trend towards logistics out-sourcing among many companies. This shift away from in-house to out-sourced logistics operations is part of a wider desire to manage logistics activities in a different way in response to new pressures in supply chains. Many organizations require ever-more responsive and reliable distribution systems that increasingly operate internationally or even globally. Often, these organizations identify logistics as falling outside their core competencies. They do not feel they have the necessary expertise to achieve the level of logistics service that they require in-house. Even if they were capable of providing these services themselves, they do not want to have to make the substantial investments that would be required in personnel and equipment. Instead, they turn to logistics third-party providers that specialize in providing these services.

Schary (1994) has highlighted the importance of transaction costs in the decision over whether to out-source logistics functions, since these costs emphasize the role and cost of co-ordination. As he suggests: "The decision whether to perform logistics functions internally depends on the combination of negotiation, management (including co-ordination) and operation. The decision whether to use internal functions or outside organizations to perform part of the functions of the supply chain is determined by the combination of costs and the limits of management skill and specific knowledge. Using outside contractors in transport and warehousing provides examples of these decisions. Transport operations involve operating costs, which are often lower for external suppliers." It is not surprising that if the costs of co-ordination and operation for outside agencies are low, there is a tendency to out-source them, because this conserves assets and provides flexibility.

Another major factor that has had an important bearing on the extent of logistics out-sourcing is, of course, freight transport regulation, especially quantity restrictions. In countries where regulation is, or has been, strict this serves to limit haulage capacity. In these circumstances, in-house freight transport operations may be the preferred choice of transport service users. Meanwhile, third-party logistics services tend to be more widely used in countries that have deregulated freight transport markets.

Evidence exists of the out-sourcing of other logistics activities, in addition to freight transport, at an increasing rate by organizations. A study by the European Logistics Association (1997) showed that European companies expect to out-source a growing range of logistics activities in future. While the 1998 report by the Holland International Distribution Council noted: "Transportation is already widely out-sourced and companies are now also increasingly out-sourcing other activities such as warehousing and value adding operations. In this respect, recent industry surveys suggest European companies have advanced further along the path of using third-party logistics providers than North American companies and that the European market is quite mature." Indeed, a study comparing the out-sourcing of logistics activities in Europe and the U.S.A. indicated that logistics out-sourcing is more common in Europe, with 76% of the European companies surveyed using third-party logistics providers compared with 58% of U.S. companies (Vanderbroeck, 1996).

A survey conducted in 1996 estimated that the third-party logistics market in the European Union was worth approximately $32 billion, which represented just less than a quarter of total logistics expenditure. This study found that out-sourcing varied significantly between countries. For example, British companies spent about 34% of their logistics budget with third-party providers, compared with only 11% and 13% in the cases of Greek and Italian companies, respectively (Marketline International, 1997).

6. Advantages of logistics out-sourcing

Since the late 1970s the quality of logistics service providers has improved significantly and the range of services they offer has greatly expanded. Many factors have encouraged manufacturers, wholesalers, and retailers to out-source some or all of their logistics activities to logistics services providers. These include:

(1) The standards of logistics service providers have risen and their efficiency has improved greatly. The specialist management and knowledge-based skills and operational experience offered by third-party operators may result in improved services at lower costs.

(2) Gaining access to the latest technology and equipment employed by the third-party provider.

(3) Financial conditions in the 1980s encouraged firms to concentrate capital investment in their core competencies and to pay for ancillary activities, such as distribution, on a current cost basis.

(4) There is one less business variable to worry about. By subcontracting a portion of the business and ensuring that acceptable standards are built into the contract, senior management can focus their attention on their core competencies.

(5) There are potential cost reductions, for example:
    (a) shared use may give better utilization of vehicles and warehouses, leading to lower unit costs due to the consolidation of different customers' demands;
    (b) the specialization of the contractor may allow volume buying of vehicles, warehouses, and mechanical handling equipment and systems;
    (c) the labour costs of a third-party operator may be lower; and
    (d) third-party companies may exist on a lower return on capital than that expected of major manufacturing and retailing companies.

(6) There is increased flexibility in terms of short-term changes in locations, fleet mix, warehouse types, and staffing levels. This allows retailers and

manufacturers to be more responsive as market or customer needs change (e.g., during seasonal peaks).

(7) The need for investment in new equipment and premises is avoided.
(8) There has been a proliferation of regulations relating to vehicle operations and product handling.
(9) There has been a rapid rate of technological change.
(10) Internal industrial relations problems can be overcome.
(11) There are external back-up systems in the event of strikes.

The larger grocery retailers in the U.K. were among the first to realize the benefits of out-sourcing part of their logistics activities. The growth of centralization during the 1970s and 1980s stimulated the demand for third-party logistics services in the sector and established many of the logistics service providers as major companies in their own right. The close, long-standing relationships that have developed between the retailers and the logistics companies is argued to be one of the main reasons why the retail distribution activity of the major multiples is so effective (Sheldon, 1998).

7. Disadvantages of logistics out-sourcing

Although the dominant trend since the 1980s has been towards the out-sourcing of road transport and other logistics activities, many companies still choose to operate either all or part of their logistics operations in-house. The decision to retain logistics activities in-house has not been widely addressed in much of the literature. However, Fernie (1990) has highlighted the main reasons for keeping some in-house logistics activities:

(1) *Cost issues*:
   (a) operations at cost plus could be run more cheaply in-house, assuming other variables remain equal, because the third-party logistics company needs to make a profit on its operations;
   (b) switching costs will be incurred by contracting out (e.g., redundancy costs, asset disposals or write-offs);
   (c) monitoring and control of costs is easier when the distribution function remains in-house (good information systems and clearly agreed service standards with the third-party operator may overcome this issue, but most of the large multiple retailers still retain an in-house presence to provide a benchmark of costs and operations for their contracted distribution operations);
   (d) the cost of monitoring the performance of the logistics can be high and is also sometimes difficult to achieve effectively; and

(e) some companies do not have the necessary information or expertise to assess which logistics providers are offering good services at competitive prices.

(2) *Control issues*: The view is that in-house logistics and distribution operations can provide the company with more control over important customer service considerations, such as delivery reliability, and a degree of compatibility with other company activities and practices. Flexibility of operations is also seen as a possible advantage of retaining an in-house distribution function, with the loyalty of the distribution operation not torn between several customers. There is also the concern that out-sourcing could result in a loss of security and that confidential information will be passed to competitors.

(3) *Economies of scale*: Many in-house operations are large enough to benefit from economies and derive similar buying power over their suppliers that the third-party specialists enjoy.

(4) *Innovation through specific expertise*: Larger or specialist in-house operators can claim to have much more expertise in particular sectors than logistics specialists. For example, distribution of frozen foods or deliveries to special delivery locations.

The last two disadvantages (economies of scale and innovation) also appeared under the advantages of using third-party services, illustrating that it is not often a simple clear-cut decision between in-house and out-sourced logistics operations. Indeed, the two types of service can be viewed as complementary and are often used in conjunction with each other. Many companies operating in-house services for the purpose of transporting goods from one place to another supplement this with contract logistics services, on either an occasional or regular basis. There are several important advantages for companies supplementing in-house operations with out-sourced operations (McKinnon, 1989):

(1) parallel logistics services can help ensure that no disruptions occur in supply of goods;

(2) geographical expansion of a company's market can often be met more economically by a contractor than in-house, especially in the early stages when sales in the new area are low;

(3) diversification into products with different handling and marketing requirements can benefit from the use of a contractor with the correct equipment and expertise;

(4) companies with significant fluctuations in sales during the year can use contractors to meet demand during the peaks and operate own account distribution throughout the year at a stable level; and

(5) partial use of contractors can help to provide the own account operator with a benchmark against which to compare the own account service.

8. Supply chain developments related to out-sourcing

The growing importance of multinational and global companies, combined with the increased emphasis on logistics management as a leading-edge technique for competitive advantage, all pose a tremendous challenge for those concerned with logistics services. Three key reactions to the challenge have important implications for the organization and purchasing of logistics services:

(1) rationalizing the vendor network,
(2) contractual agreements between carriers and shippers, and
(3) the role of quality in service provision.

8.1. Rationalizing the vendor network

Rationalizing the vendor network has spread to the logistics service sector. Manufacturers and retailers increasingly wish to concentrate a larger proportion of their logistics expenditure with fewer, but inevitably larger, logistics suppliers (this has been referred to as the concept of "one-stop shopping"). This development has been reinforced by the growing ability of large companies to make international comparisons of their logistics service suppliers and the rise in the use of management techniques such as benchmarking. The selected third-party logistics provider (or prime contractor) will, in many cases, select a limited number of subcontractors for the functional operations, such as trucking or warehousing, with control through information replacing control by ownership.

8.2. Contractual agreements between logistics providers and their customers

Relationships between logistics providers and their customers are increasingly typified by contractual agreements rather than by a series of transactions. This can lead to the logistics provider acting in much the same way as manufacturing co-makership arrangements. This more stable and long-term concept only becomes possible when firms deal with fewer providers, so there is a strong link with the trend to reducing the number of logistics suppliers. Moreover, the more sophisticated service requirements of users will mean that it becomes harder to buy services at short notice from a variety of logistics providers on the basis primarily of price. The service requirements may simply be too complicated to allow that. Long-term contracts have become more usual, spelling out the terms of service and performance. This trend has been reinforced by the growing reliance on information and communication systems, which can have the effect of tying together the logistics provider and customer.

This is beginning to result in companies that out-source logistics activities looking to develop collaborative, strategic partnerships with the third-party providers. Bowersox (1992) has identified five factors that are critical to the success of a logistics partnership:

(1) selective matching – all organizations have compatible corporate cultures and values;
(2) information sharing – partners openly share strategic and operational information;
(3) role specification – each party in the partnership is clear about the specifics of its role;
(4) ground rules – procedures and policies are clearly spelled out; and
(5) exit provisions – a method for terminating the partnership is defined.

Logistics partnerships are based on the idea of the parties working closely together to create highly competitive supply chains. In such a partnership, logistics providers will work with their client to develop and design appropriate logistics systems rather than simply performing the operations. The mutual trust and exchange of information and ideas that typifies a close working relationship can help bring about logistics benefits such as reduced inventory levels, improved delivery reliability, and enhanced customer service and quality.

Another development in the relationship between companies and their logistics providers (and other parties in the supply chain, including suppliers and customers) has been the formation of "supply chain alliances," or "supply chain associations," in which a number of companies work together in an "extended enterprise." In this approach, it is the extended enterprise that is viewed as the unit of competitive advantage rather than the individual company. An example of this approach is provided by Toyota, "which in Japan has managed to out-source a major part of its competitive advantage achievement to its network of direct and indirect suppliers. Indeed, taking productivity as a measure, 80% of Toyota's achievement of competitive advantage over its U.K. based rivals is based on the effective use of its supplier network" (Hines and Rich, 1998).

8.3. The role of quality

Despite pressure on margins, few buyers are concerned solely with price. There is an increasing emphasis on "quality" in logistics services. Although price will always be a very powerful determinant of purchasing decisions, quality has come to assume greater importance in many European and global markets. In many cases this is a consequence of the introduction of new techniques, such as just-in-time (JIT) (see Chapter 13), in manufacturing. Many companies have now embraced the concept of an integrated supply chain and believe procurement, production, and order fulfillment need to be considered as a whole. Some evidence exists for the increasing importance attached to reliability and the decreasing significance of solely cost factors. Despite these changes, if all other things are equal, then the choice between one logistics provider and another may very well be made on price. In addition, it is becoming clear that high quality in logistics services is seen as a minimum standard; service providers that cannot achieve high quality and consistent services are likely to be offered only the lowest value type of work.

The importance of good business-to-business relationships in order to promote supply chain efficiency is widely recognized. Holmes (1995) notes from a survey of logistics directors that 40% of respondents see adversarial attitudes towards customers and suppliers as an obstacle to closer supply chain integration and 50% view building mutual trust with trading partners as a key success factor. Yet he goes on to point out, when quoting from in-depth interviews, that "A ... true partnership depends on a balance of power. Without it you get one side taking advantage of the other; the whole attitude is wrong and that why partnerships get a bad name." Balance of power is interesting in the context of global logistics services, since the varying business cultures and the very different scales of operation between a global manufacturing company and its logistics service suppliers can lead to tensions.

Improved supply chain relationships can also reduce costs. According to a recent report, effective supply chain partnerships can save up to 6% of the material costs of a manufacturing company (A.T. Kearney & UMIST, 1996). However, the report also pointed out that, although many manufacturing companies claimed to have supply chain relationships with customers and suppliers, less than a third had measured the costs and benefits associated with these initiatives. Again, this highlights the opportunities that exist for logistics service providers that are prepared to adopt much more stringent measures of service level and service quality as a central part of their operation.

9. Success factors in logistics out-sourcing

Despite the trend towards increased out-sourcing, it is important to recognize that the contracting out of non-core activities does not always prove to be a successful strategy (see, e.g., Peisch, 1995). Some research has suggested that a majority of managers who have out-sourced activities (logistics and other activities) have been dissatisfied with the outcome (Lonsdale, 1999). However, rather than this being due to an inherent problem with out-sourcing itself, it is more likely to be due to one or more of the following factors:

Box 2
Factors that are frequently included in the evaluation of third-party
logistics companies

- Ability to provide suitable logistics data before, during, and after goods shipments
- Business arrangements (e.g., performance incentives, short- and long-term plans of the company, asset replacement strategy)
- Business success and development (e.g., accounts gained and lost by the company)
- Business experience and qualities (e.g., company history, quality of employees)
- Capabilities and competencies (i.e., ability to meet company's needs)
- Compatibility of technology for the required service
- Financial strength and stability
- Standards (i.e., Are they sufficiently high and are they improving?)
- Location/coverage (i.e., Does the provider's network match the requirements?)
- Management structure
- Opportunities to develop long-term relationships
- Price
- Reliability
- Reputation
- Service quality
- Speed
- Supplier certification
- Support services
- Systems flexibility and capacity

Source: Razzaque and Sheng (1998).

(1) making the wrong decision to out-source a logistics activity in the first place (i.e., the activity should have been kept in-house);
(2) poor selection of the third-party provider;
(3) poor management of the relationship with the third-party; and/or
(4) a lack of suitable performance measurement tools (i.e., methods with which to monitor the success of out-sourcing).

In approaching the decision of selecting logistics activities that could be out-sourced and the logistics companies that could perform these activities on their behalf (Box 2), companies need to address the following issues:

(1) identification of the specific logistics activities that could potentially be out-sourced;
(2) evaluation of which of these activities should be out-sourced;
(3) appraisal of the likely positive and negative effects of out-sourcing these activities on the company and its core business;

(4) specification of the level of service required from a logistics company with respect to each activity to be out-sourced;

(5) identification of the logistics companies with the necessary capabilities to provide these services; and

(6) negotiation with shortlisted logistics companies to determine the company that is best suited to achieving the required service standard at a competitive price.

10. Future trends in logistics out-sourcing

One of the factors influencing the thinking of companies providing logistics services has been the perceived desire of users to reduce the number of companies from which they purchase services. Some large corporations have pursued this "one-stop shopping" philosophy to a greater extent than others. A key question has to be the longevity of these links between provider and user. In some cases the initial optimism does not appear to be sustained. However, even where this supermarket theory of industrial buying has come unstuck there seems no reason to believe that users will wish to go back to systems where their buying power was fragmented and their ability to co-ordinate logistics across their supply network was weakened by dealing with too many service providers. This has implications for the providers in terms of their chosen strategy and will be influenced by:

(1) company culture and background (e.g., the size of the company and their ability to absorb the financial and management consequences of rapid change),

(2) customer profile (including industry, scale, main countries of operation), and

(3) customer culture (are the logistics services customers global, European, or national in their attitude to business-to-business purchasing?).

It is essential that a logistics service provider matches the needs of the customer. It is little use providing an excellent service if there is no demand for it or if companies cannot use it within their supply chain systems. At present it could be argued that some buyers are taking a short-term view and are ignoring the medium and longer term benefits of partnerships in favor of the short-term gains of lower prices. Providers of logistics services must demonstrate their ability to match their customers' expectations and ensure that they enhance and improve their ability to measure, monitor, and manage service quality.

In future we may see arrangements for providing services that are broadly similar to those prevailing in the automotive sector with tiers of suppliers. So we may see global corporations expecting that a single logistics service supplier will be responsible for the overall management of their supply chains, and probably providing a significant part of the physical infrastructure to achieve this. However, the first-tier supplier will be responsible for a second tier. The key feature will be the acceptance of full responsibility by the first-tier company, and in this respect it will be different to a simple subcontracting relationship. Managing this relationship and ensuring high and sustained quality will involve far more integration of communications and information systems than we can see at present. It will be interesting to see whether there is acceptance of this concept among manufacturers who are familiar with it in the context of component and subassembly supply, but may be more concerned with a perceived lack of accountability when an intangible service is being handled in this way.

It also seems likely that we will continue to see rationalization among logistics service providers. Whether this involves mainly companies providing similar services merging or forming alliances, or whether it involves organizations with complementary skills and service ranges, remains to be seen. What is clear is that the scale of investment required to build some of the more geographically diverse networks is too great in relation to the likely financial reward, and this is becoming a significant deterrent to companies going it alone.

The next few years promise further changes. Many manufacturing and retailing companies face continued difficult trading conditions. As a result, some may favor shorter and more price-driven relationships with service providers. However, there are alternative strategies, since in some cases efficiency gains are best achieved through longer term relationships between companies. Indeed, in an increasingly competitive environment, long-term relationships and mutual trust will become more important if we are to see greater efficiency gains in supply chain management. Logistics service companies will have to play their part in this by earning the trust of users through their own enhancements of measurement monitoring of service performance, and by being more innovative in their thinking.

References

A.T. Kearney and UMIST (1996) *Profit from partnerships*. Corby: The Institute of Logistics.
Bowersox, D. (1992) *Logistical excellence*. Burlington: Digital Press.
Dun & Bradstreet (1998) Press release (http://www.dnb.com/newsview/0798news1.htm).
European Logistics Association (1997) *Towards the 21st century: Trends and strategies in European logistics*. Brussels: European Logistics Association.
Fernie, J. (1990) *Retail distribution management*. London: Kogan Page.
Hines, P. and N. Rich (1998) "Outsourcing competitive advantage: the use of supplier associations", *International Journal of Physical Distribution and Logistics Management,* 28(7):524–546.
Holland International Distribution Council. (1998) *World-wide logistics: The future of supply chain services*. The Hague: Holland International Distribution Council.
Holmes, G. (1995) "Supply chain management: Europe's new competitive battleground", KPMG/ EIU, London, report.

Lonsdale, C. (1999) "Effectively managing vertical supply relationships: A risk management model for outsourcing", *Supply Chain Management*, 4(4):176–183.

Marketline International (1997) *EU logistics*. London: Marketline International.

McKinnon, A. (1989) *Physical distribution systems*. London: Routledge.

McKinnon, A. (1999) "The outsourcing of logistical activities", in: D. Waters, ed., *Global logistics and distribution planning: Strategies for management*. London: Kogan Page.

M. Corbett & Associates (1999) *The 1999 Outsourcing trends report*. New York: M. Corbett & Associates Ltd.

Peisch, R, (1995) "When outsourcing goes awry", *Harvard Business Review*, May/June, 24–37.

Quinn, J. and F. Hilmer (1994) "Strategic outsourcing", *Sloan Management Review*, 35:43–55.

Rabinovich, E., R. Windle, M. Dresner and T. Corsi (1999) "Outsourcing and integrated logistics functions: An examination of industry practices", *International Journal of Physical Distribution and Logistics Management*, 29(6):353–373.

Razzaque, M. and C. Sheng (1998) "Outsourcing of logistics functions: A literature survey", *International Journal of Physical Distribution and Logistics Management*, 28(2):89–107.

Roberts, D. (1999) "Sorry there's virtually nobody here", *The Daily Telegraph*, 21 August, 27.

Schary, P. (1994) "Organizing for logistics", in: J. Cooper, ed., *Logistics and distribution planning: Strategies for management*. London: Kogan Page.

Sheldon, D. (1998) *Retail distribution 1998*. Watford: Institute of Grocery Distribution.

The Outsourcing Institute and Dun & Bradstreet (1997) "The outsourcing index" (http://www.outsourcing.com).

Vanderbroeck, M. (1996) Paper presented at Global Supply Chain Management, the 1st World Logistics Conference, London.

Part 4

CUSTOMER SERVICE

Chapter 17

SERVICES MARKETING

ROHIT VERMA
DePaul University, Chicago

1. Introduction

It is now well known that services constitute the biggest section of the economy in many developed nations (Fitzsimmons and Fitzsimmons, 2000). According to the 1999 Statistical Yearbook of the United Nations, the service sector provides over 80% of the employment in the U.S.A. and over 70% of the employment in Canada, Japan, France, Israel, and Australia, when the agriculture, mining, and manufacturing sectors show negligible or negative growth. Because of rapid developments in information technology (IT), globalization, changing customer needs/preferences, and changes in the relative wealth of the developed and newly developing economies, the effective design of service systems continues to increase in importance (see Chapter 5). Some even argue that several developed nations have moved beyond the *service economy* to an *experience economy*, and therefore the ability to design effective systems for creating *desired* customer experiences will increasingly become an order winner (Pine and Gilmore, 1998).

Recognizing the need to improve the understanding of the services industry, this chapter provides an overview of services marketing concepts. Because services are inherently multifunctional in nature, operations, marketing, technology, and human issues are intimately connected to each other. Within this context, transportation services play the role of a *key enabler*, by facilitating the required and necessary movement of goods and people to satisfy the needs of the marketplace (e.g., delivery of mail-order merchandise to homes; mass rapid transport systems in urban areas). Many of the conveniences desired by the citizens of the service/experience economy cannot be fulfilled without the development of an efficient transportation system, and hence transportation and logistics services are growing at a rate faster than the growth of the entire service sector. For example, during the 1990s, while cumulative employment growth in the U.S.A. was 18%, the total service sector employment increased by 22% and transportation services employment increased by 26% (U.S. Bureau of Labor

Handbook of Logistics and Supply Chain Management, Edited by A.M. Brewer et al.
© 2001, Elsevier Science Ltd

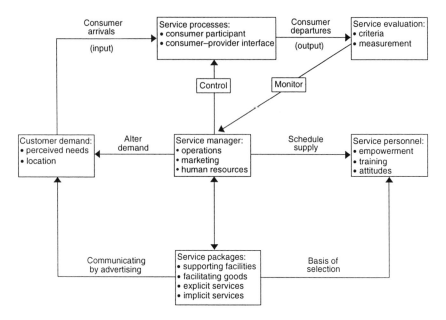

Figure 1. The open system view of services.

Statistics, 2000). Within the transportation services sector, employment in the trucking and air transportation services increased by 29% and 27%, respectively.

Transportation services, whether operating in a business-to-business (B2B) or a business-to-consumers (B2C) environments, must deal closely with various members of the supply chain (e.g., manufacturers, distributors, retailers, customers) and simultaneously satisfy the needs of many participants. Therefore, to manage effectively in this environment, a transportation service provider needs to understand the interactions between the essential elements of the service delivery systems.

The multifunctional role of a service manager can be illustrated by recognizing that a service provider operates in an "open system" connected by customers, employees, processes, and product-service package offerings (Figure 1). For example, the management of a air-transportation services company needs to consider the interrelationships between the service packages offered (pricing, features), personnel (pilots, air crew, ground crew), customer demand patterns (timing, volume, mix), and service processes (routes, check-in process, baggage handling process) in order to gain a good evaluation from their customers.

The remaining sections of this chapter provide an overview of the above and other marketing and management issues related to the service systems, with specific reference to the transportation and logistics industry.

2. Services environment

As mentioned earlier, the services sector is going through a period of massive growth and many changes. Furthermore, according to Pine and Gilmore (1998) many nations have moved to an experience economy, where human labor is increasingly used for artistic, creative, and intellectual reasons, and the standard of living is measured by quality of life in terms of health, education, and recreation. Several factors have contributed to and emerged as important services marketing issues in the changing face of post-industrial economies. This section summarizes the major trends.

2.1. Technological innovations

It will not be an exaggeration say that just as machines changed an agricultural economy into an industrial economy, today's IT is transforming an industrial economy into a service-driven economy. The availability of computers and global communication technologies (e.g., the internet and mobile voice/data communications networks) is creating limitless opportunities for designing, developing, and marketing new services. The role of technology in services in fact goes far beyond hardware and machines. It also includes innovative integrative systems, such as automated health testing and the airline industry's real-time pricing (yield) management systems. Technology can also facilitate the re-engineering of such activities as delivery of information, order taking, billing and payment, and redesign of the entire service delivery systems. It can also help in automating the "routine" efficiency-driven tasks and free up valuable human resources for selling or introducing other value-added services to consumers.

2.2. Global competition, regulatory changes, and increased privatization

Many services that were formerly purely domestic now have to compete with services from all around the world. For example, a customer service call placed by a consumer in Argentina may be routed to an operator located in Ireland, India, or Mexico, depending on the call volume, time of the day, language requirements, availability of the calling circuits, and nature of the call. In addition, the move towards deregulation of services, such as air transportation and tele-communications, around the world is giving rise to many opportunities for providing new services and/or for market share gain. Similarly, privatization of government-run business in many centrally planned nations has opened up several new opportunities for service marketers.

2.3. Franchising and service supply chains

Respected national brand names such as McDonalds, Citibank, and Hertz & Avis car rentals, among others, have spread far beyond their national boundaries. In some case such service chains are entirely company owned, whereas others are co-owned by multiple investors who manage or control specific parts of the business. Specifically, franchising involves licensing a product or service to be sold by independent entrepreneurs according to tightly specified specifications. The franchiser often takes the responsibility for marketing, customer relations, product development, and managing the supply chains, whereas individual franchise owners run the day-to-day operations of the specific store. Over the years franchising, has become a very popular approach for service capacity expansion and capital growth, and offers many possibilities for innovative service marketers.

2.4. Changing social trends

As the baby-boomers mature, the percentage of older people in many developed nations is increasing rapidly. The aging of a population creates opportunities for retired people to take part-time employment and also creates demands for various services needed by senior citizens (e.g., home delivery of goods and services). In addition, because of the increase in life-expectancy, there is an increased need for healthcare, transportation, and leisure services for older members of the community. Another significant social trend is that two-income families are fast replacing the traditional family structure of a working husband and a homemaker wife. The new two-income families are creating demands for services such as day care, preschool, "eating" out, precooked "gourmet" foods, home delivery of services, and family vacation services. For two-income families time is at a premium, and therefore they are willing to pay for services that give them more free time. In addition, a number of two-income families have an increased level of disposable income, leading to an increased demand for leisure, entertainment, and tourism services. The numbers of single adults and single-parent families are also increasing, creating the need for specific services such as organized recreational activities. The above and other social trends offer many opportunities for innovative service marketers.

2.5. Summary

The above and other trends in the services environment are creating a need for many innovative new services, such as web-based supermarkets (e.g., EthnicGrocer.com, WebVan.com), and have led to an increase in the demand for many existing services (e.g., restaurants, recreational facilities). Many new services, such as e-retailers, which require the movement of goods or people, need the assistance of an efficient distribution network, and therefore transportation services are becoming an essential part of the service package delivered to the customer. In other words, transportation plays a very significant role in customers' evaluation and choice of services. Hence understanding key services marketing concepts is essential for the effective design and development transportation and logistics operations. The next section of this chapter provides an overview of basic services marketing and management concepts.

3. Understanding service delivery systems

The service sector comprises a wide range of industries, such as entertainment, food services, healthcare, financial services, transportation and distribution services, education, and professional services. This diversity makes it difficult to make useful generalizations concerning the management of all service organizations. However, many underlying characteristics are similar across services and are often very different from those in other sectors of the economy (e.g., manufacturing, mining, agriculture) (Cook et al., 1999).

A service firm, such as an international airline, creates value for its customers without or with relatively little transformation of materials. On the other hand, a manufacturer transforms metals, polymers, and energy into a product such as a sports utility vehicle. Customers are rarely involved in the production process of the majority of manufactured goods (e.g., automobiles), but are very likely to become an essential part of the service-delivery process (e.g., urban mass rapid transit systems). The majority of services cannot be inventoried, as can products in manufacturing (an unsold airline seat is "lost" as soon as the airplane leaves the gate). Some of the above other commonly accepted characteristics of services are (Cook et al., 1999):

(1) services are intangible,
(2) the customer is a participant in the service-delivery process,
(3) generally services are produced and consumed simultaneously,
(4) services have a relatively higher variability in operational inputs and outputs,
(5) services generally have time-perishable capacity,
(6) site selection in services is directed by the location of customers,
(7) services in general are very labor intensive, and
(8) it is relatively difficult to identify appropriate measures of service output.

The above dimensions of services are fairly general and a lot of variability exists among different service firms, even within the same industry (Verma and Young,

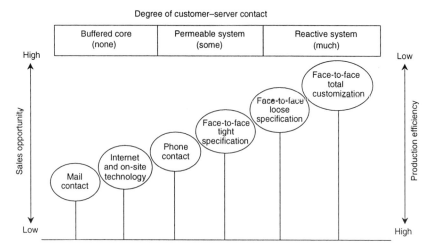

Figure 2. Customer contact in services.

2000). Hence to understand unique managerial issues associated with different types of services or to identify commonalities among different firms, close attention should be paid to the underlying features of services. Some of the commonly associated dimensions or services are described below.

3.1. Customer contact

The term "customer contact" refers to the physical presence of the customer in the service system (Chase and Tansik, 1983). According to Chase (1978), the degree of customer contact determines the sales opportunity, production efficiency, and design of the service delivery systems (Figure 2). For example, low-contact services (e.g., a package delivery company such as FedEx) are designed around efficiency considerations such as speed and utilization. FedEx, for example, has undertaken significant investment of capital, equipment, and IT to gain the maximum possible production efficiency (e.g., a large hub in Memphis, U.S.A., to minimize package-sorting times). High-contact services, on the other hand, have a relatively lower level of production efficiency, because the physical presence of the customer increases the variation in the service delivery process (e.g., even large airport terminals can only handle a very small number of the passengers compared with the number of packages handled at the FedEx hub). On the positive side, high-contact services have more opportunity to sell additional services to their customers than do low-contact services (Chase, 1978). For example, a package delivery company can rarely sell any additional services to its customers after the package has been dropped in a mailbox, whereas a passengers airline terminal has the opportunity to sell additional services (e.g., food, gifts, insurance) to its customers.

3.2. Tangibility

Customers' preferences and their evaluation of a service also depend on the relative tangibility of the product–service package purchased. As the degree of tangibility decreases, customers' ability to able to see, touch, or feel the purchased product–service bundle decreases. For example, the services provided by a passenger airline can be considered more tangible than the services provided by an international logistics consulting firm. Customers, in general, rely more on the brand name (reputation) when selecting intangible services (Fitzsimmons and Fitzsimmons, 2000). The degree of tangibility not only shapes the nature of the service delivery, but also affects the role of employees and the design of the service system.

3.3. Customization

An important marketing decision is whether all customers should receive the same service or whether service features or processes should be customized or adapted to meet individual requirements. Due to advances in computing technology, corporations are able to collect specific information about different users of their services and are able to customize their services according to the specific needs of their customers. For example, the Dell Corporation provides customized websites for each of its large corporate clients based on their own preferences for computer systems they are interested in purchasing; Amazon.com tracks and recommends the books, CDs, and DVDs its users might be interested in purchasing, based on their past choices.

3.4. Recipients of service process

Some services, such as air transportation, are directed towards customers themselves, whereas other services are directed towards things (e.g. cargo, package-delivery services). The design and marketing of the services systems focused on customers are very different from those services focused on things (e.g., airline passenger terminals, cargo terminals).

3.5. *Nature of the relationship with customers*

In many services, anonymous customers purchase one or more service packages from the service provider and then disappear (e.g., a one-time user of a package-delivery company or an airline). On the other hand, by developing a formal relationship with its customers, each customer is known to the organization and all transactions are individually recorded and attributed (e.g., frequent flyers on an airline; corporate customers of large logistics companies). The service provider has an opportunity to create a loyal customer base by giving specific loyalty credits (e.g., frequent flyer miles) when the customers are not anonymous. Furthermore, loyalty information can be used to target groups of customers with specific service options and/or promotions.

3.6. *Nature of demand and supply*

Some services face a steady and predictable demand for their services, whereas others encounter significant fluctuations. Similarly, in some services it is possible to alter capacity at short notice, whereas in other cases marketing mechanisms (e.g., pricing) must be used to deal with the unpredictable nature of demand. The interrelationships between demand, capacity-constrained supply, and pricing mechanisms can be easily observed in the passenger airline, hotel, and rental car industries, where yield management (or revenue management) is practiced widely (Kimes, 1989). The service provider, such as an international airline, constantly evaluates demand patterns, available capacity, and the past trends in cancellations and overbooking when determining prices for various seat categories. Because of the perishable nature of airline seats (i.e., once the flight has departed the potential revenue from an empty seat is lost forever), offering a discount on fares to fill the remaining seats in an aircraft becomes attractive. Selling all seats at discount, however, would preclude the possibility of selling some at full price. Although airlines were the first to develop yield management concepts, this practice is now being implemented in many other services such as hotels and rental cars. According to Kimes (1989), yield management can be a very appropriate marketing tool for services:

(1) which contain relatively fixed and perishable capacity,
(2) which operate in multiple market segments,
(3) which can sell services in advance,
(4) which operate in fluctuating-demand environments, and
(5) when the marginal sales costs are much lower than the marginal capacity change costs.

For additional information about the applicability of yield management in services, the reader is referred to Kimes and Chase (1998), Metters and Vargas (1999), and Smith et al. (1992).

3.7. Mode of service delivery

When designing delivery systems, service marketers need to decide if the customers will visit the service provider (e.g., a bus stop) or whether the service provider needs to visit the customer (e.g., a taxicab). At the same time, due to recent advances in IT, it is also possible that certain types or components of a service be delivered through mail or electronic channels (e.g., downloading music and e-books instead of going to a store). As mentioned earlier, the transportation industry plays the role of key enabler when customers desire more such services (e.g., home delivery of goods purchased online).

3.8. Type of service processes

Because customers are involved in the production and delivery of services, marketers need to understand the nature of the processes to which their customers are exposed. Service processes range from simple procedures involving only a few steps, to highly complex activities such as transporting passengers on an international flight. Because of the multifunctional nature of services, many different approaches to classifying service processes have been developed. For example, Schmenner (1986) classified service processes in four categories (service factory, service shops, mass services, professional services) based on the relative labor intensity, customer contact, and customization. Lovelock and Wright (1998) also classified service processes in four categories: people processing, possession processing, mental stimulus processing, and information processing. Other researchers, such as Bowen (1990), Kellogg and Nie (1995), and Wemmerlov (1990), have also attempted to link service–products with attributes of back-end service processes. Each study cited presents a typology of either ideal service management or theoretically derived differences between services. As a result, each classification scheme provides insights into the operations, personnel, and marketing issues related to different types of service processes.

3.9. Summary

It can be easily appreciated from the above discussion of the underlying characteristics of service delivery systems that there is no one formula for applying

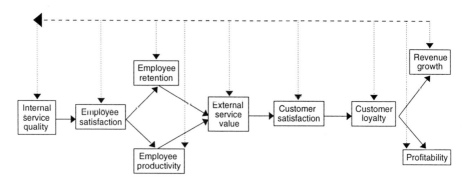

Figure 3. The service profit chain.

marketing concepts to complex transportation services firms. Therefore, innovative service providers either adapt their offerings and/or delivery process in order to better meet the needs of the markets they serve. Consequently, a number of approaches to implementing services marketing concepts have been developed. The next section in this chapter summarizes the major themes in services marketing.

4. Major themes in services marketing

It should be apparent from the preceding sections in this chapter that the service sector is going through a dynamic stage of development, presenting many exciting opportunities for marketing research and practice. Here I present only a few services marketing themes, which I believe will continue to be of great importance during the coming years. Note that this section contains only a few specific examples from any specific type of service, such as transportation. The focus is on summarizing merging themes that can be applied to a large cross-section of service firms.

4.1. Customer satisfaction and loyalty

Along with the growth in the services industry, there has been an increased interest by both academic researchers and practitioners in understanding the drivers of customer satisfaction and loyalty. Research has shown that it is significantly more expensive to recruit new customers than to generate repeat business with existing customers. Reichheld and coworkers (Reichheld, 1996; Reichheld and Schefter, 2000) have argued that for all types of services (including e-services) customer defection (and therefore retention) should be carefully monitored to assess customer satisfaction. Building on similar ideas, Heskett et al. (1994) proposed an oft-cited *service profit chain* model, which links financial performance and customer loyalty to their drivers (operations strategy, human resources issues, service features). The service profit chain model has been very influential in developing an understanding of the drivers of high performance in the services industry. Hence this model is described briefly below (Figure 3).

According to Heskett et al. (1994):

(1) *Customer loyalty drives profitability and growth.* Heskett et al. (1994) argue that a 5% increase in customer loyalty can boost profits by 25–85%.

(2) *Customer satisfaction drives customer loyalty.* The study demonstrates that "very satisfied" customers are likely to repurchase a service up to six-times more than are customers who are merely "satisfied."

(3) *Value drives customer satisfaction.* Delivering higher value service implies providing relatively more for the same cost to customers.

(4) *Employee productivity drives value.* Successfully and profitably delivering value requires building a team of employees who understand the linkages between operational productivity and customer value.

(5) *Employee satisfaction drives loyalty.* Heskett et al. found that low employee turnover is closely linked to customer satisfaction and that unsatisfied employees are three times more likely to leave the firm.

(6) *Internal service quality drives employee satisfaction.* Service workers are happiest when they are empowered to make things right for the customers and when they have the responsibility that enriches their work.

Other more quantitative approaches to linking customer satisfaction and financial performance have been developed by Rust et al. (1995) and Claes et al. (1996). Rust et al. (1995) developed a quantitative model, known as *return on quality* (ROQ), which links quality-improvement efforts with customer satisfaction, customer retention, revenue, market share, and profitability in service organizations (Figure 4). ROQ measures a ratio of changes in the net present value of profits with respect to additional quality improvement expenditures. Claes et al. (1996) linked customer loyalty and complaints with customer satisfaction, perceived value, perceived service quality, and customer expectations. This approach, also known as the *American customer satisfaction index* (ACSI), has been widely published in both the academic and the business literature (Figure 5). Although ROQ, ACSI, and the service profit chain approaches were developed by different research teams, there is a remarkable degree of agreement between their conceptual frameworks (see Figures 3 to 5), confirming the strong linkages between quality, customer satisfaction, customer loyalty, and financial performance in the services industry.

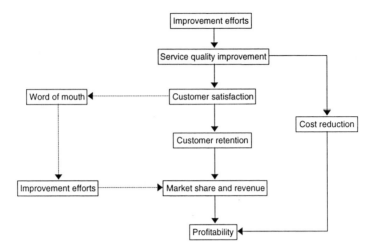

Figure 4. The return on quality approach.

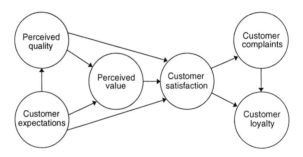

Figure 5. The American customer satisfaction index (ACSI).

A number of measurement approaches have been proposed that recognize the importance of customer satisfaction (Berry and Parasuraman, 1997). The most commonly used approaches are described below.

(1) *Transactional surveys*. Satisfaction surveys of customers following the service encounter provide feedback while service experience is still fresh in the customer's mind.

(2) *Mystery shopping*. In this approach researchers pretend be customers. The approach measures individual employee service behavior for use in coaching, training, performance evaluation, reorganization, and rewards, and is used to identify systematic strengths and weaknesses in customer contact service.

(3) *New, declining, and lost customer surveys.* Surveys that determine why customers select a service firm, reduce their purchase, or leave the firm.

(4) *Focus group interviews.* Directed questioning of a small group, usually 8–12 people (customers, potential customers, employees), provides a forum for participants to suggest improvement ideas.

(5) *One-on-one interviews.* An effective technique for achieving a deeper understanding of the customers' view of service performance.

(6) *Customer advisory panels.* A group of customers is recruited to periodically provide the firm with feedback and advice on service performance and other issues.

(7) *Service reviews.* Periodic visits with customers to discuss and assess the service relationship.

(8) *Total market surveys.* Surveys that measure customers' overall assessment of a company's service. Data are often collected from sample customers of multiple competitors (i.e., from the total market).

(9) *Employee field reporting.* Formal process for gathering, categorizing, and distributing field employee intelligence about service issues.

(10) *Service operating data capture.* A system to retain, categorize, track, and distribute key service performance operating data, such as response times, service failure rates, and service delivery costs.

(11) *Customer complaints.* These are mostly passively collected, and are used to identify the most common service failure types.

(12) *Critical incident technique.* This is a one-on-one technique to elicit details about services that particularly dissatisfy or delight customers.

One of the central underlying assumptions behind the work of Berry and Parasuraman (1997), Rust et al. (1995), Heskett et al. (1994), and Cleas et al. (1996) is that the quality of service ultimately determines value for the customer, which leads to customer loyalty and superior financial performance. Therefore, a considerable amount of research has been undertaken to link service quality with other characteristics of service delivery systems.

4.2. Service quality

The most widely used measurement instrument for service quality, known as SERVQUL, is based on the "gap" model proposed by Parasuraman et al. (1985). Parasuraman et al. conceptualized service quality as the difference between customer expectations and perceptions of service delivery. Customer perceptions are formed by personal needs, past experiences, and word of mouth. According to the model proposed by Parasuraman et al., customer perceptions of service are conceptualized as a function of service encounter or delivery, external

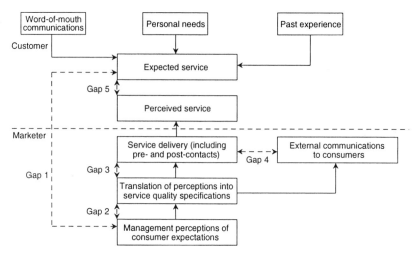

Figure 6. The service quality gap model.

communications to customers, managers' perceptions of customer expectations, and translations of managers' perceptions into service specifications (Figure 6).

In subsequent work Parasuraman et al. (1988) developed a 22-item scale, known as SERVQUAL, which measures the gap between customer expectations and customer perceptions within five dimensions of service quality:

(1) *Reliability*: the ability to perform the promised service both dependably and accurately.
(2) *Responsiveness*: the willingness to help customers and to provide prompt service.
(3) *Assurance*: the knowledge and courtesy of employees and their ability to convey trust and confidence.
(4) *Empathy*: the provision of caring, individualized attention to customers.
(5) *Tangibles*: the appearance of physical facilities, equipment, personnel, and communication materials.

In subsequent publications, Parasuraman et al. (1994) expanded the conceptualization of expectations based on further empirical studies. The enhanced version of the "gap" model includes a "zone of tolerance" within service expectations. According to Parasuraman et al. (1994), customers' expectations of service are within a zone bounded by desired service (highest expectations) and adequate service (lowest expectations). Parasuraman et al. (1994) also introduced a modified version of the SERVQUAL instrument, which measures the five service quality dimensions along desired service, adequate service, and perceived service scales.

Since the publication of the SERVQUAL instrument, a number of studies have applied the technique to a large number of industries. However, a number of studies have criticized the use of SERVQUAL as a general instrument for measuring service quality. Criticisms include both statistical and theoretical reasoning (e.g., Carman, 1990; Cronin and Taylor, 1992). At the same time, SERVQUAL in its original or modified format continues to be used in both academic and applied studies. For example, Soteriou and Chase (1998) demonstrated that perceived service quality as measured by SERVQUAL is linked to the degree of customer contact. Oppewal and Vriens (2000) presented an alternative approach for measuring service quality, using integrated conjoint experiments based on the five SERVQUAL dimensions.

With the recent exposition in electronic and information intensive services, I believe that we will continue to see many more applications, modifications, and enhancements of SERVQUAL. On the other hand, there has been a significant amount of research on the design and development of high-quality services. This stream of research is reviewed in the following section.

4.3. New service design and development

In goods-based industries, new product development has been widely studied (e.g., Wind and Mahajan, 1997). Given the inherent differences between the production of goods and services (e.g., the role of customer contact, intangibility, demand heterogeneity), the application of new product development models to services might not suffice in adequately describing how new services are optimally developed (Bitran and Pedrosa, 1998). According to a recent review article by Johnson et al. (2000), the new service development research undertaken so far has been largely descriptive rather than prescriptive. A brief review of the research is given in this section.

Johnson et al. (2000) have defined new service as "an offering not previously available to customers that results from the addition of offerings, radical changes in the service delivery process, or incremental changes to existing service packages that customers perceive as being new." According to Lovelock (1984), new service development can be divided into two categories: radical innovations and incremental innovations. *Radical innovations* include services for as yet undefined markets, such as many of the internet and IT based services (e.g., wireless web-based logistics, inventory tracking services). They also include start-up businesses in markets already served by existing services, and businesses providing radically new services to existing markets. *Incremental innovations* include service-line extensions (e.g., adding menu items in restaurants, adding new flight routes), service improvements (e.g., designing better web interfaces), and service style changes (modest forms of visible changes that have an impact on customer perceptions, emotions, and attitudes).

According to Ramaswamy (1996), service design consists of four interrelated components: service product design, facilities design, operations process design, and customer service process design. *Service product design* refers to the design of physical attributes; *facilities design* refers to the design of the physical layout of the facilities (including virtual facilities) where the service is delivered; *operations process design* refers to the activities that are needed to deliver or maintain a service; and *customer service process design* pertains to the interactions between the customer and the service provider. According to Chase et al. (1998), the three dominant approaches to service design are: the *production line approach* (e.g., McDonalds); the *self-service approach* (e.g., ATM machines); and the *personal attention approach* (e.g., Nordstrom). Ramaswamy (1996) stated that customers should be involved in all stages of the design process and that the design specifications should be developed with customer preferences in mind. He also stated that a multifunctional team should be involved in the design process and that the developed designed should be tested in "real" markets and not laboratories. Building on these concepts, Ramaswamy (1996) developed an eight-stage model:

(1) *Stage 1 – defining service design attributes.* This stage involves: identifying key customers for the service; determining the needs that the customers expect the service to fulfill; prioritizing the needs in order of importance; specifying the attributes required by a service that meets these needs; creating quantitative measures for measuring design attributes; establishing the relationships between needs and attributes; and determining the most important design attributes.

(2) *Stage 2 – specifying performance standards.* This stage involves: identifying the customers' desired performance level for each attribute; analyzing the performance of competitors; determining the relationship between performance and customer satisfaction; and specifying design performance standards for each attribute.

(3) *Stage 3 – generating and evaluating design concepts.* This stage involves: defining the key functions needed to provide the service; assembly of the key functions into processes; documentation of process flows; creation of multiple design concepts for service; and evaluation and selection of a service concept.

(4) *Stage 4 – developing design details.* This stage involves: converting selected service concepts into process components; generating, evaluating, and selecting the best design alternative for each component; testing the performance of the overall service design; making the necessary modifications; and specifying detailed functional requirements.

(5) *Stage 5 – implementing the design.* This stage involves developing and implementing service construction, pilot testing, roll-out, and communication plans.

(6) *Stage 6 – measuring performance.* This stage involves: selecting and measuring the performance of key attributes relative to established standards; measuring the capability of attributes and the efficiency of processes; identifying the root cause of poorly performing attributes, if any; and developing corrective actions, as required.

(7) *Stage 7 – assessing customer satisfaction.* This stage involves assessing customer satisfaction with respect to customer expectations and competition, and with respect to past performance of service, and identifying opportunities for improvement.

(8) *Stage 8 – improving performance.* This stage involves estimating: the relationships between financial objectives; overall customer satisfaction; service attributes; process attributes; and developing process improvement initiatives.

Although no detailed empirical work has been done to assess the approach presented above, the framework is valuable in developing a general understanding of various aspects of the service design process. To my knowledge, no other such elaborate conceptual framework for services design has been presented elsewhere. However, in recent years there has been a considerable amount of work done on developing tools or techniques for use in the service design activity, such as structured analysis and design (Congram and Epelman, 1995), function analysis (Berkley 1996), and discrete choice models (Verma et al., 2000; Pullman et al., 2001a,b).

4.4. Operational and human issues in services marketing

As mentioned earlier in this chapter, the interdisciplinary nature of services often makes it difficult to separate functional boundaries, and many operational issues directly impact on marketing practice and research. Here some operational and human issues of interest to service marketers are briefly described.

Operations strategy

Defined as the long-term pattern and priorities of decision-making, operations strategy in service is often inseparable from corporate and marketing strategies. Since the service delivery system is the main business of a service firm, alignment between the market positioning of service and products, priorities of marketing and operations functions, and an understanding of departmental trade-offs and constraints are essential. For example, Chase and Hayes (1991) have shown that the stages of the service strategy (available for service, journeyman, distinctive competence achieved, world class service delivery) are a function of customer preferences, service quality, technology, and workforce and management practices.

Demand and capacity management

Many services inherently contain highly fluctuating demand patterns. Although it is possible to stabilize certain aspects of demand with marketing mechanisms (promoting off-peak demand and discouraging peak demand periods), much can be gained by effectively using operational mechanisms, such as remote processing, use of automated technology, and use of flexible labor scheduling techniques. As mentioned earlier, revenue management (or yield management) is an approach for pricing capacity-constrained and time-perishable services (e.g., airline seats, hotel rooms, telecommunication packages). Prices are adjusted based on the demand rates, the supply of capacity at a given time, and the anticipated cancellation rates. Large-scale probabilistic optimization heuristics, which consider both the marketing and the operational variables, are used in price adjustments.

Management of waiting lines

Stochastic customer demand, capacity, and service productivity and the inability to use inventory buffers result in waiting times. Operations researchers have been interested for the last several decades in developing analytical models of queue configurations and their linkages on customer waiting times. However, there is also a psychological side to waiting. For example, preprocess waits are often perceived to be longer than in-process waits. In addition, perceived waiting times are often not linearly related to actual waiting times. It has also been suggested that service waits are impacted by environmental factors (e.g., music) or culture (e.g., Asian, western). For example, in a recent study, Pullman et al. (2000a,b) demonstrated that economic utilities for waiting times were not the same for Japanese-, Spanish-, and English-speaking customers at an international airport.

Service encounter

Most services are characterized as an encounter (or a series of encounters) between a service provider and a customer. This interaction, also known as the "moment of truth," is the time when the customer evaluates and forms opinions about service performance. The characteristics of a service encounter are determined by the customer, the service-providing employee, and the service organization. Recognizing the above interrelationship, service encounters have been of interest to service marketers, operations management, human resources, and organizational behavior researchers (e.g., Tansik, 1990; Price et al., 1995).

Service recovery and error proofing

Service recovery is a firm's response to failures in its delivery system. Even the best-in-class services fail sometimes, because service delivery systems are characterized by the simultaneous production and consumption, and because of the inclusion of customers in the production process. Good service recovery systems provide a firm with a second opportunity to "get things right" and win back market share. However, historically, most service firms have paid little attention to developing an effective service-recovery or error proofing system. Recently, Stewart and Chase (1999) found that substantial portions of service failures are a result of human error in the delivery process. Given that competition is increasing at an increasing rate in many service industries, the issue of error proofing and service recovery might be of critical importance to service marketers in the future.

5. Concluding remarks

Services provide many opportunities and challenges to managers and researchers alike. This chapter has provided an overview of the conceptual basis, trends, and major themes in services marketing. Many examples from transportation and logistics services are embedded throughout the chapter to illustrate the applicability of the concepts presented in these specific industries. The dynamic and evolving nature of services continue to present an increasing number of unanswered questions. Along these lines this chapter should, at the best, be considered as a primer on the exciting science and art of services marketing.

References

Berkley, B.J. (1996) "Designing services with function analysis", *Hospitality Research Journal*, 20(1), 73–100.
Bitran, G. and L. Pedrosa (1998) "A structured product development perspective for service operations", *European Management Journal*, 16(2), 169–189.
Berry, L.L. and A. Parasuraman (1997) "Listening to the customer – the concept of a service-quality information system", *Sloan Management Review*, Spring, 65–76.
Bowen, J. (1990) "Development of a taxonomy of services to gain strategic marketing insights", *Journal of the Academy of Marketing Science*, 18(1):43–49.
Carman, J.M. (1990) "Consumer perceptions of service quality: An assessment of the SERVQUAL dimensions", *Journal of Retailing*, 66(Spring):127–139.

Chase, R.B. (1978) "Managing service demand at the point of delivery", *Academy of Management Review*, 10:66–76.

Chase, R.B. and R.H. Hayes (1991) "Beefing up operations in service firms", *Sloan Management Review*, Autumn, 15–26.

Chase, R.B., N.J. Aquilano and F.R. Jacobs (1998) *Production and operations management: manufacturing and services*, 8th edn. New York: McGraw Hill.

Claes, F., M.D. Johnson, E.W. Anderson, J. Cha and B.E. Bryant (1996) "The American customer satisfaction index: Nature, purpose and findings", *Journal of Marketing*, 60(October):7–18.

Congram, C. and D.A. Epelman (1995) "How to describe your service: An invitation to the structured analysis and design technique", *International Journal of Service Industry Management*, 6(2):6–23.

Cook, D.P., C. Goh and C.H. Chung (1999) "Service typologies: A state of the art survey", *Production and Operations Management*, 8(3):318–338.

Cronin, Jr., J. and S.A. Taylor (1992) "SERVPERF versus SERVQUAL: Reconciling performance-based and perceptions-minus-expectations measurement of service quality", *Journal of Marketing*, 58(January):55–68.

Fitzsimmons, J.A. and M.J. Fitzsimmons (2000) *Service management: Operations, strategy, and information technology*. New York: McGraw-Hill.

Heskett, J.L., T.O. Jones, G.W. Loveman, W.E. Sasser, Jr. and L.A. Schlesinger (1994) "Putting the service profit chain to work", *Harvard Business Review*, March/April, 164–174.

Johnson, S.P., L.J. Menor, A.V. Roth and R.B. Chase (2000) "A critical evaluation of the new service innovation and service design", in: J.A. Fitzsimmons and M.J. Fitzsimmons, eds., *New service development: Creating memorable experiences*. Thousand Oaks, CA: Sage.

Kellogg, D. and W. Nie (1995) "A framework for strategic service management", *Journal of Operations Management*, 13(4):1734–1749.

Kimes, S. (1989) "Yield management: A tool for capacity-constrained service firms", *Journal of Operations Management*, 8(4):348–363.

Kimes, S. and R.B. Chase (1998) "The strategic levers of yield management", *Journal of Service Research*, 1(2):156–166.

Lovelock, C.H. (1984) "Developing and implementing new services", in: W.R. George and C.E. Marshall, eds., *Designing new services*. Chicago, IL: American Marketing Association.

Lovelock, C. and L. Wright (1998) *Principles of service marketing and management*. Upper Saddle River, NJ: Prentice Hall.

Metters, R. and V. Vargas (1999) "Yield management for the nonprofit sector", *Journal of Service Research*, 1(3):215–226.

Oppewal, H. and M. Vriens (2000) "Measuring perceived service quality using integrated conjoint experiments", *International Journal of Bank Marketing*, 18(4):154–169.

Parasuraman, A., V.A. Zeithaml and L.L. Berry (1985) "A conceptual model of service quality and its implications for future research", *Journal of Marketing*, 45:41–50.

Parasuraman, A., V.A. Zeithaml and L.L. Berry (1988) "SERVQUAL: A multiple-item scale for measuring consumer perceptions of service quality", *Journal of Retailing*, 64(1):12–40.

Parasuraman, A., V.A. Zeithaml and L.L. Berry (1994) "Moving forward in service quality research: Measuring different customer-expectation levels, comparing alternative scales, and examining the performance-behavioral intentions link", Marketing Science Institute, Cambridge, Working Paper No. 94–114.

Pine II, B.J. and J.H. Gilmore (1998) "Welcome to the experience economy", *Harvard Business Review*, July/August, 97–105.

Price, L.L., E.J. Arnould and P. Tierney (1995) "Going to extremes: Managing service encounters and assessing provider performance", *Journal of Marketing*, 59(2):83–97.

Pullman, M.E., R. Verma and J.C. Goodale (2001a) "Service capacity design with an integrated market utility-based method", in: J.A. Fitzsimmons and M.J. Fitzsimmons, eds., *New service development: Creating memorable experiences*. Thousand Oaks, CA: Sage.

Pullman, M.E., R. Verma and J.C. Goodale (2001b) "Service design and operations strategy formulation for multicultural markets", *Journal of Operations Management*, in press.

Ramaswamy, R. (1996) *Design and management of service processes: Keeping customers for life*. Reading, MA: Addison-Wesley.

Reichheld, F.F. (1996) "Learning from customer defections", *Harvard Business Review*, March/April, 56–69.

Reichheld, F.F. and P. Schefter (2000) "e-Loyalty: Your secret weapon to the web", *Harvard Business Review*, July/August, 105–113.

Rust, R., A.J. Zahorik and T.L. Keiningham (1995) "Return on quality (ROQ): Making service quality financially accountable", *Journal of Marketing*, 59(1):58–70.

Schmenner, R.W. (1986) "How can service business survive and prosper?", *Sloan Management Review*, 27(3):21–32.

Smith, B.C., J.F. Leimkuhler and R.M. Darrow (1992) "Yield management at American Airlines", *Interfaces*, 22(1):8–31.

Soteriou, A. and R.B. Chase (1998) "Linking the customer contact model to service quality", *Journal of Operations Management*, 16:495–508.

Stewart, D.M. and R.B. Chase (1999) "The impact of human error on delivering service quality", *Production and Operations Management*, 8(3):240–263.

Tansik, D.A. (1990) "Balance in service system design", *Journal of Business Research*, 20:55–61.

U.S. Bureau of Labor Statistics (2000) http://stats.bls.gov.

Verma, R. and S.T. Young (2000) "Configurations of low contact services", *Journal of Operations Management*, 18:643–661.

Verma, R., G.M. Thompson, W.L. Moore and J.J. Louviere (2000) "Effective design of products and services: An approach based on the integration of marketing and operations management decisions", De Paul University, working paper.

Wemmerlov, U. (1990) "A taxonomy for service process and its implications for system design", *International Journal of Service Industry Management*, 1(3):20–40.

Wind, J. and V. Mahajan (1997) "Issues and opportunities in new product development: An introduction to the special issue", *Journal of Marketing Research*, 34:1–12.

Chapter 18

UNDERSTANDING AND PREDICTING CUSTOMER CHOICES

HARMEN OPPEWAL
University of Surrey, Guildford

HARRY TIMMERMANS
Technical University of Eindhoven

1. Introduction

The demands for transportation and logistics are derived demands. Persons and goods are transported because they want or need to be somewhere else, not because there is a want or need for transportation per se. Even the demand for activities that at first sight seem exceptions to this rule, such as recreational travel, can in most cases be decomposed or explained from motives that are not transportation based as such. A person who enjoys traveling may do so mostly because the travel facilitates are away from the obligations of home or work and because it allows new sights and experiences, not because there is an inherent need for traveling. For this person travel is one, and perhaps even the only, feasible way of fulfilling a need for recreation. An understanding of how customers respond to travel characteristics, and their underlying logistics arrangements, therefore requires an understanding of customer travel motives and decision-making in the context of customers' needs to participate in activities such as work, shopping, and leisure consumption, and even their need for a place to live.

The pursuit of a better understanding of customer choice and responsiveness has led to at least three types of research and research literatures. The first focuses on the *explanation* of individual consumer behavior and provides general insights into the types of customer choice behavior, its determinants, and its underlying processes. Explanation in this literature is largely attained by conceptually describing the decision-making processes that lead to the choice outcomes of interest and testing general hypotheses in laboratory or field settings. The analysis typically takes place with analysis of variance and factor analytic methods, with an increasing use of structural equation models. The second type of research

Handbook of Logistics and Supply Chain Management, Edited by A.M. Brewer et al.
© 2001, Elsevier Science Ltd

concentrates on the *prediction* of customer responses from utility maximization principles. Customer responses are assumed to be a function of characteristics of the customer and his or her choice situation (e.g., a change in fares). This approach typically uses formalized conceptualizations to derive quantitative predictions of market shares of, for example, transport modes for the new circumstances. This takes the form of quantitative models that relate predictor variables representing transport and other relevant characteristics to choice behavior or choice probabilities. Procedures for building these predictive models include large-scale field surveys and regression-based methodologies, but can also include controlled experiments such as are currently used in the stated-preference paradigm. The third type is the rapidly increasing stream of formalized qualitative models that aim to predict choice behavior without assuming utility maximization. This approach builds *computer representations* of decision rules that underlie decision-making in real-world contexts. These models, or representations, take the form of decision tables, production systems, or other types of systems, such as neural nets.

The aim of this chapter is to outline these three approaches and demonstrate how they each contribute to a better understanding of customer choices and responsiveness. The chapter gives a brief introduction to the conceptual framework that underlies much of the research, highlighting some of the major concepts and key models in each area. To illustrate these concepts and models we refer to examples, noting that we do not claim these examples to be the only possible ones. It is beyond the scope of this chapter to review the extant literature on customer responsiveness for different travel and transport modes, such as rail, road, air, water, and surface, or, for that matter, responsiveness in the context of work, leisure, and residential choice. For such reviews the reader is referred to the relevant chapters in Handbook 1 in this series (Hensher and Button, 2000) and to recent research volumes such as the one by de Dias Ortúzar et al. (1998). We also refer to these sources for more in-depth and technical reviews of aspects of demand modeling, in addition to the much referenced book by Ben-Akiva and Lerman (1985) on discrete-choice modeling. We do not specifically elaborate on the determinants of service quality and the relation between perceived service quality and choice, but refer to Chapter 17 of this Handbook for a review of services marketing. The general literature on services marketing readily applies to transport and logistics because, in the end, the products that they supply are bundles of services.

The structure of this chapter is as follows. We first look at conceptual frameworks for understanding customer choices. The following section focuses on outlining predictive models of customer choice and responsiveness. Next follows a section on qualitative, rule-based models. The chapter concludes with a summary and discussion.

2. Customer behavior explained: The consumer behavior paradigm

The paradigm that is dominant in the consumer behavior discipline incorporates choice as a manifestation of a decision-making process that results from a drive for need fulfillment or goal accomplishment (e.g., Solomon et al., 1999). Choice also includes a voluntary aspect, where the person in question has different options to choose from and can be held responsible for his or her decisions. Theories of choice have been applied to a wide range of problems, including very mundane choices where it could be argued that, in effect, decision-makers have very little choice. The decision to travel to work by car can be considered a free choice, but for some people it seems to be the only possible way to get to work, at least in the short term. In the longer term a limited availability of public transport could result in a decision to move residence.

Consumer decision-making is typically conceived as a series of steps that starts with problem recognition, followed by a search for information, the identification of options or alternatives, the evaluation of these alternatives, and finally the selection of one of these alternatives. This, however, is not the final step, as post-decision behavior is also important – humans have a need to justify their choices and, hence, they continue to search for information even after they have made their decision. In addition, as a result of the decision and its further implementation, a new need may arise that requires a new round through the decision-making process.

Although the steps outlined seem universal, the extent to which they are taken with consideration and deliberation depends on various factors, including the newness of the decision, the time available to make a choice, and the risk and consequences of making a wrong choice. These latter factors are often assumed to determine the involvement of the decision-maker with the decision process. Low involvement leads to more automated, routine, or habitual processing of information and the acceptance of more simple rules for the selection of alternatives – these rules are called "heuristics." Some authors claim that this type of decision-making would be less rational, while others (e.g., Payne et al, 1983) have rightfully argued that decision-makers usually need to allocate time and effort across different tasks. It can therefore be very rational to not allocate too much time and effort to less important tasks. Few commuters will consider their travel option each morning. Instead they will settle in with their daily routine of driving to work, and only reconsider this decision when there is sufficient change in circumstances (e.g., increasing delays or petrol prices).

When the decision becomes more important to the person, the decision process becomes more problem oriented. The decision-maker will put more effort into the search for information and will, for example, more actively search for external information instead of relying on his or her own memory. In addition, a larger number of alternatives will be considered. Similarly, the evaluation of the alternatives can be assumed to be less dependent on one or a few determinants, the decision-maker being more inclined to enter a mode of compensatory decision-making, where a trade-off is made between all the positive and negative features of the alternatives.

Expertise and experience with the decision task also determine the extent to which the decision-maker is reliant on external information and the type of information that is most suitable. Customers with less experience typically put more effort into their search for information, although there are also indications that if the level of experience is very low the search effort is also small. This implies, for example, that information about public transport improvements in a suburb may easily reach current users in the area but not the infrequent and inexperienced users who are the target group for the improved services.

The above steps in decision-making do not occur in a vacuum. The way they are carried out largely depends on the decision context. This context consists of situational factors, such as the social and physical surroundings, as well as of the available time and the decision-maker's mood and attitude. Time is valued differently depending on which activity the person is involved in. For example, a certain amount of time spent waiting in a car can be evaluated quite differently from the same amount spent waiting for the last bus to come. Mood is a result of the interplay of pleasure and arousal, arousal depending on temporary physiological and psychological factors. Attitude is a longer term disposition to respond to an object. Attitudes are assumed to explain behavior to some extent, but there has been much debate about this in the literature (see, e.g., East, 1997). Attitudes, for example, do not well predict behavior if the attitude measure is not compatible with the behavior. A person who has a positive attitude towards traveling by rail in general may have a negative attitude towards taking the train during rush hours. Also, in some cases attitudes will follow and not precede behavior. For example, positive initial experience with a new piece of technology such as an electronic diary may result in a more positive attitude towards new personal electronic devices in general.

Although, in principle, virtually all decisions can be described as a series of steps, not all decisions are taken with the same amount of deliberation. External cues may lead a person to change his or her plans on an impulse. On the other hand, a person can be loyal to a certain item, person, place, or transport mode, indicating that the person is less willing to switch to another alternative than would be expected from the characteristics of this choice option. Most scholars perceive loyalty as a form of commitment to the alternative, where repeat behavior reflects a conscious decision to continue this behavior (Solomon et al., 1999). Others, however, have objected to this by pointing out that other reasons, such as habitual behavior or a lack of availability of other alternatives, may underlie the repeated choice of one alternative (East, 1997). A good way of testing for loyalty is to observe switching behavior when items become temporarily unavailable. What substitutes will be chosen, and how likely are people to return to their initial choice? Out-of-service or out-of-stock situations can clearly be used as opportunities to measure loyalty.

The social context also influences individual decision-making. People want to "look good" in the eyes of their peers or of other reference groups. In addition, according to the theory of social comparison (Festinger, 1954), people tend to feel at their best when they do slightly better than their peers. If a person drives a car to work, it represents an opportunity to demonstrate to colleagues what kind of car can be afforded. Social comparison theory predicts that the type of car that is preferred is a car that is more prestigious than the ones that colleagues drive, but not too much more prestigious. Reference groups can be either formal or informal, and a person can be a member of the group or aspire to become a member. Adolescents who, because of their age, are not yet allowed to drive a car can very much aspire to drive one and have older friends as their main reference group, the behavior of whom they observe and copy more than any other. In this sense, reference groups can exert substantial social power and induce conformity. People also want to do well in the eyes of themselves (self-perception) and tend to engage in behaviors that help maintain a positive image of themselves (ego defence). For example, they tend to put more value to dimensions of behavior on which they perform well and ignore information that threatens their positive self-esteem. Someone who drives to work in an old second-hand car apparently does not find it important to drive a fancy car. At the same time, however, the fact that he drives this type of car will make it increasingly unimportant for this person to "perform" on this possible dimension of social comparison. Thus, the causal relation between perceived importance of an attribute and choice behavior is multidirectional. It is also important to note that these behavioral phenomena are not only relevant at the moment of decision-making, but also affect post hoc decision-making behavior. A typical example is when someone reads car advertisements after a new car has been bought, mostly to get confirmatory evidence that the right choice was made.

The psychological decision processes described above are the micro-mechanisms that drive customers' reactions to changes in their social or physical environment. However, it is often not necessary to describe the processes at this level of detail. The development of an improved local transport system may, for example, only require knowledge about the extent to which different customer groups or segments will change their travel behavior. That is, instead of reducing transport behavior research to the mere application of psychology, it may suffice, and may even be more worthwhile, to describe how patterns of customer reaction vary with consumer socio-demographic and economic characteristics such as age, household type, and income, or how they are affected by demographic and societal trends (e.g., Lee-Gosselin and Pas, 1997). Will an increased punctuality of trains lead to a reduction in car use? Which segments will be the first to adopt electrical cars? The answers to such questions will typically assume the micro-mechanisms above to be in operation, but they do not require their detailed description for separate individuals. Instead, the focus will be on prediction of behavioral aggregates.

3. The prediction of customer choices from utility maximization

The complexity of decision-making behavior and the multitude of factors affecting decision-making make it very difficult to predict decision outcomes for an individual decision-maker in a specific task environment. Studies on individual decision-making therefore mostly focus on *testing* specific ideas about decision-making processes and the effect of factors such as information availability, time pressure, and task structure. However, in transportation and logistics, the focus is often not on individual persons but on the behavior of aggregates, such as the total demand for a certain transport mode or the effects of the availability, or non-availability, of a new product on the market shares of existing products. Predictive models of decision-making are therefore often estimated and applied at the aggregate level. The theory underlying these models, however, is typically based on concepts of individual decision-making, with the assumption that decisions are driven by utility maximizing behavior.

The focus in model building is on making predictions to help solve specific and applied problems, rather than on testing general theory. For example, a model could be developed to predict the effects of petrol price changes on mode choice, without any intention of developing and testing a general theory of how automobile use depends on petrol prices. The building of such predictive models takes an engineering approach. First, the relevant problem is defined; that is, the questions to be answered are explicitly mentioned and a statement is made about why and for whom these questions need answering. Next, a conceptual model is built to summarize the insights gained from previous studies, literature, and general theory regarding the main factors that can be assumed to be of influence on the outcomes of interest. This model should include the variables that are of managerial interest, for example, because they represent the key decision variables in formulating a new policy. However, the model should also include the other factors of influence, as otherwise the model results will be too dependent on specific circumstances and the uncontrolled factors may lead to spurious results. Next, a mathematical framework or set of equations is used to specify the functional relations between the explanatory and dependent variables. For example, the assumption that a consumer is less likely to choose a shopping destination at a greater distance can be expressed as a negative power function, $p(i, j) = d_{ij}^{-\beta}$, where $p(i)$ is the probability of individual i choosing alternative j, d_{ij} represents the distance between i and j, and β is a parameter to be estimated that represents the sensitivity to distance of consumer i. Existing theory may offer guidelines on how to specify these relations. In many situations, however, the modeler will have to make additional, and often ad hoc, assumptions. In those cases it is common practice to rely on standard functional forms such as additive functions. A next step, which often involves many arbitrary decisions by the modeler, is how to measure and operationalize the main constructs in the model. Distance can be measured in distance "as the crow flies" or as the actual distance traveled by car in meters, or it can be measured in minutes of travel time. Clearly, the precise nature of the relations and the parameter values to be estimated will depend on all these decisions.

3.1. Discrete-choice models

The most common type of customer response models in transportation are discrete-choice models. In general, discrete-choice models allow one to predict probabilities of choices between discrete-choice outcomes. Examples include the probability of choosing the car versus public transport, or the probability of choosing destination A versus destination B or C. Discrete-choice models serve to predict the probability that an individual will choose a particular discrete-choice alternative as a function of the attributes that describe the alternative, the socio-economic characteristics of the individual, and the characteristics of the choice context. It results in predictions of market choice and consumer demand, and when linked with external criteria provides indicators of feasibility and competitive impact.

There are several introductions to discrete-choice modeling approaches (e.g., Ben-Akiva and Lerman, 1985; Bhat, 2000; Koppelman and Sethi, 2000). Our current aim is to give a brief explanation of random utility theory (McFadden, 1974), which is a theory that emerged from developments in econometrics and mathematical psychology. It provides the conceptual underpinning for a range of model types that all aim to predict individual choice behavior from utility maximization principles. Random utility theory assumes that the utility U_j of an alternative j consists of a fixed or deterministic component V_j and a random or stochastic component ε_j:

$$U_j = V_j + \varepsilon_j. \tag{1}$$

The deterministic part of U_j is a function of the variables that describe the decision alternative j and, possibly, of other alternatives, the decision-maker, or the decision task or context, and is typically assumed to be of the following form:

$$V_j = \sum_k \beta_k x_{jk}, \tag{2}$$

where the β_k are parameters to be estimated and x_{jk} are coded predictor values.

The stochastic component is used to capture the influence of variables that were not, or not correctly, included in the functional specification of V_j and of a possible inherent randomness of the utility of an alternative. For example, if mood has an effect on utility, this effect could be specified in the function that predicts V_j or it could go unmeasured, adding to the variability of ε_j. Assuming a utility value for each available alternative, random utility theory predicts that the alternative with the highest utility will be chosen. Given this, and given the statistical distribution that is assumed for the stochastic component ε_j, different mathematical models can be derived that predict choice probabilities from the available alternatives and their characteristics.

The most common assumption about the stochastic component is that it is identically and independently distributed for different alternatives and that it follows a Gumbel distribution. This assumption leads to the well-known multinomial logit model. The multinomial logit model takes the following form:

$$P_{j|A} = \frac{\exp(U_j)}{\sum_{j' \in A} \exp(U_{j'})} = \frac{\exp\left(\sum_k \beta_k x_{jk}\right)}{\sum_{j' \in A} \exp\left(\sum_k \beta_k x_{j'k}\right)}. \tag{3}$$

This equation means that the probability that alternative j is chosen from choice set A is equal to j's share in the sum of the exponentiated utilities of all the alternatives in A. The multinomial logit has the property that it will predict that the introduction of any new choice alternative will attract a market share from existing choice alternatives in direct proportion to their utility. This property is directed related to the so-called independence from irrelevant alternatives (IIA) property, which can be derived as follows:

$$\frac{P_{j|j \in A}}{P_{j'|j' \in A}} = \frac{\exp(U_j)/\sum_{j'' \in A} \exp(U_{j''})}{\exp(U_{j'})/\sum_{j'' \in A} \exp(U_{j''})} = \frac{\exp(U_j)}{\exp(U_{j'})}. \tag{4}$$

Thus, the odds for any two choice alternatives are dependent only on these two alternatives, and are independent of the availability and characteristics of any other alternatives in the set. Non-availability and similarity among choice alternatives do not impact the choice probabilities predicted by the model. Clearly this property is unrealistic in many transport applications, as illustrated by the "red bus, blue bus" problem. Suppose a person is equally likely to choose the car or a bus, the latter of which happens to be red. Then, according to the multinomial logit model, the addition of an almost identical alternative, for example a blue bus, would still result in a reduction in share of the existing modes, including the red bus, each in proportion to the utility they represent.

To avoid this IIA property, the assumption of identically and independently distributed error terms may be relaxed. By assuming that the errors are normally distributed and correlated across choice alternatives the probit model can be derived. This model, however, is more difficult to estimate than the logit model, especially in cases with many alternatives. The application value of the probit model is therefore limited, although the possibilities for application have improved with the recent development of estimation methods based on simulation.

Another approach to circumventing the IIA property is to include predictors representing the existence and/or attributes of competing choice alternatives in the choice model. This results in the mother, or universal, logit model (Train and McFadden, 1976). These models include "cross-effects" parameters γ that measure the effects of the availability of certain alternatives, and their characteristics, on the utilities of other alternatives:

$$V_j = \sum_k \beta_k x_{jk} + \sum_{j'} \sum_k \gamma_k x_{j'k}. \tag{5}$$

This approach was used by, for example, Anderson et al. (1992) to measure not only the effects of transport mode attributes on mode choice but also the effects of the availability of specific modes on the choice shares of other modes. The universal logit allowed them to test and measure the differential effects of car and public transport on transport mode shares. Their findings supported the general finding that modes of public transport compete more heavily among each other than they do with the car. A similar approach can be taken to investigate the competition among certain travel destinations or of items in retail assortments.

A third way of developing non-IIA choice models is to specify a nested logit model. The modeler specifies a tree structure that groups alternatives to reflect the extent to which they are assumed to share unobserved attributes. For example, modes of public transport may share characteristics, such as perceived lack of comfort and flexibility, that do not apply to car travel. A nested logit model would account for this by specifying a tree structure in which first the choice of car versus public transport is modeled as a function of car and generic public transport attributes, such as travel time and travel cost. A next level would specify how the choice of bus or train within the group of public transport modes depends on attributes that differentiate between these modes of transport. By placing similar alternatives in the same "nest" and estimating a separate weight per nest, the differences in similarity between alternatives can thus be accounted for.

The nested logit choice model can be expressed as a product of marginal and conditional choice probabilities for the levels in the tree that define the choice alternatives. In our example, the probability of choosing rail transport is the product of the unconditional probability of choosing "public transport" (versus private transport) and the conditional probability of choosing "rail" given that a choice for public transport was made:

$$P(\text{rail}) = P(\text{rail} | \text{pub})P(\text{pub}). \tag{6}$$

The unconditional probability $P(\text{pub})$ depends on the utility derived from attributes x_1 that differentiate public transport modes from private transport and the utility of the best alternative in the subset of public transport alternatives, which are described by attributes x_2:

$$P(\text{pub}) = \frac{\exp(V_{x_1} + \max(U_{x_2}))}{\sum \exp(V_{x_1'} + \max(U_{x_2'}))}. \tag{7}$$

In the nested logit model this latter maximum can be expressed as a weighted summation across all public transport modes, and is called the "inclusive value." The weight $1/\mu$ incorporated in the inclusive value component is a parameter that should have a value between zero and one to be consistent globally with utility maximization:

$$\max(U_{x_2}) = \frac{1}{\mu} \sum \exp(V_{x_2})\mu. \tag{8}$$

The probability $P(\text{rail} | \text{pub})$ of choosing rail, given that a choice for public transport has been made, is in fact a multinomial logit model defined on the conditional choices (i.e., the available public transport modes):

$$P_i(\text{rail} | \text{pub}) = \frac{\exp(V_{x_2})\mu}{\sum \exp(V_{x_2'})\mu}. \tag{9}$$

Note that the utilities in this model are multiplied by a scale value μ. This scale value was also present in the previous "upper nest" model, but was set to unity, as is the case in any "stand-alone" logit model.

An illustrative application of the nested logit model is the study by Hensher and King (2001), who used the nested logit model to combine data from different sources and develop a model that predicts choice of car park in central business district areas.

3.2. Choice sets

Although the violation of the IIA property has drawn the most attention in the literature, there are also other limitations of discrete-choice models that should not be ignored. First, the application of discrete-choice models assumes that for each decision-maker the choice set is known. In some cases this is a reasonable assumption. For example, when modeling the choice between car and public transport it seems reasonable to assume that a respondent knows that there is, in principle, the possibility of choosing either alternative. However, if we start differentiating between different public transport alternatives it may already be the case, for example, that some respondents do not know about a particular bus service in their area. Even though it is available, for these people the bus does in fact not occur as an alternative in their choice sets. In addition, even if all alternatives have been identified, the information about these alternatives may be distorted or incomplete. For example, people who take up residence in a suburb have been found to be not so well informed about the accessibility of their new neighborhoods (Raux and Andan, 1997).

Parameter estimates become inconsistent when choice sets are mis-specified (Stopher, 1980). Several attempts have been made to model choice set construction and composition as a separate step in the decision process. In marketing, terms such as "awareness set" and "consideration set" are used to distinguish between the various ways in which an alternative can enter the choice set. However, there is only limited evidence that there is a real gain in modeling choice set construction, the argument being that measures of awareness and consideration are not sufficiently discriminative from measures of utility (Horowitz and Louviere, 1995). Several interesting attempts have nevertheless been made to model choice sets (see, e.g., Roberts and Nedungadi, 1995).

3.3. Choice constraints

One other limitation of discrete-choice models is the dependence of the choice sets and/or the utilities of the alternatives on the specific circumstances or choice contexts. Circumstances can include changes in life-cycle stages, which have a large impact on the use of transport (Raux and Andan, 1997), on task-related variables, such as the time available for task completion, or on the decision-maker's choice history. Circumstances may limit the availability of choice alternatives to the extent that it becomes more worthwhile to focus on modeling the occurrence of constraints for different population segments than on modeling the final choices that people within these segments make.

Choice constraints are an important element in time–space geography as developed by Hägerstrand (1970). Hägerstrand identified three types of constraint:

(1) *capability constraints*, which refer to physical constraints (e.g., humans physically need a minimum number of hours to sleep and eat);
(2) *coupling constraints*, which reflect the fact that one has to be with particular people at the same location at the same time, or use particular materials to perform certain activities; and
(3) *authority constraints*, which relate to the institutional context of laws, rules, and other regulations (e.g., shop acts regulate the opening hours of stores, dictating when shopping activities can be performed).

Attempts to accommodate these constraints in the choice modeling framework are illustrated by Stemerding et al. (1999), who modeled the effects of weather conditions and time of departure on choice of transport mode and destination for a recreation trip. Their study also accommodated effects of transport attributes such as car parking facilities and public transport connections.

3.4. Group decision-making

A final factor to mention here that is easily overlooked in modeling studies is the influence of the composition of the decision-making unit. Many consumer decisions regarding transport are made in the context of household decision making where multiple individuals have input. To predict these decisions one needs to know the utilities of the alternatives for the group as a whole, or at least for the key decision-maker, or it requires knowledge of the utilities that the separate individual attach to the alternatives. In addition, it requires insight into how these utilities are combined into a group utility and decision during the group's decision-making process. For a recent investigation into this topic, see Molin et al. (1999), who found that, when asked individually, parents find their own travel time to work the most important, but when asked jointly as a group agree that the children's travel time to school is the most important.

4. Computational process models to describe and predict choice behavior

It has been indicated in the previous section that there is a tendency in discrete-choice modeling to incorporate increasingly more complex types of behavior, such as availability effects, contextual effects, and similarity. It is fair to say, however, that most commonly used models in transportation and logistics involve a linear utility function. This means they implicitly assume a compensatory decision-making process; that is, a low evaluation of one attribute may, at least partially, be compensated by high evaluation scores on one or more other attributes. Alternative behavioral principles that were discussed in the beginning of this chapter are, however, more difficult or even impossible to represent in mathematical equations of the type that underlie discrete-choice models.

The assumption of utility-maximizing behavior, which is characteristic of discrete-choice models, has therefore been criticized by some scholars who have argued that individuals do not necessarily arrive at "optimal" choices, but rather use heuristics that may be context dependent. This position is supported by the results of behavioral decision theory (e.g., Kahneman et al., 1982), which states that various kinds of contexts influence the heuristics used and hence the outcome of decision processes. Consequently, there is an increasing line of research that attempts to find and apply different representations of consumer decision-making and choice behavior. Many of these models involve Boolean logic as opposed to algebraic functions. Consequently, context-dependent decision-making, non-compensatory decision-making, functional equivalence, and complex feedback and feed-forward structures are relatively easy to build.

An early formalism to represent such choice heuristics is the use of a so-called "production system." A production system is a set of IF, THEN, ..., ELSE rules that specify which decision will be made as a function of a set of conditions. Consumers are typically assumed to collect and constantly update their information about the (transportation or logistics) environment, mediated through their interaction and experience with the environment, and to process this information to form evaluations and develop imperfect and limited cognitive representations that are stored in their long-term memories. A subset of this imperfect information about the environment and sets of heuristics, called short-term memory, is used for every-day decision-making. Thus, in its most basic form, a production system is a set of IF–THEN rules, a short-term memory (STM), and a mechanism for controlling which rule to apply in a given context. IF–THEN rules take the following form:

IF (condition = X) THEN (perform action Y).

The STM represents the current state of the system and takes a form similar to that of a condition in the IF–THEN rules. The control mechanism matches the STM against various production rules and chooses the rule the condition side of which best matches the STM so that its action can be executed. The action chosen generally leads to changes in the state of the system, and consequently in the contents of the STM. The basic cycle of matching the rules against the STM and carrying out the most appropriate action continues to occur in an iterative manner until the system ceases operation in some terminal state.

Thus, the modeling approach assumes that activity patterns may be represented in terms of a sequence of state–action (SA) pairs in which the state S represents a possible state of both the individual and the decision-making context, while the action A represents the decision taken in that state. An individual's state at any time may be represented in terms of a conjunction of values x_i over a set of n state variables $\{X_i,\ i = 1,\ n\}$. Hence,

$$(X_1 = x_1) \wedge (X_2 = x_2) \wedge \dots \wedge (X_n = x_n) :: \rangle \ \mathbf{A}. \tag{10}$$

Each distinct action of an individual may be expressed as an appropriate production rule in terms of a simple logical language known as "disjunctive normal form" (DNF). This represents the conditions under which it is permissible to take each of the feasible actions. The DNF rules are particularly simple examples of IF–THEN rules comprising a general production system.

Different formalisms can be used to represent these rules. Recently, the decision-table formalism has been used a lot because it has some clear advantages in terms of verifying the exhaustiveness, exclusiveness, and consistency of the decision rules. The decision-table formalism has some advantageous properties. First, because they are exclusive, consistent and complete, decision tables return a response for every possible case within the domain of interest. This behavior is not guaranteed by traditional production systems. Second, it is capable of representing various types of interactions between variables.

Traditionally, the rules of these computational process models were derived from think-aloud tasks or other knowledge-elicitation techniques. Experts were prompted to elicit the relevant rules. Decision-making processes were simulated using such expert knowledge. More recently, however, there has been an increase in the use of inductive learning algorithms and neural networks to derive the logical rules from empirical data. An example is given in Arentze et al. (2001), who developed an algorithm for scheduling problems. The algorithm starts with an empty schedule, and successively adds activities. More formally, let P be the program to be implemented on a given day by a particular individual, $R_k \subseteq AP$ be the subset of activities not yet scheduled at the kth step of the scheduling process, and $S_k \subseteq AP$ be the complementary subset of activities that has already been scheduled. Initially, $R_k = AP$ and $S_k = \varnothing$. A scheduling decision sd_k consists of two steps: (1) selecting an activity $a \in R_k$, and (2) selecting a position for a in the current schedule S_k. The system includes for both steps a set of decision rules that represent scheduling constraints and preferences. The rules are organized in a hierarchy. From high to low order, the rules successively reduce the set of options (activities or positions) until the most preferred option is uniquely identified or until all rules have been applied. In the case of indifference, options are randomly selected. Constraints have the highest priority. They represent constraints on feasible schedule positions for a given activity. Preference rules, in contrast, represent preferences, which may vary between individuals. The objective of the learning algorithm is to find the subset of preference rules that minimizes the deviation between predicted scheduling decisions and observed choices.

The estimation problem can be conceptualized as a problem of training a rule-based system based on examples (in the form of observed schedules). In the proposed method, action alternatives within decision tables are conceptualized as competing rule sets (which have the same condition, but different actions). An error parameter is attached to each rule (*jk* cells of each decision table). This parameter indicates the past success of the rule in reproducing observed cases. Each time a case is to be processed, the system selects a rule from each competing rule set, as a probabilistic function of error-values. After processing a case, the system updates the error-values of the rules that were used, based on feedback in terms of the similarity measure. This algorithm describes an incremental learning process, in which feedback information is accumulated in the error values of rules.

The performance of the system will depend on the specification of the rule-selection function, the specification of the similarity measure, and the method of updating rule error values.

At least two methods of updating error values can be identified. In the first method, the error value of a rule is expressed as the average error value across all the cases for which the rule delivered output. In the above "average-error" method, the past performance of a rule keeps exerting influence on the current rule value throughout training. Although the effect of current performance on the running average becomes weaker and weaker as training proceeds, the system exhibits a complete memory of past cases. In the second method, the system still keeps a record of past performance, but the weight of past cases on current error values declines as training proceeds.

Full-fledged examples of such rule-based systems in transportation and logistics research are still rare. The most elaborated rule-based model is Albatross, an activity-based model of transport demand (Arentze and Timmermans, 2000). Using the decision-table formalism, decision heuristics are induced from an activity diary. These rules describe behavior (which activities are conducted, where, when, for how long, and with whom, and the transport mode involved) as a function of personal characteristics, context, and the history of the activity schedule. The model performed well; there was evidence of better predictive ability compared to a conventional discrete-choice model.

5. Summary and discussion

We have discussed three very different approaches to the study of customer choice and responsiveness, and the question that seems pertinent is: What is the value of each approach, and to which extent are the approaches commensurable? While, the approaches all share a recognition that customer choice involves a series of steps, and the explanation of choice requires that the subjectivity of the decision-maker is accommodated, they do this in very different ways. The consumer behavior discipline accommodates a wide range of different approaches, but the dominant approach is positivistic and focuses on the development of theory and the empirical testing of hypotheses derived from this theory. The theory is broad and verbally described, and does not easily allow one to make precise predictions in applied contexts. Hypotheses are tested in, preferably, controlled laboratory situations, where the first goal is to ensure internal and construct validity of results. Establishing the external validity is also key, however, because the goal is to develop (more or less) general theory about customer behavior, and the research results themselves have little inherent value, especially if they are based on the much criticized habit of testing theory on student samples only. The aim is to explain behavior in order to build understanding in the sense of the researcher (and research clients) gaining intellectual control over the phenomenon in question, such that the range of possible phenomena to expect is better understood and new solutions to problems is more easily discovered or devised.

This is in stark contrast with discrete-choice modeling, which strictly operates within a utility maximizing framework and leaves this elegant framework largely unquestioned. The focus is on the development of models that allow the measurement and prediction of customer choices and sensitivities in applied contexts. The operationalization and specification of the model are key issues here, in addition to the proper estimation of model parameters and the use of statistical theory to avoid acceptance of unlikely results. In specifying the model the aim is to include the most important determinants of the choice behavior under study plus the variables that the research clients are interested in as potential policy instruments. "Explanation" in this context means being able to predict a sufficient amount of variation in the dependent variable with a model specification that includes the most relevant attributes. It thus means that a correspondence is established between the data at hand and, in the end, the utility maximizing principles as expressed in a mathematical model. Generalizability of results is not a first issue because it is accepted that each new application may require at least a new calibration of an existing model, and often even involves the development of a totally new model specification in order to allow prediction of the required level of precision. Proper assessment of the external validity of models is nevertheless regarded as useful, as are meta-studies that analyze and summarize the results from various previous studies. One case in which external validity is especially important is where stated-preference methods are used to measure preferences in controlled hypothetical circumstances. The application of these measures and models to the "real" world clearly assumes that the results transfer to this external domain, which is an assumption that too often remains untested.

The computational process models combine some of the features of the previous two approaches. The ingredients are intuitively appealing ideas about the rules that people apply during decision-making. This would suggest that the models themselves also lead to an improved understanding. However, the multitude of rules and the large number of possible interactions between them suggest that the models as such do not summarize data in an enlightening way. The researcher and research client have to accept what the data and rules that were put into the model have generated, and thus explain this possible outcome. Of the three approaches, the computational process models are the least parsimonious and the most difficult to generalize. However, as said, they do have the advantage of being able to accommodate interactions between a large number of variables producing precise predictions. Systematic variation of the variables, their interactions, and the sequence of rule applications can be used to test the sensitivity of a result, and can be a basis for generalizations.

The three approaches represent three different ways of studying customer choice and responsiveness. The usefulness of each depends on the type of problem that needs to be solved. The consumer behavior discipline is most useful if the aim is to gain a general understanding of the reasons for and ways in which people may respond to, for example, a new service and who will be the first to adopt the service. The consumer behavior discipline brings a not so well organized, but often still useful, toolkit with a range of concepts and "half-way" theories that can help to better describe and understand customer behavior. If the goal is to obtain precise predictions of, for example, changes in market shares of transport modes after introducing a new service, then discrete-choice models may prove most fruitful. Furthermore, if the first goal is to investigate more closely a wide range of possible changes in circumstances, such as changes in shop opening hours on customer behavior, then computational process models may prove most valuable.

We hope this review will have helped the reader understand that there is more than one way of explaining and predicting customer choice and that she or he will be stimulated to use, and explore further, each of these different types of research.

References

Anderson, D., A. Borgers, D. Ettema and H. Timmermans (1992) "Estimating availability effects in travel choice modelling: A stated choice approach", *Transportation Research Record*, 1357:51–70.

Arentze, T.A. and H.J.P. Timmermans (2000) *ALBATROSS: A learning-based transportation oriented simulation system*. Eindhoven: EIRASS.

Arentze, T.A., F. Hofman and H.J.P. Timmermans (2001) "Estimating a rule-based system of activity scheduling: A learning algorithm and results of computer experiments", *Transportation Research*, in press.

Ben-Akiva, M. and S. Lerman (1985) *Discrete choice analysis*. Cambridge, MA: MIT Press.

Bhat, C.R. (2000) "Flexible model structures for discrete choice analysis", in: D.A. Hensher and K.J. Button, eds., *Handbook of transport modelling*. Amsterdam: Pergamon.

de Dias Ortúzar, J., D. Hensher and S. Jara-Diaz (1998) *Travel behaviour research: Updating the state of play*. Oxford: Pergamon-Elsevier.

East, R. (1997) *Consumer behaviour: Advances and applications in marketing*. Hemel Hemstead: Prentice Hall.

Festinger, L. (1954) "A theory of social comparison processes", *Human Relations*, 7(May):117–140.

Hägerstrand, T. (1970) "What about people in regional science?", *Papers of the Regional Science Association*, 23:7–21.

Hensher, D.A. and Button, K.J., eds. (2000) *Handbook of transport modelling*. Amsterdam: Pergamon.

Hensher, D.A. and J. King (2001) "Parking demand and responsiveness to availability, pricing and location in the Sydney central business district", *Transportation Research A*, 35:177–196.

Horowitz, J.L. and J.J. Louviere (1995) "What is the role of consideration sets in choice modelling?", *International Journal of Research in Marketing*, 12:39–54.

Kahneman, D., P. Slovic and A. Tversky (1982) *Judgement under uncertainty: Heuristics and biases*. Cambridge: Cambridge University Press.

Koppleman, F.S. and Sethi, V. (2000) "Closed-form discrete-choice models", in: D.A. Hensher and K.J. Button, eds., *Handbook of transport modelling*. Amsterdam: Pergamon.

Lee-Gosselin, M.E.H. and E.I. Pas (1997) "The implications of emerging contexts for travel-behaviour research", in: P. Stopher and M. Lee-Gosselin, eds., *Understanding travel behaviour in an era of change*, pp. 1–28. Oxford: Pergamon-Elsevier.

McFadden, D. (1974) "Conditional logit analysis of qualitative choice behaviour", in: P. Zarembka, ed., *Frontiers in econometrics*, pp. 105–142. New York: Academic Press.

Molin, E., H. Oppewal and H. Timmermans (1999) "Group-based versus individual-based stated preference models of residential preferences: A comparative test", *Environment and Planning A*, 31(11):1935–1947.

Payne, J.W., J.R. Bettman and E.J. Johnston (1992) "Behavioural decision research: A constructive processing perspective", *Annual Review of Psychology*, 43:87–131.

Raux, Ch. and O. Andan (1997) "Residential mobility and daily mobility: What are the ties?", in: P. Stopher and M. Lee-Gosselin, eds., *Understanding travel behaviour in an era of change*, pp. 29–52. Oxford: Pergamon-Elsevier.

Roberts, J. and P. Nedungadi (1995) "Studying consideration in the consumer decision process: progress and challenges", *International Journal of Research in Marketing*, 12:3–7.

Solomon, M., G. Bamossy and S. Askegaard (1999) *Consumer behaviour*. Upper Saddle River, NJ: Prentice-Hall.

Stemerding, M.P., H. Oppewal and H.J.P. Timmermans (1999) "A constraints-induced model of park choice", *Leisure Sciences*, 21(2):145–159.

Stopher, P. (1980) "Captivity and choice in travel-behaviour models", *Transportation Engineering Journal of the ASCE*, 106:427–435.

Train, K. and D. McFadden (1978) "The goods/leisure trade-off and disaggregate work mode choice models", *Transportation Research A*, 12:349–353

Part 5

LOGISTICS PERFORMANCE MEASUREMENT

Chapter 19

COSTING THEORY AND PROCESSES

WAYNE K. TALLEY

Old Dominion University, Norfolk

1. Introduction

Costs incurred in the provision of transport services, both passenger and freight, by for-hire transport providers (carriers) have been classified as: accounting versus economic costs, non-shared versus shared costs, and internal versus external costs. Accountants are primarily concerned with measuring costs for financial reporting purposes. Consequently, accounting costs reflect the historical outlay of funds to resources (e.g., labor and capital). Alternatively, economists are concerned with measuring costs for decision-making purposes. Economic costs reflect the opportunity costs of resources in their current use (i.e., the earnings that resources would have received if utilized in their next best alternative). Resource prices in competitive resource markets reflect the opportunity costs of resources in these markets.

Non-shared (shared) costs are costs that can (cannot) be traced to a particular passenger or freight shipment. For example, if a single freight shipment is transported in a vehicle from location A to location B, the entire cost of the vehicle trip is traceable to it. If two or more shipments are transported, the cost of the vehicle trip is not traceable to a particular shipment, but is to be shared among the shipments. In this case, the shared cost is a common cost, since one shipment does not unavoidably result in another shipment being transported in the same vehicle. If a transport movement unavoidably results in another movement occurring, the cost that these movements share is a joint cost. For instance, a joint cost arises when a front-haul vehicle trip necessarily creates a back-haul vehicle trip. If a vehicle transports freight from location A to location B and then must return to location A, the vehicle cost of the round-trip is a joint cost to be shared between the front-haul and back-haul trips.

Internal (external) costs generated by transport carriers in the provision of transport services are borne (not borne) by these providers and therefore enter (do not enter) into their decision-making processes. Examples of external costs include air, noise, water, and esthetic pollution costs of transport services.

Handbook of Logistics and Supply Chain Management, Edited by A.M. Brewer et al.
© 2001, Elsevier Science Ltd

What are the costs incurred by transport carriers in the provision of particular transport services? This chapter addresses this question. Costs are restricted to those borne by carriers. The next section discusses costing transport service in theory. Topics include carrier resource, production, and cost functions, as well as allocating shared costs. Then, costing transport carrier service in practice is discussed. Specifically, individual-carrier and multicarrier costing processes (or methods) are discussed. The former utilize data of a single carrier, whereas the latter utilize data of two or more carriers for determining carrier costs in the provision of particular transport services. The discussion of individual-carrier costing processes include cost-center, resource, and time-series statistical cost function costing processes, whereas the discussion of multicarrier costing processes include cross-section statistical cost function, cross-section statistical resource function, and regulatory costing processes.

2. Costing transport service in theory

Costs that transport carriers incur in the provision of transport service reflect: the resources utilized and their amounts; whether there are production and cost efficiencies in the provision of service; whether costs are long-run or short-run costs; and how shared costs are allocated among individual freight shipments and passengers.

Five general types of resources are employed by transport carriers: labor, energy, way, terminal, and vehicle resources. Labor resources include workers directly involved in the physical movement of freight and passengers, maintenance of facilities, and management of operations. Energy resources include gasoline and diesel fuels, electricity, and natural energy (e.g., wind and water currents). The way is the path over which the transport carrier operates, and consists of the right of way that is supplied by nature plus added physical facilities. For example, a highway consist of the land area right of way and added pavement, traffic lights, and signs. Terminals are physical structures, which serve as activity centers where freight shipments are loaded and unloaded and passengers embark and disembark from vehicles. Vehicles may either be power units, such as railroad locomotives and tugboats, or carrying units, such as rail cars and barges.

2.1. Resource function

The amounts of resources employed by a transport carrier will depend on:

    (1) the amounts and types of freight shipments and passengers to be transported, and

(2) the quality of the service to be provided.

A carrier varies the quality of its service by varying the levels of its operating options, and thus the means by which it can differentiate its service (Talley, 1988). Carrier operating options include:

(1) speed of movement,
(2) frequency of service,
(3) reliability of service,
(4) accessibility of service, and
(5) susceptibility to loss, damage and injury.

The greater the speed of movement, the higher the quality of service to the user, since shipments and passengers will arrive at their destinations within a shorter time period. Frequency of service is how often the service is provided within a stated time period. The greater the frequency, the higher the quality of service, since the service will more likely be available when desired by shippers and passengers.

Reliability of service is the degree to which shipments and passengers arrive at their destinations at their scheduled times. Reliability may be measured by the variability between scheduled and actual times of arrival. The more reliable the service, the higher the quality of the service (e.g., receivers of freight shipments can reduce their inventories and thus their inventory costs). Accessibility of service is the spatial convenience of transport service. The more accessible the service, the higher its quality, since the user would incur less time in accessing the service. Susceptibility to loss, damage, and injury is the likelihood of loss and damage to shipments and injury to passengers. The lower this likelihood, the higher the quality of the service, since costs associated with loss, damage, and injury would be less.

If no excess capacity (i.e., no underutilization) exists for resources employed by a transport carrier, improvements in quality of service and/or increases in the amounts of shipments and passengers to be transported will require that the carrier employ greater amounts of resources. The relationship between the minimum amount of a given resource to be employed by a carrier and the levels of its operating options and amounts of shipments and passengers to be transported has been called by Talley (1988) the "carrier's resource function."

2.2. Production function

Whereas operating options are generally under the control of the carrier, freight shipments and passengers to be transported are not. For a transport service to exist, both the user and the provider must have participated in its provision, the user providing himself or his freight to be transported and the provider providing the resources to move them from one place to another. If either party is unwilling to participate, a transport service will not occur. The most commonly used measure of freight (passenger) transportation service is the ton-mile (passenger-mile). A ton-mile is one ton of freight moved one mile, whereas a passenger-mile is one passenger moved one mile.

A carrier's production function describes the maximum amount of transport service that it can provide from the employment of given levels of resources. If the carrier adheres to its production function in the provision of service, it is said to be technically efficient. If the resources of the production function are increased by the same percentage and transport service increases by a greater percentage, the carrier exhibits increasing returns to scale. If the transport service increases by the same (lesser) percentage, the carrier exhibits constant (decreasing) returns to scale in the provision of service.

2.3. Cost function

A carrier is cost efficient if the cost it incurs in the provision of transport service is the minimum cost, given the prices of resources. If so, it is also must be technically efficient. If not, but becoming so, the same resources can now provide greater service at the same cost or the same service at less cost.

A long-run cost function for a carrier expresses the relationship between the minimum cost to be incurred and the level of transport service provided and resource prices. Long-run marginal cost is the addition to long-run cost attributable to providing an additional unit of transport service. The long run is a period of time that is sufficiently long so that the carrier can vary the amounts of all resources utilized. In the short run, one or more resources are fixed.

Short-run total costs are the sum of short-run variable and fixed costs. Short-run variable costs vary with the level of service provided. Fixed costs are the costs associated with resources that cannot be varied, and therefore remain fixed regardless of the level of service. The relationship between the minimum variable cost to be incurred in the short run by a carrier and the level of transport service provided, resource prices (except those of fixed resources), and the level(s) of the fixed resource(s) is its short-run variable cost function. Short-run marginal cost is the addition to short-run variable cost attributable to providing an additional unit of transport service. Short-run total costs are used by transport carriers for making pricing decisions, whereas long-run costs are used for making investment decisions.

A carrier exhibits economies of scale if its long-run unit costs (long-run costs divided by the amount of service provided) decline as service increases. The basic rationale for economies of scale is resource specialization (i.e., as the level (or scale) of service increases, the opportunity for resource specialization increases). For example, vehicles that are more efficient in performing a limited set of tasks can be substituted for less efficient all-purpose vehicles, and workers that are more efficient in performing a small number of related tasks can be substituted for lower skilled but more versatile workers. A carrier exhibits diseconomies of scale if its long-run unit costs rise as service increases. Diseconomies may arise from increasingly complex problems of co-ordination and control faced by the carrier management. If a carrier exhibits constant returns to scale, its long-run unit costs remain constant as service increases. Whether economies of scale exist in a transport industry is important for assessing competition among carriers of different sizes within the industry. For a discussion of returns to scale in the U.S. trucking industry, see Xu et al. (1994).

A complementary long-run concept to economies of scale for a multiservice carrier is economies of scope; that is, the cost advantage to a carrier in providing two or more services as against specializing in the provision of these services as single-service carriers. Specifically, economies of scope exist when a carrier can provide given amounts of two or more services more cheaply than if each amount of a given service were provided by a single-service carrier. Economies of scope for a multiservice carrier may arise when its services share resources. Whether economies of scope exist is important for evaluating whether the costs in the provision of transport services will be lower for multiservice or single-service carriers. For a discussion of economies of scope in the U.S. airline industry, see Keeler and Formby (1994).

A carrier in the short run exhibits economies of density if its short-run total unit costs decline as the utilization of a fixed resource, or resource density, increases. A study by Harris (1977) found that U.S. railroads exhibit economies of density, where density was measured as revenue ton-miles of rail service per mile of fixed railroad track.

2.4. Allocating shared costs

Three basic problems arise in allocating cost among transport movements that share the cost:

(1) identifying the shared cost,
(2) identifying the transport movements, and
(3) selecting the rule for allocating the shared cost among the transportation movements.

The selection of a rule is often argued to be an arbitrary decision, since cost shares cannot be unambiguously identified with individual transport movements. Alternatively, it is also argued that if the rule satisfies (previously) agreed upon rule-selection criteria, then its selection is not arbitrary even though there may be more than one rule that satisfies these criteria (i.e., the rule is neither an arbitrarily selected nor a unique rule).

Criteria (Talley, 1983) for selecting rules to allocate common cost among transport movements include:

(1) they provide core allocations,
(2) they detect cost inefficiency,
(3) they provide fair allocations, and
(4) they generate low computation costs.

Common cost allocations are core allocations if:

(1) the cost allocation to an individual movement is no greater than the corresponding cost to be incurred in providing resource capacity exclusively for that movement (*condition of individual rationality*);
(2) the sum of the cost allocations are equal to the common cost (*condition of group rationality*); and
(3) the sum of the cost allocations assigned to any subgroup of the transport movements is no greater than the cost to be incurred exclusively for this subgroup (*condition of coalition rationality*).

Rules that allocate common costs based upon the costs of exclusive resource capacities (found in the condition of individual rationality) have been referred to as "alternative capacity rules" (Talley, 1988). By doing so, their cost allocations can be investigated for core allocations. One such rule is the *Moriarity rule*:

$$f_k = \frac{C_k}{\sum_n C_k},$$ (1)

where f_k is the fraction of the common cost to be allocated to the kth sharer (or transport movement) of this cost, C_k is the exclusive capacity cost to be incurred by the kth sharer, and n is the total number of sharers of the common cost. If C_n is the common cost, then $f_k C_n$ is its cost allocation to the kth sharer.

If the cost shares of a given resource capacity (e.g., a given vehicle or terminal) assigned to transport movements that share this capacity are core cost allocations, there is no incentive for carrier management to change this resource capacity, since capacity cost cannot be lowered from such a change. Thus, management can detect whether resource capacities are cost efficient by investigating whether core cost allocations exist for these capacities. Alternatively, management can detect whether a resource capacity shared by transport movements is cost inefficient by investigating if at least one of the core allocation rationality conditions is violated in determining cost allocations. The fairness criterion for selecting a common cost allocation rule is concerned with whether the cost allocations place an unfair burden on any transport movement. Even if the allocation rule satisfies the efficiency and fairness criteria, the computation costs incurred in applying the rule may be prohibitive.

As opposed to common costs, core cost allocations for joint cost allocations cannot be determined, since the movements that incur these costs do not occur independently of one another. Criteria (Talley, 1986, 1988) for selecting a joint-cost allocation rule include fairness and low computation costs, as well as whether the rule adheres to a certain transport pricing criterion (e.g., marginal cost pricing).

3. Costing transport service in practice

Costing processes (or methods) for determining carrier costs in practice in the provision of particular transport services include individual- and multicarrier costing processes. The former utilize data of a single carrier, whereas the latter utilize data of several carriers.

3.1. Individual-carrier costing processes

Individual-carrier costing processes that are often used by transport carriers include cost-center, resource, and time-series statistical cost function costing processes.

Cost-center costing

A cost center is the center to which costs are accumulated. In costing transport carrier service, this center is a service measure. The cost-center costing process assigns system-wide carrier expenses to various service measures. System-wide costs assigned to a given service measure are then divided by the system-wide total for that measure. These unit cost coefficients are then used to determine (or estimate for future time periods) costs incurred (to be incurred) by the carrier in provision of particular transport services.

The cost-center costing process has been used by both freight and passenger carriers. One of the first cost-center costing processes to appear in the literature for transport carriers is that developed by Ferreri (1969). The Ferreri process determines the operating cost of a bus service route. The *Ferreri equation* for determining this cost is:

$$C_i = C_m M_i + C_h H_i + C_r R_i + C_v V_i, \qquad (2)$$

where C_i is the carrier operating cost for the ith bus route, C_m is the average system-related vehicle-mile cost (per vehicle-mile of service), M_i is the vehicle-miles of service provided on the ith bus route, C_h is the average system-related vehicle-hour cost (per vehicle-hour of service), H_i is the vehicle-hours of service provided on the ith bus route, C_r is average system-related revenue-passenger cost (per dollar of passenger revenue), R_i is passenger revenue on the ith bus route, C_v is average system-related peak-vehicle cost (per peak vehicle), and V_i is the number of peak vehicles utilized on the ith bus route.

By substituting values (for a given time period) for the average system-related costs and values for the four service measures for the given bus route, eq. (2) can be solved for an estimate of the operating cost for the bus route. If the number of vehicles on the route is fixed, the equation describes a short-run cost function.

A basic problem in applying the Ferreri cost-center costing process is in determining average system-related costs. The latter are determined by assigning each carrier operating expense to one of four service measures (vehicle-mile of service, vehicle-hour of service, dollar of passenger revenue, peak vehicle). However, if an expense is related to more than one service measure, a problem arises (which is not addressed by the process) in how to allocate this shared expense among the service measures. The Ferreri cost process has also been criticized for: having constant unit costs regardless of service levels; using passenger revenue and the number of peak vehicles as service measures; and holding constant the quality of service (by implicitly assuming homogenous service measures).

Resource costing

The resource costing process seeks to determine the minimum amounts of resources required to provide a given level of carrier service, and then multiplies these amounts by their corresponding resource prices (or unit costs). These products are then summed to obtain the cost of the service level. The process may utilize either an engineering or a statistical resource function.

An engineering resource function determines the minimum amount of a resource required by a carrier based on laws of physics. A simple example of such a function is found in a study by DeSalvo (1969) for determining the number of gallons of diesel fuel g to be consumed per hour by a rail diesel engine generating horsepower HP in providing rail freight line-haul service:

$$g = 0.073 \text{ HP}. \tag{3}$$

This relationship is based on: 139 000 BTU in a gallon of diesel fuel and 2545 BTU in one horsepower per hour.

There are several advantages in using engineering resource functions in resource costing processes: First, these functions are subject to little or no error in determining minimum resource requirements, since they are based on laws of physics. Second, these functions may not only include levels of service and carrier operating options (unless they remain constant), but also engineering characteristics of the resources (e.g. horsepower as in eq. (3)). A disadvantage of engineering resource functions is that their resources may not be market resources (i.e., resources for which market prices exist); if resources are not market resources, market prices (needed in determining resource costs) will not be available.

A statistical resource function is an estimate of an assumed stochastic resource function utilizing carrier historical data. A stochastic resource function has a random resource-dependent variable expressed as a function of explanatory variables and a random error term. The statistical technique of regression analysis is usually used in the estimation. An advantage of this function in resource costing processes is that the resource is a market resource (since historical data are available) and therefore can be priced by the market. A disadvantage is that the amount of the resource determined from the function (by substituting values for its explanatory variables and solving) is not necessarily the minimum resource requirement.

Time-series statistical cost-function costing

A time-series statistical cost function is an estimate of an assumed stochastic cost function utilizing time-series historical data. The latter are historical data for a given carrier over many time periods. A stochastic cost function has a random cost-dependent variable expressed as a function of explanatory variables and a random error term. When estimated with time-series data, the function is an estimate of a short-run variable cost function (see the previous discussion for the explanatory variables of this function). The estimation is usually performed using regression analysis. Estimates of carrier short-run variable costs are found by substituting in the estimated function values for the explanatory variables, and then solving.

Specific functional forms of statistical cost functions may be either non-flexible or flexible. Non-flexible cost functions place restrictions, implicit in their functional forms, on the underlying production function of the carrier. Examples include linear and log–linear (e.g., Cobb–Douglas) cost functions. Alternatively, flexible cost functions place less restrictions than do non-flexible cost functions on underlying production technologies, and thus are less restrictive for allowing the data via estimation to reveal carrier production properties. A flexible cost function that has been used extensively in carrier costing is the translog cost function, where cost is expressed as a second-order Taylor's series expansion of explanatory variables. An example of a time-series statistical translog cost function for a railroad is given in Braeutigam et al. (1984).

A major advantage of statistical cost functions is that they have a strong theoretical base (e.g., cost function theory suggests explanatory variables and interpretation of estimated coefficients). A disadvantage is that the estimation technique, regression analysis, involves assumptions that may be violated in estimations and, if so, may result in biased estimates of cost function coefficients. For a recent survey of transport statistical cost functions, see Oum and Waters (1996).

3.2. Multicarrier costing processes

Multicarrier costing processes for costing transport carrier services include cross-section statistical cost function, cross-section statistical resource function, and regulatory costing processes. Cross-section statistical cost functions are analogous to time-series statistical cost functions, except that they use cross-section data and are estimates of long-run cost functions. Cross-section data are historical data for two or more carriers for the same time period. (By comparison, panel data are historical data for the same carriers for many time periods.) An implicit assumption of cross-section statistical cost functions is that the carriers have identical long-run cost functions. Each carrier can obtain long-run cost estimates by substituting in the estimated cost function its values for the explanatory variables, and then solving.

The cross-section statistical resource function costing process is a resource function costing process for which multicarrier (or cross-section) data are used to obtain statistical resource functions. These functions, in turn, can be used by each carrier (by substituting in the function its values for explanatory variables, and then solving) to obtain estimates of its resource requirements; cost estimates are then determined by multiplying these resource estimates by market resource prices.

One of the first studies to present cross-section statistical resource functions for transport carriers was that by Borts (1952). He used historical data of 76 railroads for the year 1948 to estimate (using regression analysis) the parameters of rail freight line-haul labor and fuel resource functions. Borts obtained the following linear statistical resource functions for man-hours employed Rh and net tons of fuel consumed Rf in rail freight line-haul service:

$$\text{Rh} = -1\,860\,000 + 7.5Q2 + 29.3Q3 - 26.4Q4 - 3.5R4 + 238.9R5, \qquad (4a)$$

$$\text{Rf} = -24\,800 + 0.092Q2 + 2.02Q3 + 0.019Q4 + 25.7R4 + 28.5R5, \qquad (4b)$$

where Q2 is the number of carloads of freight transported by a given railroad, Q3 is the number of loaded freight-car-miles for a given railroad, Q4 is the number of empty freight-car-miles for a given railroad, R4 is thousands of pounds of total tractive capacity of freight locomotives for a given railroad, and R5 is the number of miles of mainline track for a given railroad.

Multicarrier costing processes have also been used by government agencies for obtaining cost estimates for the transport carriers that they regulate. These regulatory costing processes use data from the carriers that the agency regulates to determine unit costs for various expense categories. The unit costs are then used by the agency in their regulatory deliberations (e.g., in approval or denial of price increases by carriers).

One example of a regulatory costing process is Highway Form B (HFB), a truck-carrier regulatory costing procedure of the U.S. Interstate Commerce Commission (ICC). The HFB has three sequential phases (Talley, 1988): phase I assigns collected ICC truck-carrier expenses to four major expense categories (line-haul, pick up and delivery, billing and collection, terminal platform); phase II computes system-wide unit costs for these categories; and phase III adjusts the unit costs in phase II when shipment characteristics differ from shipment averages.

References

Borts, G.H. (1952) "Production relations in the railway industry", *Econometrica*, 20:71–79.

Braeutigam, R.R., A.F. Daughety and M.A. Turnquist (1984) "A firm specific analysis of economies of density in the U.S. railroad industry", *Journal of Industrial Economics*, 33:3–20.

DeSalvo, J.S. (1969) "A process function for rail line haul operations", *Journal of Transport Economics and Policy*, 3:3–27.

Ferreri, M.G. (1969) "Development of a transit cost allocation formula", *Highway Research Record*, 285:1–9.

Harris, R.G. (1977) "Economies of traffic density in the rail freight industry", *The Bell Journal of Economics*, 8:556–564.

Keeler, J.P. and J.P. Formby (1994) "Cost economies and consolidation in the U.S. airline industry", *International Journal of Transport Economics*, 21:21–45.

Oum, T.H. and W.G. Waters II (1996) "A survey of recent developments in transportation cost function research", *The Logistics and Transportation Review*, 32:423–463.

Talley, W.K. (1983) "Fully allocated costing in U.S. regulated transportation industries", *Transportation Research*, 17B:319–331.

Talley, W.K. (1986) "A rule for allocating joint truck-carrier costs", *Transportation Research*, 20B:49–57.

Talley, W.K. (1988) *Transport carrier costing*. New York: Gordon & Breach.

Xu, K., R. Windle, C. Grimm and T. Corsi (1994) "Re-evaluating returns to scale in transport", *Journal of Transport Economics and Policy*, 28:275–286.

Chapter 20

BENCHMARKING AND PERFORMANCE MEASUREMENT: THE ROLE IN QUALITY MANAGEMENT

DAVID GILLEN

University of California – Berkeley and Wilfrid Laurier University, Waterloo

1. Introduction

Benchmarking represents an evolving component of modern management practice as part of the total quality management (TQM) operation. Not unsurprisingly, the shift to TQM coincided with the development of management strategies driven by the objective to satisfy customers rather than simply minimize cost and insure supply.* This fundamental paradigm shift, which shaped other industries in the 1980s, is now affecting the role and management of transportation service providers and, more recently, those that supply infrastructure services, such as airports. Airport benchmarking, for example, can provide management with comprehensive data and a consistent analytical methodology so they can utilize the power of benchmarking for operational decisions and long-term strategic planning (Ashford and Moore, 1992). Throughout this chapter, airports are used as an example of the benchmarking process, but the concept is applicable across all public and private modes of transportation and to other industries as well (Talvitie, 1999).

TQM is a business philosophy as well as a set of guiding principles founded on the notion of customer satisfaction. It involves *designing* the organization to satisfy customers every day. It generally comprises two strands: careful design of the product or service and ensuring that the organizations system can consistently produce the design. Quality management is both a cause and a consequence of the

---

*Improving productivity and efficiency are two factors that drove business decisions up to the mid-1980s and led to organizational structures that were very much an engineering-driven set of functional relationships. The traditional value chain in an airport, for example, ran from airport marketing to airport operations and maintenance, terminal and cargo management, and concession management. As we shift to a consumer-driven business strategy, the value chain leads to an organization that is structured on the basis of customer value-adding activities, marketing, quality management for airside and landside, and commercial management.

Handbook of Logistics and Supply Chain Management, Edited by A.M. Brewer et al.
© 2001, Elsevier Science Ltd

structural and behavioral changes that have characterized a majority of sectors in the economy in the last 15 years.

This chapter examines the management practice of the integration of new quality management techniques with the adoption of a customer orientation and a shift away from functional management. An additional theme of this chapter is the development of performance measures and the comparison of these measures with those regarded as the "best in the business." This process is formally known as benchmarking, and is a central part of quality management.

The key proposition in benchmarking is that "what gets measured gets managed," and that without measurement there will be no sense of the incremental gains or costs of undertaking quality improvement (Rolstades et al., 1995a,b). Without quality improvement in the new economic environment, the success of a firm, even for traditional commodities, is questionable. In the absence of performance measures and benchmarking, management cannot defend itself against criticism, quantify shortcomings, articulate its accomplishments, or develop sustainable strategies in the modern world of broadening competitive markets for goods and services, including transportation. Indeed, measurement will identify the "enablers" or drivers, the processes and practices that make the best-in-class performance possible. There must therefore be a (dynamic) set of measures that track the performance of the integration of the firm's activities and provide the input into any benchmarking exercise, which identifies the reasons for the excellence. TQM, benchmarking, and performance measures are an important part of the larger organizational framework. They are integrated methods of achieving the strategic goals of the firm.

Benchmarking identifies centers of excellence within an organization, and superior business practices and processes used by other organizations and competitors. The aim is to increase revenue, reduce costs, and deliver quality to customers. A benchmark will affect every level of the organization, from senior management responsible for strategic planning, to operations management, to finance and marketing. To be effective, benchmarking must be integrated with management tools and performance measures that link management to the economic performance of the firm. To accomplish this, any benchmark must provide consistency and take account of the diversity in airport scale, structure, and operational activity, in order to ensure the comparisons are reasonable and accurate (Doganis and Graham, 1995).

2. Total quality management and competitive strategy

Quality management requires attention to the whole production and delivery system (Gillen and Noori, 1995). The modern consumer, passenger, or shipper wants and expects value for money – the right products, latest technology, high

Box 1
Customer paradigm shift – from seller to buyer power

Old	New
• Productivity and quality conflict	• Increasing productivity with improved quality
• Quality defined by conformity to some standard	• Quality is defined in terms of meeting user needs.
• Quality measured by degree of non-conformance	• Quality is measured by continuous user satisfaction
• Achieve quality with inspection	• Quality is determined by product design and achieved via effective controls
• Some defects allowed	• Defects are prevented through process controls
• Quality is a separate function	• Quality is part of every function
• Poor quality is blamed on a specific input (labor)	• Poor quality is a management responsibility
• Supplier relationships are short term and cost oriented	• Supplier relationships are long term and quality oriented
• Inspection equals quality	• Quality is continuous improvement
• Focus is short-run profits	• Focus is long-term survival (discounted net present value)

quality, and the "right" price. This does not mean that costs should be minimized, or quality maximized, but rather that the sum of firm and customer costs should be minimized. For example, airlines have traditionally viewed airport services as cost centres. A modern approach would see airlines and airports as strategic partners. These are the new demands, which illustrate the need for the introduction of TQM into the transportation sector, particularly the infrastructure services, and that the perceived trade-off between price and quality has shifted.

The fundamental differences between the old and new view of quality in the strategy of the firm are illustrated in Box 1. In this perspective an investment in quality can add value for the customer. This value creates a margin that a supplier can exploit in whole or in part. In addition, by investing in "quality," TQM can, and generally does, generate cost reductions for producers (reduced waste, increased productivity, improved efficiency) and for the customer (reduced operating and transactions costs). There is also the opportunity to increase revenue, since the demand for the product has increased because of the higher quality.*

Box 1 also reflects the essential components of quality management. The shift from a supply to a customer focus requires a clear vision of the objectives and a

---

*By adding quality, customers see value in the product, and this value is rent. The rents can be shared, in some distribution, between customers and producers. It is more cost effective, and therefore important, to retain a satisfied customer than it is to compete for new ones.

consistency of purpose for service. Adding customer value cannot be accomplished at any cost nor with a reliance on the old sources of cost economies, such as scale and utilization economies. The process must be redesigned to minimize system costs by examining each process and the alternatives available (e.g., make versus buy) as well as the integration of the processes, which may minimize the cost of a particular process but may not minimize overall costs.

A customer orientation is human capital intensive, and this requires an investment in human resources. Motivation is going to come from a commitment to the organization and contracts that provide incentives with a reward structure that is positive. This means eliminating targets or quotas as the motivation for production. Finally, quality management involves a continuum across the vertical chain of production and distribution. This means each stage of the production and distribution process will have to buy into the transformation to quality management (suppliers, labor, management, customers). In order to accomplish quality management we therefore need an information system that provides the initiative for change, a platform for continuous improvement, and the basis of communication.

3. Benchmarking and performance measures

3.1. Benchmarking

What is benchmarking? It is a process of continuous measurement, internally and externally, and comparison against established external best practice leaders to obtain knowledge to improve performance. What is the role of benchmarking? "It is an instrument for providing a reference point." A simple straightforward definition such as this makes clear the specific role of benchmarking as a means of creating attention and momentum for change. It is a process of internal performance measurement and external monitoring of industry leaders. The definition also identifies the linkages between benchmarking as an instrument and its relationship to or within other quality management methods. Benchmarking should be broadly based and not limited to technical areas (e.g., engineering, finance). It should also be considered to be part of continuous improvement to ensure that internal processes are designed and functioning to achieve "market" success; "best practice" refers to the best way of meeting customers' needs while minimizing system costs (Rolstadas, 1995b).

Any firm, whether in the transportation sector or not, could be a benchmark for a particular process or product for a transportation firm, regardless of how large it is. It is also possible that businesses outside the industry might provide the standard, or best in class. A famous example of corporate benchmarking occurred with Xerox. This firm had a very good distribution system but, because distribution was a major component of overhead costs, management sought savings and quality improvements. The benchmarking process led Xerox to look at the process of distribution in other industries besides the copier industry, and L.L. Bean, a mail-order firm, was identified as the best for warehousing and distribution. The case study of their operation identified the enablers, which included greater use of computer-directed activity, including bar-coding to assist automation, items being arranged by turnover rate, continuous service to incoming orders, and incentives to employees for higher productivity and order-filling accuracy. Xerox was able to take this knowledge and apply it in their context to improve their performance.

There is a clear difference between a performance measure and a benchmark. The former provides a continuing measure of productivity, cost efficiency, operating excellence, or level of quality and service delivery. A benchmark, on the other hand, is a point of reference or target. A Benchmark can refer to a core functional area (e.g., production), a support area (e.g., finance), a business process (e.g., product design), a subfunction (e.g., billing), or even a specific task (e.g., receipt recording). Once the benchmark has been identified, the performance measure evaluates progress in achieving it.

In order to benchmark it is necessary to fully understand the process or activity and how well it performs. This means that the firm (airport) must first undertake an intensive internal analysis using quantifiable measures such as throughput, product quality, customer satisfaction, equipment set-up times, and availability of inputs. From this point the analysis should move to a more detailed study of how an area operates. This raises questions of: what exactly is the nature of the work (tasks); what are the constraints the area is operating under; and what are the strengths and weaknesses? It is at this point performance measures become the vehicle for translating the information gained from benchmarking into improved or superior performance.

The purpose of the measures is to supply the firm and the area management with indicators that will assist them in providing a superior service to their customers, and to do so in a cost-efficient way while pursuing their mission objectives.* This move to performance evaluation, productivity assessment, and benchmarking by large numbers of firms across a wide variety of industries is recognition of the fact that the link between a firm's performance-measuring system and its actual performance is strong. What gets measured gets managed. However, measurement is not sufficient for success – the results must be utilized through benchmarking. To evaluate the extent to which an organization recognizes this principle, the manager should ask:

---

*The area could be defined as functional, operational, or product.

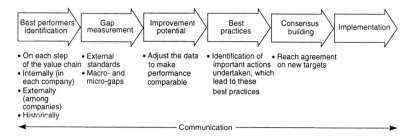

Figure 1. The steps in benchmarking.

(1) How can the ongoing efforts to meet a firm's goals be evaluated?
(2) How can people be encouraged to improve the firm's performance?

As illustrated below, benchmarking can be broken down into a number of steps. In Figure 1 we have illustrated six steps. The first requirement is to identify the best performers, and those processes set the standard. Once a point of reference has been chosen, it is necessary to measure the difference between the standard and the firm's performance. In the process of measurement, comparability must be maintained. Transitory effects must be considered, as must the differences in the context. This gap can be converted into an objective. However, in a given planning period, this objective need not necessarily be to close the gap in one step. Thus the benchmark objective need not be to move immediately and directly to the standard of the best performer, but a shift in performance by an amount that is challenging yet attainable. Trying to move to best of class in one move may frustrate the entire process.

Once the differences in performance between the firm and the standard have been measured and the benchmark set, the next step is to discover the enablers of the firm. The enablers, or drivers, are those activities that move the firm from its present state towards best in class. The decision of how much of the gap to close over a given time period will be determined very much by the enablers and whether the benchmarking firm decides it must make fundamental changes to its processes and organization. What remains is to implement the change. This is a non-trivial exercise, since change always threatens some people and firms. Throughout the benchmark process, communication among the participants is essential to success.

3.2. Performance measures

Because airports and other forms of transportation infrastructure have traditionally been perceived as public utilities, planning and engineering have influenced management practices more than modern business methods. In other words, a supply-side outlook was reflected in the view that bringing a homogeneous commodity (seat-miles, vehicle-miles, ton-miles, available runway capacity) to market was what was important. As a result, traditional management was operations oriented and would divide the firm, an airport for example, into subsystems (operations, administration and leases, concessions) and then into functional areas. This approach, however, ignores the customer orientation of new business management. Specifically, the perspective of "airports as a business" and the introduction of new management tools such as TQM and benchmarking has been garnered from a competitive strategy perspective in which the objective is to meet customer needs and add value to customer services.

As airport ownership and organization are evolving from purely public ownership and management, the importance of developing competitive strategies has taken on increasing importance (Doganis and Graham, 1987). The focal point of these strategies is the customer, which include passengers and cargo agents, the carriers, and other airport operations. The fundamental change is one from a functional management orientation at the airport to a customer-driven service in which the airport is not considered a cost center for an airline but a strategic partner in supplying inputs. It is, therefore, essential to place the exercise of performance measurement into the broader context of competitive strategy.

The selection of a process to measure and to compare with best practice must be related to a strategy that is directed at the core businesses of the firm. The vision or mission of the firm (airport) must drive the selection of what to measure and how to measure on a continuing basis (Doganis, 1983). This is a key contribution of TQM – describing a set of guiding principles which focuses on careful product or service design, and assurance that the design can be produced consistently for those products or services which the firm (airport) views as its primary business. However, this is an interactive process, meaning that dynamic vision will force new strategies, which will in turn drive the need to develop new performance indicators.

4. Linking performance measures, enablers, and strategy

To summarize what has been described thus far, benchmarking is a process of moving the firm in a mode of continuous improvement. Once the firm has identified the standard or best in class of a particular process or service, it establishes a benchmark, or target, for the firm to achieve in a given time period. As improvements take place and strategies change, the benchmark is continuously adjusted. In order to successfully complete the benchmarking the firm needs information both on how well they are doing and on how the improvements can be achieved. This requires the development of performance measures and the identification of the enablers. Enablers, which are practices, processes, or methods, are the means by which superior performance is achieved. Benchmarks provide the level of excellence, while enablers identify the reasons behind the success and how the process was implemented.

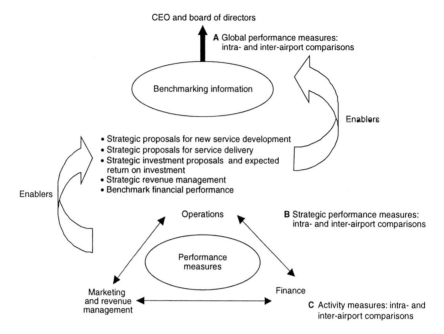

Figure 2. The integrated tiers of benchmarking and performance measurement.

The mechanism identified for linking strategic, managerial, operations, and customer views is illustrated in Figure 2. In setting out the core indicators, a hierarchical relationship is been established, running from micro-based activity measures and to global airport measures. The core indicators can be thought of as a subset of measures used at the global level (A), which summarize the core activities of the firm. For an airport this would include operational, financial, and costing indices for aircraft operations, numbers of passengers and cargo, and community accessibility.

Using the framework illustrated in Figure 2, the most basic micro-level performance indicators for each activity in each business area in the transportation firm are identified as activity measures (C). At this level the measures define activities in terms of their physical (productive) capacity that is supplied at a given time and level of utilization. In and of themselves these are not of strategic value, but they do identify the enablers, the drivers of performance.* For example, the maintenance schedule for baggage belts is an operational issue but will be influenced by the importance of hubbing at the airport.

---

*In this sense they are similar to activity-based costing measures.

At level B in Figure 2, the strategic level, performance measures are identified that managers of the customer-driven management areas could find valuable in running their area, in integrating with other operational areas, and in collectively providing information for senior management policy and direction. These measures, continuing with the airport example, encompass revenue generation, financing, and operational and cost indicators, which are of strategic importance both to the area and to the whole airport. For example, at the operational level of parking system downtime, available capacity and turnover rate of vehicles (in available spaces) can be identified as activity measures of capacity supplied and used. At the strategic level, the revenue per space and revenue per vehicle are income performance measures that have value, since they not only indicate how well the parking function area is generating revenue, but they may also indicate a need to institute changes in either the level or the structure of parking fees. These measures are also of value at the global level, discussed below, because parking revenue is one component of total airport revenue. An increase in parking revenue could mean less would be required of other areas, such as carrier terminal fees, which would reduce the entry costs of new carriers to the airport; if expanding the number of carriers was a desirable strategy of the airport managers.

At the global level (A in Figure 2), performance measures are generally compared across airports, but can certainly be compared over time for a given airport. At this level, it is the airport chief executive officer (CEO) and the board of directors who have collectively identified the mission of the airport and the set of indicators that will be benchmarked to serve as a measure of how well the airport is doing. The global level performance measures provide information on the success, or lack thereof, of the strategies pursued. At this level the measure can refer to a process (e.g., the number of passengers per gate) or an aggregate measure for the total airport (e.g., the rate of return on assets for the total airport). The strategic performance measures provide the foundation for changes to the global indicators, and the activity level measures supply the enablers for the strategic-level measures.

For example, it is commonplace to compare, across airports, the aggregate number of operations, operations per runway, or operations per meter of runway. This provides a measure of runway capital productivity, a coarse measure of service accessibility to the community and a measure of the success of the strategies pursued by the airport CEO. However, the manager of airside facilities (runways, taxiways, ramps, parking aprons) would also focus on the processes by which the number of hourly or daily flights could be allocated or increased;* available capital (square meters of airside capacity), usage rate of the capital,

---

*We can set aside for the moment that the airport CEO, in conjunction with the directors, may wish to pursue alternative strategies of maximizing the number of carriers and flights that use the airport, or may wish to focus on a key client, such as a hub carrier, and develop airport strategy in conjunction with the carrier's strategy.

availability of capacity, air traffic control operations, airline scheduling decisions, and pricing of airside services. These latter measures are strategic-level measures, and are under the purview of the airside management. This management must examine the activity measures to determine how they might affect the amount of airside capacity available. This may include changes in maintenance or snow-clearing procedures or priorities, or repair or maintenance programs.

For each level of a performance measure the categories are financial, cost efficiency (or expenses), marketing, revenue generation, and operations. The purpose in defining these categories is to emphasize the concept of the measurement, which can be applied across the different airport activities. For example, one might consider the activity of moving aircraft from the landing runway to the gate and assess the airport's performance in terms of the financial return, the revenue from landing fees (and hence the rate structure), the number of operations per use of airside facility (productivity), and simply the average time from landing to gate (Gillen and Cooper, 1995; Gillen and Stang, 1996). Alternatively, one can assess this activity for only one performance dimension but consider it as an operational activity, as part of airside management, and as a comparison across airports.

In constructing and reporting performance measures the following are important. First, each measure should be reported in three forms; the (arithmetic) mean, the median, and the standard deviation. The reasoning is that a simple mean, which is commonly used, provides no information on the variability of the data. The median is not subject to the influence of extreme values in the data. The standard deviation is a measure of data dispersion around the measure of central tendency, the mean. A high standard deviation is a signal that the data are widely dispersed and have a broad range. The information contained in any one piece of data will therefore be lower than when the standard deviation is lower.*

Second, data should be provided for a time series, since reporting for only one year can be misleading, as the data point may be an outlier and contain significant transitory influences. Furthermore, the use of a time series is essential to track progress over time.

Third, all data should be expressed in real terms. All financial data, costs, and revenues should be adjusted by a price index to remove the bias of inflation, particularly if intercountry comparisons are made. Which index to use is open to question, since a capital price index may be appropriate in one circumstance while a wholesale price index may be more appropriate in another. Nonetheless, comparisons in constant or real dollar terms are essential.

Finally, the context in which the data are collected is important to understand their level and structure. Significant economic and institutional events need to be

---

*In a normal distribution 68% of the data are plus or minus one standard deviation from the mean, and 98% of the data values are plus or minus two standard deviations from the mean.

recorded, since these can result in swings in the data which have no relationship to what strategy has been followed or how productive some inputs were.

The process of developing and implementing performance measurement and benchmarking is an evolutionary process. It is an exercise in identifying the information needs and developing information systems that will generate the data in a friendly, usable, and broadly accessible format. Initially, the exercise is important in evaluating the transition to a new internal organization, while later it can be used in evaluating and implementing strategies. The performance measures also form the basis of a report card to the community to show what has been done with the transportation resources.* This should be provided on an ongoing basis to track the evolution of strategies and operations.

A useful and effective performance measure should be judged on the following criteria:

(1) Does the measure track performance and offer insight into the sources of change?
(2) Do the measures provide ex ante guidance rather than simple description?
(3) What is the value added of the measure?
(4) Can the measure be realistically quantified?
(5) Is the measure complementary to other indices? Are they determined jointly?
(6) Can the measure be calculated with readily accessible data on a continual basis?
(7) Are task or subfunction indices easily aggregated?

5. Implementing benchmarking and performance measurement

Benchmarking can be applied at any level within the firm, whether to projects, profit centers, or processes. Clearly, the data requirements will differ, as will the commitment of resources. An internal benchmarking exercise is a valuable first start in learning the process. Benchmarking can also be external to the firm, where activities in other industries or firms within the same industry may provide the best-in-class model.

The benchmarking excise should include the following steps. First, develop a model for placing benchmarking in the wider context. Any activity that is to be benchmarked is part of an integrated set of processes and is affected by the vertical

---

*This is similar to the idea of the balanced scorecard (Kaplan and Norton, 1992), which organizes financial, and non-financial performance measures around four perspectives: customer, internal, financial, and innovation. It is motivated by the objective of multiple-perspective approach, which overcomes incompleteness when measures are one-dimensional.

linkages between processes. This context cannot be ignored in the benchmarking exercise or in using the results. Second, identify participants in the benchmarking activity. Training is required for the benchmarking team. Knowledge of the process of benchmarking, overview training, and the process is needed to provide guidance. Since benchmarking is not an isolated exercise but is part of a management philosophy, the benchmarking team must extend beyond those who have responsibility for what is being benchmarked. It should include members from senior management. Third, identify key issues related to implementation of organizational changes. To use and benefit from benchmarking, a cultural change is required. Benchmarking is, in essence, using a form of market to check on the activities of a firm that are not normally judged by an external market. In addition, having world-class performance in one or a few processes, but not all, will not improve the position of the firm. World class in all areas is needed.

6. Summary

Performance measures are the information base for strategies to meet customer needs, improve productivity and enhance competitiveness. Quality management, in turn, is founded on performance measurement systems linked to bench-marking. Performance indicators (measures) are the basic building blocks in the evolution and provide the fundamental inputs to the analytical process by which the enablers are identified.

Management needs to establish yardsticks to track improvements and to see if stakeholder requirements are being met. Managers are dealing with a continuous improvement process and they need to ensure consistency with strategies. While the principles of quality management and continuous improvement should be encapsulated in each performance measure, the system of measures must provide practicable management tools. As a result they should possess several important properties. First, they should measure and not evaluate. Performance measures are the input to the evaluation. Second, they should not place excessive emphasis on precision, since the aim is to discover how and if improvement is occurring. Thirdly, any activity requires multiple indicators, since no one measure tells the whole story. Furthermore, the range of measures should include an assessment of all inputs and not focus exclusively on labor. Fourth, the measures should blend quantitative and qualitative elements. Not all activities are easily reduced to numbers, and scales and indexes may need to be created. Finally, incentives should not be tied to standards, but rather to the improvement relative to the established benchmark.

Standards act as targets, and the theme of quality management is continuous improvement. The metrics need to cover the following dimensions of business activity:

(1) *achievement metrics* – direct indicators including financial operating, costing measures;
(2) *diagnostic metrics* – indirect measures that assess critical success factors, such as customer satisfaction, product quality and reliability, cost efficiency, and flexibility; and
(3) *competence metrics* – measures that describe how well prepared the airport is for the future, such as investment in product development, diversification, and training.

Benchmarking is a powerful yet quite simple management tool, since it is transparent and is accessible to front-line workers as well as CEOs. It motivates and provides direction for action. However, it must be integrated into the management of the firm. To be effective, benchmarking requires a combination of detailed, and sometimes difficult, analysis in conjunction with determined leadership that is capable of setting targets and implementing effective programs.

Benchmarking has been undertaken in a large number of industries where firms have made a change to quality management principles and practices. There are a number of key lessons that emerge from their experience. First and foremost there must be commitment from the senior levels in the organization. If these upper management levels do not sponsor the benchmarking exercise there will be a lack of credibility, commitment, and support. Second, the benchmarking team must reflect the purpose of the exercise. This means it must reflect excellence from the chairperson on down. The chairperson must be a champion and provide the leadership throughout the process. Thirdly, planning is essential. This means a carefully laid out process, which is documented at every step.

The benefits of benchmarking to an airport or highway authority or port can be far reaching. The airport industry, for example, is a reflection of the airline industry; it is dynamic and rapidly changing. This change within the industry and the role aviation is playing in the new economy, with its emphasis on e-commerce and supply chains, has created the need for airports to analyze their operating performance and evaluate their competitive strengths. Benchmarking can serve as a critical management tool in this pursuit. As more pressure is brought to bear on the airport to meet customer needs, benchmarking can be used to identify centers of excellence within an organization and superior business practices used by competitors and other organizations. This information can then be used to increase performance and facilitate change. Benchmarking will prove valuable at each level of the organization, from senior managers, through operations, marketing and finance, and product development.

References

Ashford, N. and C. Moore (1992) *Airport finance*. New York: Van Nostrand Reinhold.
Doganis, R. (1983) "Airport economics and management", in: *World Conference on Transport*

Research: Research for transport policies in a changing world. Hamburg: SNV Studiengesellschaft Nahverkehr.

Doganis, R. and A. Graham (1987) "Airport management: The role of performance indicators", Transport Studies Group, Cranfield University, Research Report No. 13.

Doganis, R. and A. Graham (1995) "The economic performance of European airports", Department of Air Transport, Cranfield University.

Gillen, D. and D. Cooper (1995) "Measuring airport efficiency and effectiveness in the California aviation system", Institute for Transportation Studies, University of California, Berkeley, Research Report UCB-ITS-RR-95-67.

Gillen, D. and D. Stang (1996) "Performance measurement: The keystone in benchmarking and total quality management. A performance measurement system for Pearson International Airport", Airports Division, Transport Canada, Ottawa, Report.

Gillen, D. and H. Noori (1995) "Performance measuring matrix for capturing the impact of AMT", *International Journal of Productivity Research*, 33(7):2037–2048.

Kaplan, R.S. and D.P. Norton (1992) "The balanced scorecard – measures that drive performance", *Harvard Business Review*, January/February, 71–79.

Rolstadas, A. (1995a) *Performance management: A business process benchmarking approach*. London: Chapman Hall.

Rolstadas, A., ed. (1995b) *Benchmarking: theory and practice*. London: Chapman Hall.

Talvitie, A. (1999) "Performance indicators for the road sector", *Transportation* 26:5–30.

Chapter 21

GREEN LOGISTICS

JEAN-PAUL RODRIGUE
Hofstra University, Hempstead

BRIAN SLACK
Concordia University, Montreal

CLAUDE COMTOIS
Université de Montréal

1. Introduction

The two words that make up the title of this chapter are each charged with meaning, but combined they form a phrase that is particularly evocative. "Logistics" is at the heart of modern transport systems. As has been demonstrated, the term implies a degree of organization and control over freight movements that only modern technology could have brought into being. It has become one of he most important developments in the transportation industry. "Greenness" has become a code word for a range of environmental concerns, and is usually considered positively. It is employed to suggest compatibility with the environment, and thus, like "logistics," is something that is beneficial. When put together the two words suggest an environmentally friendly and efficient transport and distribution system. The term "green logistics" has wide appeal, and is seen by many as eminently desirable. However, as we explore the concept and its applications in greater detail, a great many paradoxes and inconsistencies arise, which suggest that its application may be more difficult than might have been expected on first encounter.

In this chapter we begin by considering how the term has been developed and applied in the transportation industry. Although there has been much debate about green logistics over the last 10 years or so, the transportation industry has developed very narrow and specific interests. When the broader interpretations are attempted, it will be shown that there are basic inconsistencies between the goals and objectives of "logistics" and "greenness." We conclude this chapter by exploring how these paradoxes might be resolved.

Handbook of Logistics and Supply Chain Management, Edited by A.M. Brewer et al.

2. Development and application of green logistics

In common with many other areas of human endeavor, "greenness" became a catchword in the transportation industry in the late 1980s and early 1990s. It grew out of the growing awareness of environmental problems, and in particular with well-publicized issues such as acid rain and global warming. The World Commission on Environment and Development (1987), with its establishment of environmental sustainability as a goal for international action, gave green issues a significant boost in political and economic arenas. The transportation industry is a major contributor to environmental degradation (Banister and Button, 1993; Whitelegg, 1993). The developing field of logistics was seen by many as an opportunity for the transportation industry to present a more environmentally friendly face. During the early 1990s, there was an outpouring of studies, reports, and opinion pieces suggesting how the environment could be incorporated in the logistics industry (Muller, 1991; Tanja, 1991; Murphy et al., 1994). It was reported that the 1990s would be "the decade of the environment" (Kirkpatrick, 1990).

As we look back on that decade we can observe that interest in the environment by the logistics industry manifested itself most clearly in terms of exploiting new market opportunities. While traditional logistics seeks to organize forward distribution, that is the transport, warehousing, packaging, and inventory management from the producer to the consumer, environmental considerations opened up markets for recycling and disposal, and led to the entire new subsector of reverse logistics. This reverse distribution involves the transport of waste and the movement of used materials. While the term "reverse logistics" is widely used, other names have been applied, such as "reverse distribution," "reverse-flow logistics," and "green logistics" (Byrne and Deeb, 1993).

Inserting logistics into recycling and the disposal of waste materials of all kinds, including toxic and hazardous goods, has become a major new market. There are several variants. An important segment is customer driven, where domestic waste is set aside by home-dwellers for recycling. This has achieved wide popularity in many communities. A second type is where non-recyclable waste, including hazardous materials, is transported for disposal to designated sites. As landfills close to urban areas become scarce, waste has to be transported over greater distances to disposal centers. A different approach is where reverse distribution is a continuous embedded process, in which the organization (manufacturer or distributor) takes responsibility for the delivery of new products as well as their take-back. For example, BMW is designing a vehicle the parts of which will be entirely recyclable (Giuntini and Andel, 1995).

The way in which the logistics industry has responded to the environmental imperatives is not unexpected, given its commercial and economic imperatives. However, the fact that it has virtually overlooked significant issues such as

Table 1
The number of articles with an environmental orientation published between 1997 and 1998

Journal	% of articles
International Journal of Physical Distribution and Logistics Management	1.7
Logistics Spectrum	1.2
Logistics Focus	4.8

pollution, congestion, and resource depletion means that the logistics industry is still not very "green." This conclusion is borne out by published surveys. Murphy et al. (1994) asked members of the Council for Logistics Management what were the most important environmental issues relating to logistics operations. The two leading issues selected were hazardous-waste disposal and solid-waste disposal. Two-thirds of respondents identified these as being of "great" or "maximum" importance. The least important issues identified were congestion and land use, two elements usually considered of central importance by environmentalists. When asked to identify the future impact of environmental issues on logistical functions, again waste disposal and packaging were chosen as leading factors. Customer service, inventory control, production scheduling, all of which are key logistical elements, were seen to have negligible environmental implications.

By the end of the 1990s, much of the hyperbole and interest in the environment by the logistics industry had been spent. A count of the number of articles with an environmental orientation in three journals between 1997 and 1998 revealed that they represented an insignificant proportion of all articles (Table 1). Most of the articles that were identified as having an environmental content dealt with hazardous-waste transport issues.

This suggests that at the beginning of the 21st century the logistics industry in general is still a long way from being considered green. Reverse logistics has been its major environmental preoccupation. While this is an important step, recycling being one of the important elements in sustainability, many other environmentally significant considerations remain largely unaddressed. Are the achievements of transport logistics compatible with the environment?

3. The green paradoxes of logistics in transport systems

If the basic characteristics of logistical systems are analyzed, several inconsistencies with regard to environmental compatibility become evident. Five basic paradoxes are discussed below, and are summarized in Table 2.

3.1. Costs

The purpose of logistics is to reduce costs, notably transport costs. In addition, economies of time and improvements in service reliability, including flexibility, are further objectives. Corporations involved in the physical distribution of freight are highly supportive of strategies that enable them to cut transport costs in the present competitive environment. However, the cost-saving strategies pursued by logistics operators are often at variance with environmental considerations. Environmental costs are often externalized. This means that the benefits of logistics are realized by the users (and eventually by the consumer if the benefits are shared along the supply chain), but the environment assumes a wide variety of burdens and costs. Society in general, and many individuals in particular, are becoming less willing to accept these costs, and pressure is increasingly being put on governments and corporations to include greater environmental considerations in their activities.

Although there is a clear trend for governments, at least in their policy guidelines, to make the users pay the full costs of using the infrastructures, logistical activities have largely escaped these initiatives. The focus of much environmental policy is on private cars (emission controls, gas mixtures, pricing).

Table 2
Paradoxes of green logistics

Dimension	Outcome	Paradox
Costs	Reduction of costs through improvements in packaging and reduction of waste. Benefits are derived by the distributors	Environmental costs are often externalized
Time/flexibility	JIT and door-to-door delivery provide flexible and efficient physical distribution systems	Extended production, distribution and retailing structures consuming more space and more energy and producing more emissions (CO_2, particulates, NO_x, etc.)
Network	Increasing system-wide efficiency of the distribution system through network changes (hub-and-spoke structure)	Concentration of environmental impacts next to major hubs and along corridors
Reliability	Reliable and on-time distribution of freight and passengers	The modes used, trucking and air transportation, are the least environmentally efficient
Warehousing	Reducing the needs for warehousing facilities	Inventory shifted in part to roads (or in containers), contributing to congestion and space consumption

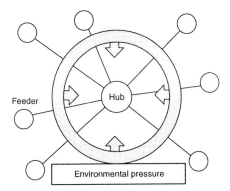

Figure 1. The hub-and-spoke network and the environment.

While there are increasingly strict regulations being applied to air transport (noise, emissions), the degree of control over trucking, rail, and maritime modes is less. For example, diesel fuel is significantly cheaper than gasoline in many jurisdictions, despite the negative environmental implications of the diesel engine. Yet trucks contribute, on average, seven times more per vehicle-kilometer to nitrogen oxides emissions than cars, and 17 times more particulate matter.

The external costs of transport has been the subject of extensive research. Recent estimates in Europe suggest that annual costs amount to between ECU 32 and 56 billion (1996). Cooper et al. (1998) estimated the costs in Britain as ECU 7 billion, or twice the amount collected by vehicle taxation.

The hub-and-spoke structure (Figure 1) has characterized the reorganization of transportation networks for the past 20 years, notably for air and rail and maritime freight transportation. It has reduced costs and improved efficiently through the consolidation of freight and passengers at hubs. Despite the cost savings in many cases, the flows, modes, and terminals that are used by pursuing logistical integration are the least sustainable and environmentally friendly. The hub-and-spoke structure concentrates traffic at a relatively small number of terminals. This concentration exacerbates local environmental problems, such as noise, air pollution, and traffic congestion.

In addition, the hub structures of logistical systems result in a land take that is exceptional. Airports, seaports, and rail terminals are among the largest consumers of land in urban areas. For many airports and seaports the costs of development are so high that they require subsidies from local, regional, and national governments. The dredging of channels in ports, the provision of sites, and operating expenses are rarely completely reflected in user costs. In the U.S.A., for example, local dredging costs were nominally to come out of a harbor improvement tax, but this has been ruled unconstitutional and channel maintenance remains under the authority of the U.S. Corps of Army Engineers. In Europe, national and regional

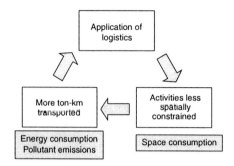

Figure 2. The vicious circle of environmental logistics.

government subsidies are used to assist infrastructure and superstructure provision. The trend in logistics towards hub formation is clearly not green.

The actors involved in logistical operations have a strong bias to perceive green logistics as a means of internalizing cost savings, while avoiding the issue of external costs. As pointed out earlier, a survey among the managers of logistical activities pointed out that the top environmental priority is reducing packaging and waste (Murphy et al., 1994). Managers were also strongly against any type of governmental regulation pertaining to the environmental impacts of logistics. These observations support the paradoxical relationship between logistics and the environment that reducing costs does not necessarily reduce environmental impacts.

3.2. Time and speed

In logistics, time is often the essence. By reducing the time of flows, the speed of the distribution system, and consequently its efficiency, is increased. This is achieved in the main by using the most polluting and least energy efficient transportation modes. The significant increase in air freight and trucking is partially the result of time constraints imposed by logistical activities. The time constraints are themselves the result of an increasing flexibility of industrial production systems and of the retailing sector. Logistics offers door-to-door services, mostly coupled with just-in-time (JIT) strategies (see Chapter 13). Other modes cannot satisfy as effectively the requirements that such a situation creates. This leads to a vicious circle (Figure 2). The more physical distribution through logistics is efficient, the less production, distribution, and retailing activities are constrained by distance. In turn, this structure involves a higher usage of logistics and more ton-kilometers of freight transported. There is overwhelming evidence for an increase in truck traffic and a growth in the average length of haul (Cooper et al., 1998) and, although McKinnon (1998) has suggested that JIT is not *greatly* increasing road freight volumes (italics added), it cannot be considered a green solution. The more door-to-door and JIT strategies are applied, the greater the negative environmental consequences of the traffic they create.

3.3. Reliability

At the heart of logistics is the overriding importance of service reliability. Its success is based on the ability to deliver freight on time with the least threat of breakage or damage. Logistics providers often realize these objectives by utilizing the modes that are perceived to be the most reliable. The least-polluting modes are generally regarded as being the least reliable in terms of on-time delivery, lack of breakage, and safety. Ships and railways have inherited a reputation for poor customer satisfaction, and the logistics industry is built around air and truck shipments, which are the two least environmentally friendly modes.

3.4. Warehousing

Logistics is an important factor promoting globalization and international flows of commerce. Modern logistics systems economies are based on the reduction of inventories, as the speed and reliability of deliveries removes the need to store and stockpile. Consequently, a reduction in warehousing demands is one of the advantages of logistics. This means, however, that inventories have been transferred to a certain degree to the transport system, especially the roads. This has been confirmed empirically. In a survey of 87 large British firms cited by McKinnon (1998), there had been a 39% reduction in the number of warehouses and one-third of the firms indicated an increased amount of truck traffic, although the increase was thought to be small in most cases. Inventories are actually in transit, contributing still further to congestion and pollution. The environment and society, not the logistical operators, are assuming the external costs.

Not all sectors exhibit this trend, however. In some industrial sectors, such as computers, there is a growing trend for vertical disintegration of the manufacturing process, in which extra links are added to the logistical chain. Intermediate plants, where some assembly is undertaken, have been added between the manufacturer and consumer. While facilitating the customizing of the product for the consumer, it adds an additional external movement of products in the production line.

3.5. e-Commerce

The explosion of the information highway has led to new dimensions in retailing. One of the most dynamic markets is e-commerce. In 1998, business-to-businesses transactions undertaken through e-commerce accounted for $43 billion, while business-to-consumer transactions accounted for $8 billion. In 1999, e-commerce boomed, reaching $150 billion, 80% of which was between businesses. These numbers are expected to reach $1.3 trillion and $108 billion, respectively, by 2003. In 1999, the computer manufacturer and distributor Dell sold $15 million worth of computers a day strictly from orders placed on its website. This is made possible by an integrated supply chain with data interchange between suppliers, assembly lines, and freight forwarders. Even if for the online customers there is an appearance of a movement-free transaction, the distribution that online transactions create may consume more energy than other retail activities. The distribution activities that have benefited the most from e-commerce are parcel-shipping companies such as UPS and Federal Express, which rely solely on trucking and air transportation. Information technologies related to e-commerce applied to logistics can obviously have positive impacts. The National Transportation Exchange (NTE) is an example where freight distribution resources can be pooled and where users can bid through a website to use capacities that would have otherwise been empty return travel. So, once again, the situation may be seen as paradoxical.

The consequences of e-commerce on green logistics are little understood, but some trends can be identified. As e-commerce becomes more accepted and used, it is changing physical distribution systems. The standard retailing supply chain coupled with the process of economies of scale (larger stores, shopping malls) is being challenged by a new structure. The new system relies on large warehouses located outside metropolitan areas, from where large numbers of small parcels are shipped by vans and trucks to separate online buyers. This disaggregates retailing distribution, and reverses the trend towards consolidation that had earlier characterized retailing. In the traditional system, the shopper bore the costs of moving the goods from the store to the home, but with e-commerce this segment of the supply chain has to be integrated in the freight distribution process. The result potentially involves more packaging and more ton-kilometers of freight transported, especially in urban areas. Traditional distribution systems are thus ill fitted to answer the logistical needs of e-commerce.

How green are logistics when the consequences of its application, even if efficient and cost effective, have led to solutions that may not be environmentally appropriate?

4. Discussion and evaluation

Our overview suggests that green logistics is still a long way from being achieved. The environment is not a major preoccupation or priority in the industry itself. The exception is where reverse distribution has opened up new market possibilities based on growing societal concerns over waste disposal and recycling.

Here the environmental benefits are derived rather than direct. The transportation industry itself does not present a greener face, indeed in a literal sense reverse logistics adds further to the traffic load. The manufacturers and domestic waste producers are the ones achieving the environmental credit.

It is not a question of whether or not the logistics industry will have to present a greener face. Pressures are mounting from a number of directions that are moving all actors and sectors in the economy in the direction of increasing regard for the environment. In some sectors this is already manifest, while in others, such as the logistics industry, it is latent. The issue is when and in what form it will be realized. Three scenarios are presented and discussed:

(1) a top-down approach, where "greenness" is imposed on the logistics industry by government policies;
(2) a bottom-up approach, where environmental improvements come from the industry itself; and
(3) a compromise between government and the industry, notably through certification.

While not mutually exclusive, each of the above scenarios presents different approaches and implications.

The first scenario is that government action will force a green agenda on the industry, in a top-down approach. Although this appeared as the least desirable outcome from the survey of logistics managers (Murphy et al., 1994), it is already evident that government intervention and legislation are reaching ever more directly over environmental issues. In Europe there is a growing interest in charging for external costs, as the European Union moves towards a "fair and efficient" pricing policy. Cooper et al. (1998) estimate that this could bring about a rise of 20–25% in transport costs. While there is some evidence that price elasticities are low in the logistics industry, around –0.1 (Bleijenberg 1998), the extent of the impact is more likely to be determined by how quickly the tax is applied. A sharp increase in costs could have a more serious impact than a more gradual, phased-in tax. In North America there is a growing interest in road pricing, with the reappearance of tolls on new highways and bridges built by the private sector, and by congestion pricing (Fielding, 1995). As yet there have been no studies of the effects on the logistics industry, but higher road costs are a clear outcome of policy intervention.

Pricing is only one aspect of government intervention. Legislation controlling the movement of hazardous goods, reducing packaging waste, stipulating the recycled content of products, and the mandatory collection and recycling of products are already evident in most jurisdictions. Indeed, it is such legislation that has given rise to the reverse logistics industry. Truck safety, driver education, and limits on drivers' time at the wheel are among many types of government action with a potential to impact the logistics industry.

A difficulty with government intervention is that the outcomes are often unpredictable, and in an industry as complex as logistics many could be unexpected and unwanted. Environmentally inspired policies may impact on freight and passenger traffic differentially, just as different modes may experience widely variable results of a common regulation. Issues concerning the greenness of logistics extend beyond transport regulations. The siting of terminals and warehouses is crucial to moving the industry towards the goal of sustainability, yet these are often under the land-use and zoning control of lower levels of government, whose environmental interests may be at variance with national and international bodies.

If a top-down approach appears inevitable, in some respects at least, a bottom-up solution would be the industry preference. Its leaders oppose leaving the future direction to be shaped by government action. There are several ways a bottom-up approach might come about. As demonstrated by the example of reverse logistics, these occur when the business interests of the industry match the imperatives of the environment. One such match is the concern of the logistics industry with empty moves. McKinnon (1998) reports that improvements in fleet management and freight distribution in Britain between 1983 and 1993 reduced the proportion of empty moves by 11%, which, other things being equal, cut CO_2 emissions by 720 000 ton/year. With the growing sophistication of fleet management and IT control over scheduling and routing, further gains are achievable.

Less predictable, but with a much greater potential impact on the greenness of the industry, are possible attitudinal changes within and without logistics. These changes are comparable to that which has already occurred in recycling. There has emerged striking public support for domestic recycling. Although this has been mandated to some degree in some jurisdictions, the mantra of the three Rs (reduce consumption, reuse, recycle) has achieved unparalleled popularity. This has been extended by some firms in successfully marketing their compliance and adoption of green strategies. Firms have found that by advertising their friendliness towards the environment and their compliance with environmental standards, they can obtain an edge in the marketplace over their competitors. A comparable situation has been investigated in the context of the logistics industry by Enarsson (1998). He argued that purchasing departments become a critical point in the move towards applying green logistics. Traditionally, price and quality characteristics formed the basis of choice, but because preservation of the environment is seen as desirable in general, greenness can become a competitive advantage. Ultimately, pressure from within the industry can lead to greater environmental awareness and respect. Companies that stand apart will lose out because purchasers will demand environmental compliance.

Somewhere between the bottom-up and top-down approaches are the moves that are being implemented with environmental management systems. Although governments are involved to varying degrees, a number of voluntary systems are in place, notably ISO 14000 and EMAS (Environmental Management and Audit System). In these systems firms receive certification on the basis of establishing an environmental quality control tailored to that firm, and the setting up of environmental monitoring and accounting procedures. Obtaining certification is seen as evidence of the firm's commitment to the environment, and is frequently used as a public relations, marketing, and government relations advantage. Decisions to proceed with a request for certification have to come from the highest decision-making levels of corporations, and involve a top to bottom assessment of operations. This represents a fundamental commitment of the corporation to engage in environmental assessment and audit that represent a significant modification of traditional practices, in which efficiency, quality, and cost evaluations prevailed. So far, there has been no research into the compliance of logistics firms with ISO 14000, although several large corporations with in-house logistics operations, such as Volvo, have been studied (Enarsson, 1998).

5. Blueprint for green logistics

A healthy environment is critical for efficient transport and, through its capacity to open markets and promote economic growth, transport is essential for effective and lasting environmental management. However, the growing internationalization of trade has broadened the concept of logistics to global logistics. Globalization and global logistics have, in many instances, harmed the environment by encouraging governments and firms to compete on the international market by lowering environmental standards in certain countries, while maintaining higher standards in rich countries. As a result, there has been increasing support for environmental initiatives undertaken at the international level and a growing reliance on local communities to address environmental problems, as the underlying environmental issues differ between and within countries. Therefore, the successful implementation of green logistics must come from the complex interplay of both global and local environmental governance. Indeed, the most important policy recommendations, implementations, and operationalization of green logistics that would work occur at the local level. Obviously, international trade is not more harmful to the environment than regional or local trade, but proper assessment of green logistics must be integrated. Most scenarios on the future of world trade and freight transport rest on the increase in energy consumption due to multimodal infrastructures. Therefore, there is a need to promote a regional approach to green logistics. The idea is not for smaller and more frequent shipments, which would result in more trips by smaller vehicles, but rather to reduce the number of trips. One objective might be to minimize movements through land-use policies that reduce the level and geographical separation of industrial activities. While the extent to which regions contribute to logistics is unclear, their role is often crucial to enable decisive and effective action to protect the environment. Since the conflict between the economic significance of logistics and the impact on the environment is first and foremost a political topic, green logistics will most effectively be implemented in settings with strong institutional factors responsible for enforcing and monitoring environmental sustainability. Therefore, further government intervention promoting greater environmental regulation appears inevitable.

References

Banister, D. and K. Button, eds. (1993) *Transport, the environment, and sustainable development.* London: E. & F.N. Spon.

Bleijenberg, A. (1998) *Freight transport in Europe: In search of sustainability.* Delft: Centre for Energy Conservation and Environmental Technology.

Byrne, P. and A. Deeb (1993) "Logistics must meet the 'green' challenge", *Transportation and Distribution*, February, 33–35.

Cooper, J., I. Black and M. Peters (1998) "Creating the sustainable supply chain: Modelling the key relationships", in: D. Banister, ed., *Transport policy and the environment.* London: E. & F.N. Spon.

Enarsson, L. (1998) "Evaluation of suppliers: How to consider the environment", *International Journal of Physical Distribution and Logistics Management*, 28:5–18.

Fielding, G. (1995) "Congestion pricing and the future of transit", *Journal of Transport Geography*, 3:239–246.

Giuntini, R. and T.J. Andel (1995) "Advance with reverse logistics", *Transportation and Distribution*, February, 73–76.

Kirkpatrick, D. (1990) "Environmentalism: The new crusade", *Fortune,* 12 February, 44–51.

McKinnon, A. (1998) "Logistical restructuring, freight traffic growth and the environment", in: D. Banister, ed., *Transport policy and the environment.* London: E. & F.N. Spon.

Muller, E.W. (1990) "The greening of logistics", *Distribution*, January, 27–34.

Murphy, P., R.F. Poist and C.D. Braunschweig (1994) "Management of environmental issues in logistics: Current status and future potential", *Transportation Journal*, 34:48–56.

Tanja, P.T. (1991) "A decrease in energy use by logistics: A realistic opportunity?", in: Freight transport and the environment, European Conference of Ministers of Transport, Brussels, pp. 151–165.

Whitelegg, J. (1993) *Transport for a sustainable future: The case for Europe.* London: Bellhaven.

World Commission on Environment and Development (1987) *Our common future.* Oxford: Oxfrod University Press.

Part 6

ORGANIZATIONAL LOGISTICS

Chapter 22

ORGANIZATIONAL LOGISTICS: DEFINITION, COMPONENTS, AND APPROACHES

ANN M. BREWER
University of Sydney

1. Introduction

Organizational logistics represents the intersection of the two broad disciplinary traditions of organizational theory and strategic supply chain management. The concept of organizational logistics comes in the wake of the last 20–30 years of a cognitive move away from traditional bureaucracy, an enduring characteristic of work organizations to more innovative structural forms. This period of organizational history has been obsessed with corporate culture, out-sourcing, and downsizing (Peters and Waterman, 1982; Brewer, 1995; Bennis, 1999; Quinn, 1999). These preoccupations have given rise to numerous managerial artifacts, including continuous performance improvement, benchmarking, and re-engineering, and intellectual ones such as organizational learning, core competence, and the rise of the network organization. The problem is that there has been little integration between the managerial and intellectual arenas.

Until recently, organizations have paid negligible attention to strategic logistics management, including information and marketing channels beyond physical distribution requirements, and have focused even less on supply chain management (i.e., the management of value-adding relationships among buyers, suppliers, and customers). Part of the explanation for this is that organizational applications have not been considered part of the study of logistics and supply chain management. In the coming decade, if value-adding relationships are to be realized there will have to be a noted movement towards integration of the supply chain with much of what has been learned by organizational theorists in the past incorporated into logistics and supply chain management in the future.[*] Indeed, it

---

[*]It is interesting to note that in an era of integrated logistics and sophisticated technology that the primary focus of yesteryear, namely physical distribution, will probably be one of the major impediments to supply chain implementation in the future. To overcome the physical distribution impediments, transport reliability, damage avoidance, delivery flexibility, and speed need to be managed more effectively within the existing freight transport infrastructure.

Handbook of Logistics and Supply Chain Management, Edited by A.M. Brewer et al.
© *2001, Elsevier Science Ltd*

is crucial that these two distinct disciplines of organizational theory and logistics and supply chain management are no longer neatly partitioned as they have been to date.

In a pragmatic sense, organizational logistics is about an integration of key functions (i.e., a cross-functional approach), targeted towards customer service, and minimizing costs and waste (i.e., mostly dissipated effort). Organizational logistics is about decisive orchestration, not only in terms of planning and co-ordination, but also in realizing internal and external synergies across the supply chain and the diverse actors within it. In this sense, it is focused not only on building relationships within groups, such as suppliers, buyers, and customers, but also with members involved in the internal workings of the organizations engaged in the supply chain. One of the important tasks is to understand the various networks of enterprises, as well as the nature of those relationships and intersecting processes. This process is referred to as "supply chain mapping" (Christopher, 1998), and is important for understanding how to structure the enterprise, integrate logistics processes internally and externally, and manage change. Since the focus is on networks of actors and their interrelationships, the term "supply networks"* supersedes "supply chain," which emphasizes a linear relationship between the actors instead of one based on "connectedness."

Traditional channel structures are being challenged, as they are unlikely to accommodate new relationships among suppliers, buyers, and customers. Organizational logistics is about defining and facilitating pathways for realizing the opportunities for growth through creating and leveraging internal and external relationships. The pathways are structured for the purpose of realizing value in the logistics pipeline, and extending this to the supply network. In this sense the pathways are goal-oriented and attempt to integrate personal, subunit, and enterprise objectives with supply network objectives. In the past, businesses have been concerned with performance improvement. This process was referred to as "adding value" (see Chapter 8 for wider discussion of value), which emphasized performance gaps proving to be less effective, since value was treated as an appendage to business processes rather than as integral to them. This is evident when comparing adding value with the more comprehensive process of organizational logistics, which focuses on performance opportunities by developing capacity and competence concurrently within business processes, as explained below.

As a result, organizational logistics is about the ability of the enterprise to appropriate value through creating and managing the capacity of its people

---

*The term "supply network" is a substitute for "supply chain management," which has been in vogue for the last decade or so. Later in this chapter, "supply network" will be supplanted by "demand network."

working together to achieve strategic objectives. For analytical purposes, organizational work can be viewed across three broad domains in the supply network. The first domain is entrepreneurial, focusing on choices of markets to serve; the second is technological, focusing on choices of technologies for process and distribution; and the third is administrative, concentrating on relationships, processes, culture, and the integration of all domains (Miles and Snow, 1978). Each of these domains gives rise to a distinct culture, and one of the challenges of organizational logistics is to bring about and sustain a high level of cultural pluralism, both within the pipeline and within the supply network.

Today, organizational logistics is not only about the allocation of resources (e.g., values, knowledge, information, technologies, facilities) but also the creative deployment of them, including their utilization, leveraging, and development. An organizational logistics approach echoes the resource-based theory of organizations, which postulates that management administers an array of assets that need to be identified and controlled. This theory was originated in the 1950s by Penrose (1959) but, like so many organizational theories, has been popularized more recently by others such as Prahalad and Hamel (1990). A resource-based view indicates that organizations have a choice over the management (including implementation) of their resource or assets base. Nowhere is this truer than in regard to knowledge, competence, and capacity.

2. Definition

While organizational logistics is concerned with managing the multiple streams of technologies both inside the organization and throughout the supply network, the primary focus here is on the capacity of people to learn and share knowledge and information collectively. The justification for limiting the topic in this way is based on the knowledge that, while organizations are frequently adept at managing markets, products, and services successfully, they are less proficient in integrating organizational processes, relationships, and resources. The real difficulty lies in identifying competencies and searching for new opportunities where aptitude can lead to innovation in services and products, as well as markets. A strong motivational magnetism through the workforce needs to emerge. However, a motivational force can create a dissonance between strategic intent and current resources, which needs to be understood and managed. This is particularly true where enterprises are operating in more complex customer service domains. Customers are more challenging today in terms of their expectations about service, time, quality, and responsiveness. Suppliers can no longer dictate service dimensions. This means that organizations will have to become far more spontaneous and develop new innovative strategies to equip them to service the demanding customer.*

Organizational logistics is an integrative expedient of strategic capacity because of a threefold approach:

(1) the mode of organizing through the establishment of strategic goals and key performance indicators as well as structural alignments;
(2) the critical role of knowledge and learning, which includes cognitive and physical capabilities; and
(3) the contributions of people (e.g., commitment) to the strategic capacity of organizations.

In general, organizational logistics is defined as a process of organizing in terms of either the total demand network and/or enterprise members. The process of organizing constructs the context in which the organization operates. This construction takes the form of three processes. The first process is *enactment*, a creation of information and knowledge through research, development, and learning, undertaken by organizational members as well as network actors. The information and knowledge that they perceive and respond to in turn shape the core processes that the organization ultimately adopts. The second is a *selection* process, whereby the organization attends to, orders, and interprets the information. The third process is *retention*, or learning, by which the organization does something with its information and knowledge. However, how the context is constructed is often at variance with another organization's interpretation of a similar context.

3. Components of organizational logistics

3.1. Core competence

A significant component of organizational logistics is the workforce, which makes an unquestionable investment in creating enterprise and network value. The relationship between individual inputs and organizational outputs is a complex one to quantify. However, when the "valuable" organization is conjured, up a diverse set of competencies is depicted, ranging from the technical through to the interpersonal, all of which are the domain of human beings. The "valuable" workforce is also conceived as flexible; that is, one able to adjust to new

---

*For this reason, the term "supply network" will now be superseded by "demand network." A demand network is a substitute concept for both "supply network" (defined earlier in this chapter) and the former "supply chain management."

requirements in novel or complex situations and stretch itself beyond its current capability. Many workforce competencies are explicit, although countless more remain tacit (Polanyi, 1966) and idiosyncratic to individuals, groups, as well as the overall organization. Considered in aggregate terms, this potential capacity is frequently lost to the organization.

Tacit competencies provide organizations with a potential competitive advantage, more so if they become explicit. Both Winter (1987) and Prahald and Hamel (1990) argue that these distinctive competencies form a key part of the strategic portfolio of an organization and facilitate its adjustment to competitive contingencies, particularly in a demand network. Competencies, according to these authors, are not easy to imitate, providing an organization with further potential to sustain its competitive advantage. More importantly, the realization of distinctive competencies is a major source of regeneration for the organization.

A core competence is ideally linked to a key objective in the business and corporate strategy, and facilitates process integration throughout the various operational levels of the organization and demand network. A process is a bundling of functional activities that transform inputs into outputs, although it is never as simple as this in practice. In organizations where the focus is not on integration, each of these activities are delimited by their functional origin, preventing their cross-functional integration. The problem with this approach is that functional activities "add" less value strategically when treated independently than they do when intentionally packaged together for strategic purposes.

Why is this the case? It was not until towards the end of the 20th century, when a number of social conditions emerged, ranging from technological innovations and globalization, through to environmental constraints, posing new opportunities and constraints, that we observed organizational and societal interrelationships being taken into account seriously by key stakeholders. This trend has led to a reconfiguring of organizational forms that has meant an increased ambiguity and convolution of business strategies and practices, resulting in a demand for knowledge to address novel and/or complex issues. These more-or-less associated trends influence the "collective motivation" to rethink conventional management approaches to both the demand network and its members so as to find or learn alternative or more appropriate ways of doing things that are not being done within existing social relationships or arrangements. At the end of this century, this has led to the emergence of a new knowledge discipline, termed organizational logistics.* Consequently, the construction of organizational knowledge has become a core competence in itself. Prahalad and Hamel (1990) defined core competence as "the collective learning in the organization," and

---

*A term coined by Hensher in 1999. In the current author's opinion, it is a term that is not synonymous with "organizational behaviour and structure," but supplants it.

Box 1
Dimensions of inter-organizational relationships

Formalization of relationship
• Agreement binding relationship
• Structural formalization (e.g., intermediary)
• Characteristics of the parties

Intensity of relationship
• Size of resource investment in the relationship
• Frequency of interaction

Reciprocity in the relationship
• Resource reciprocity (e.g., direction of exchange)
• Mutual agreement

Standardization of relationship
• Kinds and quantities of elements exchanged

Source: based on Negandhi (1975) and Levine and White (1961).

view this integral process not only for knowledge production but also for the management of knowledge.

The realization of the importance of the creation and management of organizational knowledge has given rise to two trends. The first is the development of a demand network management, which relies on the integration of processes, referred to above, facilitating the development of broad competence, and hence strategic capability, across the network. Strategic capability is also further enhanced through the secondary trend of management appreciating the need for making greater investment in knowledge systems such as education and training, research and development, and information and telecommunications systems (Sjoholt, 1999). Since the demand network and its actors rely on a multiplicity of competencies, it is important to find ways to develop cognitive frameworks, which allow integrative structures to propagate successfully among different actors "belonging" simultaneously to more than one organizational entity. The key to enhanced strategic capability lies in changing people's behavior and performance, not so much individually but rather through inter- and intra-organizational relationships, work design, and organizational learning.

3.2. Inter-organizational relationships

These trends have given rise to new organizational relationships and cognitions, both of which are significant components of organizational logistics. For example, it is now commonplace to expect to see new partnerships being forged between suppliers and customers that are completely focused on time-based criteria. The results are partnerships reflected in enduring commitments to transact business, usually characterized by preferred supplier status being granted to those who can meet demanding criteria. Similarly, these macro-relationships are mirrored inside organizations.

Box 1 shows four dimensions of inter-organizational relationships based on the classic work by Negandhi (1975) and Levine and White (1961). Their theories place an emphasis on the interrelationship between societal conditions and the development of formal relationships between members of the demand network. All relationships implicate, either directly or indirectly, the flow and control of these dimensions. A highly intense relationship leads to a greater degree of formalization, as each member furnishes business transactions with official recognition. The sustainability of the relationship is shaped by the extent of reciprocity and standardization.

One of the major influences in shaping organizational logistics is the internal governance or forms of power and authority, which are intentionally structured to achieve outcomes by specific institutions. Power defines what counts as rationality and knowledge, and thereby what counts as reality (Flyvbjerg, 1998). Forms of power are reflected and symbolized in external arrangements, often acting as barriers. If one does not take into account power relationships, the concept of organizational logistics becomes theoretically less rigorous and of limited practical use. Without understanding forms of power, the organization would be characterized as a goal-seeking open system composed of tangible and intangible assets organized along consensus lines. In a demand network context, it is difficult to sustain this line of reasoning.

The problem lies in how to overcome barriers in order to realize organizational transformations. When managing change, it is important to focus on those aspects that contribute to diminishing power and the intensity of negative elements, and to devise innovative strategies to deal with major factors that hinder change.

Relationships are formed to create value beyond those that the parties could achieve independently of each other. Consequently, relationships with the highest anticipated value return are more likely to form and to be sustained. A value equation (based on Wahba and Lirtzman, 1972) to characterize the expected utility for inter-organizational relationships is

$$\text{RCV} = P_{s/R}V_{s/R} - P_{f/R}V_{f/R}. \tag{1}$$

RCV is the relationship of created value, where $P_{s/R}$ is the probability of success of value being created for relationship R, $V_{s/R}$ is the value to relationship R given success of value creation, $P_{f/R} = (1 - P_{s/R})$ is the probability of failure of value for relationship R, and $V_{f/R}$ is the value to relationship R given failure to create value.

In a test of their theory, Wahba and Lirtzman (1972) reported that parties joined together to create value, and those with the highest expectations for doing

Box 2
Comparison of dominant work paradigms of the 20th and 21st centuries

20th century
- Work occurs within functions and is restricted to a single enterprise and physical space
- Work is initiated by functional personnel
- Work groups are labeled as "teams" and are supervised and evaluated as part of a managerial hierarchy
- Management commits the team to projects, and defines the scope, outcomes, and deadlines
- Decision-making is top-down in the managerial hierarchy
- Management, not the team, is recognized for performance
- Planning of work and its execution are treated as separate activities, with a preoccupation with information technology

21st century
- Work is organized in cross-functional programs, and is not confined to a single enterprise focus or physical space
- A steering group signs off on programs that can be initiated by any member(s) of the demand network
- People are organized into project groups, which may include members not employed in the same organization; the customer defines the expectations, and the group supervises and evaluates these within the client brief that is set within a contract between the parties involved
- Project groups define the scope, outcomes, and deadlines in consultation with the customer or client
- Decision-making is multifocused, and structured within a broad management agreement
- Project groups are accountable to both management and customers for their performance, including success and failure
- Planning and execution of work are integrated with a greater focus on interpersonal skills (e.g., communication, team working); information technology is used as a support device (e.g., administrative)

so were more likely to achieve this. There are a number of factors that have to be taken into account in these new inter-organizational relationships. These include judicious appraisal of the history of the parties, including: operational assumptions; mode of operation; strategic thinking; formalized communication channels; the extent to which resource allocation is equitable among the parties; and an available avenue for the mediation and balance of power.

3.3. Work design

Work design today is about building, harnessing, and utilizing collective knowledge and learning. The main way to achieve this is through relationships; that is, by finding ways to ensure collaboration among actors (units, subunits, groups, individuals) to exploit, harness, and challenge each others' resources. Moreover, with various streams of information and communication technologies emerging, it is possible to consider new forms of organizations that secure workers to less prescriptive tasks and activities (e.g., rules, allocations of resources), liberating them temporally and spatially from conventional structures, to form integrative ones. Relationships in the workplace start to reflect the macro-relationships of the supply chain. Box 2 depicts the contrasting approaches to work design that have dominated the 20th century and those likely to predominate in the 21st century. One important conclusion based on Box 2 is that management in the future will need to have a far wider focus on work design than in any previous period if they are intent on achieving value management within their organization.

3.4. Organizational learning

Within the process of organizational learning, the concept of core competence recurs. While core competence is elemental to the strategic capacity of demand networks, it is the product of learning that emerges at the personal, subunit, enterprise, or combination of these. Organizational learning is a primary example of core competence that will characterize the effective organization of the future. As stated earlier, organizations that can "learn" from their social context and transform and implement solutions rapidly will realize greater potential in the future, particularly in a demand network context.

The strength of the linkage between people processes and business strategy is critical if the relationships, either inter-organizational or intra-organizational, are to benefit. As corporate and business strategies are modified, this needs to be aligned with the human resource strategy. This linkage is also what process flexibility is about, and is a further ingredient of organizational logistics.

The effectiveness of relationships in the future will be under the control of the parties that comprise them. This has always been the case, but has rarely been acknowledged by management. The simple question then becomes why not permit people to modify processes that they are involved in to improve their own effectiveness? It is not suggested that this is done on an ad hoc basis, but through an "institutionalized" process of organizational learning. Argyris (1990) presents a powerful theory to support this argument (see also Argyris, 1977, 1982, 1986, 1993; Argyris and Schon, 1978). He presupposes that there are two kinds of theories that structure individual response to learning:

(1) espoused (i.e., if–then propositions underpinning actions), and
(2) theory in use (i.e., if–then propositions that people enact).

Argyris argues that there is often a paradox between (1) and (2), of which people are not fully cognisant. The paradox replicates the contradiction between theory (e.g., education and training) and practice (e.g., actual performance). Moreover, the characteristics of (2) blind us to viewing this paradox.

Table 1
A comparison of model 1 and model 2 learning (based on Argyris' theory)

Dimensions	Model 1 learning	Model 2 learning
Value	Goal directed, with a focus on maximizing one's interests and dominance	Goal directed, with a focus on shared knowledge and what is best for the interests of sustaining relationships
Action	Advocate, evaluate, and attribute without gathering information, and collaborating with relevant parties	Advocate, evaluate, and attribute based on gathering information and collaboration with relevant parties
Outcome	Misunderstanding, need to defend one's position, aggravating error with little feedback (except negative) on relevance of values, goals, and performance of relationship	Improved understanding of context, minimizing error, and maximizing feedback on relevance of values, goals, and performance of relationship
	Minimal learning and negligible enhancement of competence	Harness tacit knowledge, thereby maximizing learning and competence

Using Argyris' framework, organizational learning is the collective uncovering and rectification of error that occurs between (1) and (2). People "choose" from two broad learning processes, either model 1 or model 2, as outlined in Table 1. Selecting model 2 in preference to model 1 results in an integrated approach, which is again the very essence of organizational logistics.

4. An approach to organizational logistics

Organizational logistics allows management choice over the development of organizational forms that not only best suit the context, including power bases, in which they operate, but also addresses current internal and external issues. The work of organizations in the 21st century requires variable, specialized resources in shifting sequences that respond to customer demands. The most appropriate organizational form is contingent upon the nature of the business and potential markets, the strategy being pursued, organizational strengths and weaknesses, leadership and culture, technology, size, and business life cycle. It is impossible, and unnecessary, to predict the exact form that organizational logistics will take in the future, as there will be many and various forms. What is likely is that an organizational logistics approach will be based on core processes by eroding rather than revering functional and authority barriers, which requires a completely different way of managerial thinking from the past.

The predicament for management will be how to define "process". Christopher (1998) argues that the value of process will be realized when the starting point is with the customer's order, and subsequently by mapping the flows of resources, including human activities, technologies, and procedures. A process-based approach allows managers to choose strategies that create differential value to customers contingent upon their demands. An organization wanting to become more demand responsive will more than likely need to redesign core processes by rethinking the following dimensions (in no order of priority):

(1) *information processes* – where and how actors acquire knowledge, information, and learning;
(2) *learning processes* – the reasoning, decision-making, planning, and information-processing activities that are involved in processes;
(3) *performance processes* – the effort, activities, tools, equipment, and devices used by actors to participate in processes;
(4) *relationship management* – the nature of the link (e.g., interdependency) with other actors required in participating in processes; and
(5) *contextual processes* – the spatial, temporal, and market contexts in which processes are conducted.

5. Conclusion

Organizational logistics is a vital component of the value equation, as it provides actors with a new lens with which to assess, create, and leverage value in emerging relationships across the demand network. It achieves this through four inter-related processes:

(1) *positioning* – the selection of appropriate strategic and processual approaches;
(2) *integration* – the co-ordination of key points on the demand network;
(3) *flexibility* – the ability to adjust to changing and unexpected market circumstances; and
(4) *performance* – developing and measuring intra- and inter-organizational exigencies and outcomes.

It is important to bear in mind that a significant dimension in the management of demand networks is the behavioral one. In any arena, whether it be intra- or inter-enterprise, it is people and their actions that are responsible for the achievement of positioning, integration, flexibility, and performance. With this in mind, "organizational logistics" may be too limiting a term to explain this fully. Conceivably, a more appropriate term is "behavioral logistics," of which organizational logistics is a subset. This provides a new debate for a new century.

References

Argyris, C. (1977) "Organisational learning and management information systems", *Accounting, Organisations and Society*, (2):13–29.

Argyris, C. (1982) "The executive mind and double loop-learning", *Organisational Dynamics*, 11(2):5–22.

Argyris, C. (1986) "Skilled incompetence", *The Harvard Business Review*, 64(5):74–79.

Argyris, C. (1990) *Overcoming organisational defences: Facilitating organisational learning*. Boston, MA: Allyn and Bacon.

Argyris, C. (1993) "Education for leading-learning", *Organisational Dynamics*, 21(3):5–17.

Argyris, C. and D. Schon (1978) *Organisational learning*. Reading, MA: Addison-Wesley.

Bennis, W. (1999) "Recreating the company", *Executive Excellence*, 16(9):5–6.

Brewer, A.M. (1995) *Change management: Strategies for Australian organisations*. Sydney: Allen and Unwin.

Christopher, M. (1998) *Logistics and supply chain management: Strategies for reducing cost and improving service*, 2nd edn. London: Pitman.

Flyvbjerg, B. (1998) *Rationality and power: Democracy in practice*. Chicago: University of Chicago Press.

Levine, J. and White, P.E. (1961) "Exchange as a conceptual framework for the study of inter-organisational relations", *Administrative Science Quarterly*, 5:583–601.

Miles, R.E. and C.C. Snow (with A.D. Meyer and H.J. Coleman) (1978) *Organizational strategy, structure and process*. New York: McGraw Hill.

Negandhi, A.R. (1975) *Interorganisational theory*. Kent, OH: Kent State University Press.

Penrose, E. (1959) *The theory of the growth of the firm*. New York: Wiley.

Peters, T.J. and R.H. Waterman (1982) *In search of excellence, lessons from America's best run companies*. New York: Harper & Row.

Polanyi, M. (1966) *The tacit dimension*. London: Routledge and Kegan Paul.

Prahalad, C. K. and G. Hamel (1990) "The core competence of the corporation", *Harvard Business Review*, May/June, 79–91.

Quinn, J.B. (1999) "Strategic outsourcing: Leveraging knowledge capabilities", *Sloan Management Review*, 40(4):9–21.

Sjoholt, P. (1999) "Skills in services: The dynamics of competence requirement in different types of advanced producer services", *Service Industry Journal*, 19(1):61–79.

Wahba, M.A. and S.I. Lirtzman (1972) "A theory of organisational coalition formations", *Human Relations*, 25:515–525.

Winter, S. (1987) "Knowledge and competence as strategic assets", in D. Teece, ed., *The competitive challenge*. Cambridge, MA: Ballinger.

Chapter 23

PUBLIC POLICY AND LOGISTICS

MARTIN DRESNER and CURTIS M. GRIMM
University of Maryland

1. Introduction

A starting point for the discussion of public policy and logistics is the vital role played by logistics in a country's economy. Logistics is key for two reasons. First, actual logistics expenditures are a major part of gross domestic product (GDP) (e.g., 10.5% of U.S. GDP in 1996). Secondly, logistics supports the movement and flow of many economic activities, facilitating the production and sale of most goods. Thus, a properly functioning logistics system is vital to an effective supply chain and strong economy (Lambert et al., 1998).

Logistics has an important economic impact on various aspects of the economy, including economic development, specialization of production, variety of goods available for consumption, and consumer prices. Starting with the discussion of Adam Smith, economic specialization and division of labor has been found to be vital to a country's economic development. The success of economic development hinges on logistics, since logistics allows firms to capitalize on comparative cost advantages in the production of goods and services. Goods can be produced in the regions with the greatest comparative advantage and then transported to other regions where demand exists. Logistics also contributes to the variety of goods available in the economy. Through an efficient logistics system, consumers can partake of a choice of products from around the world. Lower prices of goods are also a function of an effective and efficient logistics system, with benefits multiplying throughout the economic chain, when raw materials and input prices are included in the analysis. Since products can be produced in regions with greatest competitive advantage, consumers can benefit from lower prices and increased purchasing power (Coyle et al., 1996).

The importance of logistics in the economy is a prime driver of the motivation for government to establish public policies that facilitate the logistics system. An important dimension within the context of logistics public policies is government planning and promotion, particularly with regard to transportation. Included would be program funding and investment such as public ownership of transportation activities. Public policies are devised by government agencies to develop and influence transportation in the public interest. This includes the functioning of regulatory agencies that administer and implement policies developed by law-makers and the development and debate over new policies to regulate or promote transportation. The public planning process assesses future transportation needs, and is linked to programs and policies devised to meet societal needs. This may lead to changes in transportation infrastructure, such as privatization, when planning and assessment concludes that a change in ownership would better facilitate the logistics system. The move towards privatization is a major trend occurring worldwide, with the belief that a market system will more effectively allocate and provide transportation infrastructure vis à vis government ownership and provision.

Much of transport planning occurs in a context where environmental or social needs are of paramount importance. Much investment in mass transit, for example, is justified on the basis of reduction in social costs of automobile use, such as pollution, noise, safety, and other externalities.

To illustrate the role of public policy and logistics we provide three case studies. These case studies of logistics public policy issues and developments provide details on the functioning of government in policies and planning. We provide three examples of how logistics public policy can affect the logistics system and have important implications for the economy and society.

2. Case studies

2.1. Public policies regarding regulation and competitive access in the U.S. railroad industry

Over the past 25 years, the U.S.A. has deregulated large sections of the economy, including surface freight, airlines and, most recently, telecommunications and electricity. The substitution of market forces and competition for government regulation has had a profound effect on logistics costs and efficiency. Public policies regarding the role of government in regulating the transportation sector will continue to be critical in the future.

The U.S. railroad industry is distinctive in that it has always been predominantly in the private sector. Accordingly, the primary motivation for deregulation was the post-World War II financial deterioration within the industry. Prior to 1980, U.S. railroads were overburdened by an outmoded regulatory framework, and found themselves hampered by regulations that were causing them to lose more and more traffic. On the freight side, the railroads' share of intercity ton-miles fell from 65% at the end of World War II to 35% by the 1970s. Rail passenger traffic also declined precipitously in the decades following World War II; however, existing regulations made it difficult for railroads to exit the money-losing passenger service. The Rail Passenger Service Act in 1970 allowed railroads to discontinue all passenger service. Finally, following decades of very low rates of return on investment for the industry (2–4%), the bankruptcy of Penn Central focused policy-makers on the paramount importance of policy change if the rail industry were to remain in the private sector. While the initial policy response to the Penn Central bankruptcy was the creation of Conrail, a government owned corporation, the option of wholesale rail nationalization was distinctly unpalatable to most policy-makers. The stage was set for Congress to address the need for fundamental regulatory change to maintain a private sector rail freight industry.

The Railroad Revitalization and Regulatory Reform Act of 1976 (4R Act) was the first step towards deregulation. However, the Act had little effect initially because the Interstate Commerce Commission, the regulatory agency charged with its implementation, was not in favor of deregulation. However, following appointment of deregulation-minded commissioners to the Interstate Commerce Commission in the late 1970s, important administrative deregulation began and the agency was also brought into the policy debate as a strong advocate rather than opponent of deregulation. At the same time, a growing body of literature supported the need for rail deregulation.

The key piece of deregulation legislation was the Staggers Rail Act of 1980, which granted substantial new freedoms to the railroads, including virtually complete pricing flexibility (Grimm and Windle, 1999). The result was a dramatic revitalization of the U.S. railroad industry. During the 1990s the industry's return on equity averaged 10.7%.

Shippers have also benefited from deregulation. Based on the first decade of deregulation, a study by Winston et al. (1990) found that the annual benefits to shippers from lower rates and improvements in service time and reliability amounted to at least $12 billion (1999). For example, the Staggers Act allowed shippers to negotiate individual and confidential contracts, which allowed specification of service dimensions as well as rates. Coupled with similar changes occurring in the U.S. motor carrier industry, shippers were able for the first time to redesign logistics systems with optimal transportation characteristics within such systems.

Despite the success of railroad deregulation, shippers have increasingly clamored for new legislation to increase the level of competition in the rail industry. As discussed by Grimm and Winston (2000), three factors have been key in this initiative: a recent railroad merger wave, dissatisfaction with regulatory protection of captive shippers under Staggers guidelines, and developments promoting competition in railroads abroad and in other U.S. network industries.

In the post-Staggers era, the U.S. railroad industry has experienced a significant number of mergers and a sharp reduction in the number of class 1 (major) rail carriers (Grimm and Plaistow, 1999). This has given rise to increasing concerns by shippers as to the competitive effects of these mergers. Mergers have occurred in two major waves. In the early 1980s, for example, Chessie System and Seaboard Coast Line formed CSX, Norfolk and Western and Southern Railroad formed Norfolk Southern, Missouri Pacific and Western Pacific became part of Union Pacific, and the St Louis–San Francisco Railroad along with Colorado Southern and Fort Worth Denver formed part of Burlington Northern. However these consolidations, as well as subsequent mergers up to the mid-1990s, were primarily end-to-end.

The merger wave that began in the mid-1990s raised more serious issues regarding horizontal competitive effects. The Burlington Northern–Santa Fe and Union Pacific–Southern Pacific mergers left only two major railroads in the western U.S.A., while Conrail's recent absorption by Norfolk Southern and CSX left only two major railroads in the east. Moreover, Union Pacific's mismanagement of its merger with Southern Pacific and a surge in traffic led to widely publicized service disruptions. Some shippers also experienced poor service in the aftermath of the Conrail acquisition. The costly delays crystallized the problems that could arise when shippers must depend on only one railroad, this being a growing possibility as the industry consolidates. At the time of writing, there are signs of yet further rail merger activity. Burlington Northern and Canadian National Railway have proposed a merger which, if approved, could touch off another round of mergers. The U.S. Surface Transportation Board has issued a 15-month moratorium on mergers, a decision that is currently being challenged in the courts.

The second factor prompting legislation to promote competition is dissatisfaction with regulatory protection of captive shippers (those dependent on a single railroad to provide service). Shippers, and various organizations that represent them, complain that rail rates are not always reasonable and that the Surface Transportation Board's rate complaint process is time-consuming, costly, and complex. Because few shippers file complaints, which can cost from $500 000 to $3 million, few rates are found to be unreasonable. Nonetheless, the industry earns roughly 30% of its revenue under conditions that would require the Board to determine in a rate case whether a railroad is market dominant.

Policy-makers in other network industries, such as telecommunications and electricity, and in railroad industries abroad are attempting to stimulate competition to address concerns that are similar to those of the captive shippers. For example, the Telecommunications Act of 1996 requires incumbent local exchange carriers to enter into interconnection agreements with competitors to address structural monopoly at the loop level, while other countries are taking steps to separate ownership of the track from provision of service to allow competitive private sector rail service to develop. The long-standing competitive access provisions in Canada's rail regulation, which include mandatory interswitching for captive shippers if an alternative railroad is located within

30 km, also provides a model for advocates of increased rail competition. These developments are prompting calls for policy-makers to take more forceful measures to ameliorate concerns about competition in the U.S. railroad industry.

A range of policy proposals is currently under debate. Some proposals, modeled on developments in rail industries abroad, would separate ownership of infrastructure from provision of service. Another radical proposal would "unbundle" the rail system into primary terminals, secondary terminals, long haul railroads, and collection/distribution railroads. Terminals would be interchange points for two or more railroads, and would have a significant collection and distribution function. Current terminal owners, in most cases class I railroads, would be forced to spin off these assets and allow independent operations to form, but they would continue to haul freight between primary terminals.

Other proposals would provide modest changes consistent with the Staggers Act framework. One by the U.S. Department of Transportation would give the U.S. Department of Justice primary authority for approving rail mergers and would authorize the Surface Transportation Board (STB) to approve more mandatory railroad interchange of traffic. Another, the Railroad Shipper Protection Act, would require rail carriers, upon shipper request, to establish a rate between any two points on the carrier's system where traffic originates, terminates, or may be interchanged. The shipper would then be allowed to challenge the rate under maximum rate guidelines. Proposals have also surfaced which would simplify maximum rate guidelines and provide for more restrictive limits on rates for captive shippers.

For the most part, the major railroads have argued for maintenance of the status quo, characterizing any changes in rail policy as "re-regulation." Furthermore, they argue that, although their financial situation has dramatically improved since deregulation, they do not come close to earning monopoly profits. Any such re-regulation would wipe out the benefits of the Staggers Act and return the industry to the days of revenue shortfalls prior to 1980. Also, the railroads trumpet the gains from deregulation and the disappointing international experience with forced access as support for the status quo. They claim that proposed policies to address the captive shipper issue would be counterproductive and characterize differential pricing (including that based on differing competitive circumstances) as efficient and necessary – without it, the size and quality of the nation's rail network would have to shrink because carriers would have no plausible way to cover their fixed (and total) costs.

The debate regarding the future direction of rail regulatory policies is expected to continue for many years into the future. The range of actions may include radical change, incremental change within the Staggers Act, maintenance of the status quo, or even total deregulation of the rail industry, as proposed recently by Grimm and Winston (2000). Getting the policy right is of critical importance to the logistics system in North America.

2.2. European public policies

In the 1990s, the countries of the European Union (EU) undertook a series of public policy initiatives designed to create a single market within the Union. These initiatives were generally contained in the document known as the Treaty of Maastricht, which took effect on January 1, 1993, and were phased in throughout the decade. The process included, among other initiatives, the harmonization of product standards among EU countries, the standardization of public procurement procedures, the removal of administrative delays at border crossings, the elimination of restrictions on the provision of cabotage transportation services, the development of common work rules, the harmonization of the taxation systems among EU countries, and the development of a single European currency.

In general, the changes brought about under the Treaty of Maastricht are thought to have a positive effect on business costs, especially in the logistics arena. The reduction in border restrictions, combined with the harmonization of product standards and the elimination of the cabotage restrictions, have allowed some firms to consolidate their operations into fewer, "trans-continental" facilities. Firms that previously operated manufacturing or distribution facilities in each of the EU countries have been able to serve their European customers more cost-effectively by consolidating manufacturing plants and distribution centers. As one commentator indicated, "A company that manages logistics on a pan-European or regional basis, rather than country by country, stands to achieve enormous cost savings in transportation, inventory, real estate, taxes, personnel, capital equipment, and more" (Gooley, 1999). A 1997 study of over 300 European-based manufacturers found that nearly half of respondents had reduced the number of manufacturing sites they maintained, while over 90% reported that they were restructuring their distribution networks (Gooley, 1999).

Nike, the U.S.-based sportswear firm, is an example of a company that restructured its European operations following the signing of the Maastricht treaty. Using a third-party logistics provider, Menlo Logistics, Nike chose a facility in The Netherlands to serve as its European distribution center. At The Netherlands distribution center, Menlo provides a wide range of logistics activities, including customs clearance for inbound shipments, receiving, storing, picking, packing, carrier selection, routing, and tracking to support Nike's European operations (Bowman, 1999).

Gillette, the U.S.-based consumer products firm, is another example of a company that has centralized its operations in Europe. The firm reduced the number of razor blade plants in Europe from five to three, and centralized control over its purchasing, manufacturing, and marketing functions. Plant managers no longer report to national managers, but directly to the head of manufacturing in the U.S.A., thus facilitating centralized planning (Economist Intelligence Unit, 1995). According to Gillette management, scheduling production has been helped tremendously by centralized planning of production and distribution. In addition, the removal of barriers at border crossings in Europe has allowed Gillette to speed the movement of goods to customers. Whereas, it used to take 7–10 days to truck products from manufacturing facilities in the U.K. to the Gillette distribution center in Frankfurt, Germany, trucks carrying Gillette product can now make two trips in the same time (Economist Intelligence Unit, 1995). According to one Gillette manager, the increased integration in Europe has represented an enormous opportunity for the firm to extract big cost reductions from economies of scale with very little loss in operating flexibility (Economist Intelligence Unit, 1995).

Gooley (1999) notes that, although the development of pan-European operations is spreading quickly, its acceptance is far from universal. Acceptance differs substantially by industry. Whereas the chemical industry is entirely pan-European, the pharmaceutical industry still faces many local restrictions that preclude a continent-wide approach to manufacturing and distribution. Other impediments to the implementation of a continental-side supply chain include the following (Council of Logistics Management, 1993):

(1) Differences in national social and environmental legislation. A primary reason why Norway has not joined the EU is that it does not wish to alter its environmental standards to conform to those of the EU.
(2) Cultural diversity within Europe. Legislation could not eliminate the preference differences among the various EU populations that can complicate the implementation of a pan-European manufacturing or distribution policy.
(3) The principle of "subsidiarity" that allows countries to enact legislation in areas which are best handled at the local level, despite EU initiatives, and can create disharmony in the way businesses operate in the different European countries.
(4) The allowance for countries to opt out of various provisions of the Treaty of Maastricht. The U.K., for example, has chosen not to participate, at least at present, in the European monetary union.

A report by the consulting firm A.T. Kearney adds that, for some products (e.g., those low value to weight commodities with high transportation costs), a continent-wide integration will likely not work. In addition, products that require suppliers to be in close proximity to customers, and those where there are advantages to having multiple production units, for example to avoid labor disruptions, will likely not be integrated in a European continental strategy (Economist Intelligence Unit, 1995).

In summary, the EU has taken several initiatives that allow firms to harmonize operations within the Union. Firms have responded by consolidating their

European operations, thus increasing operating efficiencies. Impediments to integration, however, remain.

2.3. *Privatization of port operations in Latin America*

One of the fastest growing regions for international trade is Latin America. However, a major impediment to even stronger trade growth has been the backward state of logistics systems and transportation infrastructure at Latin American ports. From the perspective of the international shipper, the operating efficiency at a port can be a key consideration in the choice of a Latin American export market. Most shipments to a Latin American country must be off-loaded directly into that country, since inter-country rail and road links are generally poor. Thus, if a shipper wants to sell to Brazil, shipments must be made directly to Brazil, rather than through Uruguay, Argentina, or some other neighboring country.

Burkhalter (1992) outlines the problems facing Latin American ports in an analysis of what he terms a "composite" Latin American and Caribbean port. He divides his analysis into a number of categories:

(1) *Location and physical infrastructure*. The composite Latin American port is "surrounded by a city without sufficient space for expansion" and situated well up-river from the sea, necessitating a long and expensive piloted trip to port. Channels to the port are not dredged to sufficient depths to allow for the operation of modern container and bulk ships. The composite port was originally constructed at the beginning of the century for bulk and break-bulk shipments, and has not been sufficiently updated with modern container and bulk-handling equipment. Intermodal rail connections are in "an advanced state of repair," while trucking companies have an adequate road system but make use of monopolistic pricing practices and experience slow port turnaround times due to bureaucratic impediments.

(2) *Institutional environment*. Little co-ordination and few communication links exist between the various groups (e.g., dock workers, banks, truckers, customs agents, port authorities) involved at the composite Latin American port. The port is heavily overstaffed and suffers from low labor productivity, at only 20–40% the labor productivity level of ports in industrial countries.

(3) *Operations*. Goods passing through the port are generally subject to delays, sometimes for several weeks, due in large part to the lack of co-ordination at the port. In an effort to facilitate their operations, private companies, such as shipping lines and freight forwarders, hire their own non-union workers to handle their goods, while still paying for the services of the

unionized state dock workers. The lack of sophisticated equipment, combined with the inefficient labor force, results in unloading and loading times four times higher than at ports in industrial countries. The net effect of these operating inefficiencies is total port costs two to three times higher than comparable costs in industrial countries.

Perhaps the port most desperate for reform is Santos, the largest and most important port in South America. The Port of Santos, in the mid-1990s, was described as one of the most inefficient in Latin America. Moving cargo through Santos was slow and expensive, with a high frequency of piracy, the port being staffed by inefficient and overpaid cargo handlers (Coone and Katz, 1996). Much of the time, the majority of container cranes was out of commission and there was a lack of bonded warehouse space (*World Trade*, 1995). The problems at the port created delays for the shipping lines that service Santos, and resulted in the implementation of congestion fees by the shipping lines, further increasing the costs of shipping through the port. However, since the mid-1990s, the Brazilian government has taken a number of steps to privatize the operations at the Port of Santos and to otherwise increase the efficiency of the port's operations. As one commentator put it, "The dream of transforming Santos into a Latin American hub may still be distant but favorable conditions have finally been set after a long period of stagnation" (Ogier, 1998). Privatization has resulted in much needed investments in container facilities. New container cranes purchased by a private terminal operator have allowed for improved productivity and a reduction in gate-to-vessel costs by about 50% after only 3 months of private sector operations (Ogier, 1998).

Earlier privatization efforts have been successful in other Latin American countries. In Venezuela, most maritime trade is in oil, with only two ports, Porto Cabello and La Guaira, handling 75% of all non-oil cargo. Both of the ports have good physical characteristics, by Latin American or Caribbean standards, the major exception being lack of modern cargo handling equipment, and inadequate port depth at Puerto Cabello. Despite the good facilities at the ports, efficiency has traditionally been poor. Losses in Puerto Cabello in 1991 were on a par with some African countries, and in La Guaira "agency gangs" or informal system dock workers undertook 60% of cargo operations for general cargo vessels and 100% for roll-on, roll-off vessels (Burkhalter, 1992). However, in the 1990s, an important modernization effort took place at Puerto Cabello. Within a 4-year period following initial privatization efforts, the total number of employees at the port decreased from 5200 to 1640. The operating time at the port increased from 5 days per week, 8 hours per day, to 7 days per week, around the clock. The number of ships calling at Puerto Cabello nearly tripled, while the number of containers handled more than doubled. Port labor productivity rose from 650 ton per worker to 4400 ton, while the average length of time for a ship in port fell from 163 hours to 58 hours. In addition, the port authority now reports cargo losses to be practically non-existent. (Diaz-Matabalis and Dresner, 1996).

The port improvements at Cabello were achieved by means of a two-step process: In the first step, ports that were under control of the Venezuelan government, including Cabello, were transferred to local control. This transfer was part of a short-lived modernization process that started in 1989 and ended in 1993, due to political problems. Next, the government of the state of Carabobo, where Puerto Cabello is located, instituted an ambitious, well-executed project for personnel reduction and the out-sourcing of all operations to 42 private operators.

The Port of Veracruz, in Mexico, is another example of successful privatization of port operations in Latin America. Moves to privatize the port began in 1996, with a handful of private operators taking over management of the port from a government-run company. The private operators have increased channel depth from 31 feet to a minimum of 35 feet, thus allowing for the operation of larger vessels and reducing delay time for vessels in port (Wallengren, 1999). In addition, new investments pledged by the private operator will substantially increase container terminal capacity and add new bulk handling equipment and storage capacity to the port. According to one freight forwarder familiar with the operations at Veracruz, "The place is really run quite efficiently now" (Wallengren, 1999).

In summary, Latin American ports have a reputation for inefficient operations. Privatization of port facilities in Brazil, Venezuela, Mexico, and elsewhere in Latin America have resulted in new port investments, leading to increased productivity and lower operating costs.

3. Conclusion

In summary, logistics is a dynamic area within which there are many vital public policy issues and actions. The appropriate policy can greatly enhance logistics systems and functioning of the economy. For example, deregulation and micro-economic reform of transportation has revolutionized the transportation systems of many countries and greatly facilitated improvements in logistics systems and economic performance. At the same time, logistics managers must be familiar with public policy issues and changes, and work to facilitate appropriate policy. Changes in public policy have had to, and will continue to, have important impacts on companies, bringing about both threats and opportunities to existing companies. For example, deregulation in the U.S. rail system has had critical effects on optimal plant location and the degree to which firms can use logistics to gain competitive advantage. The integration of the European economies has allowed European companies to increase productivity by centralizing their production and logistics processes. Finally, port privatization in Latin America has been important in the development of international trade in that region.

References

Bowman, R.J. (1999) "Logistics in Europe: Tear up the old maps", *Global Logistics and Supply Chain Strategies*, July, 38–51.

Burkhalter, L.A. (1992) "Institutional considerations and port investments: Recent experience in Latin America and the Caribbean", in A.J. Dolman and J. van Ettinger, eds., *Ports as nodal points in a global transport system*, pp. 321–348. New York: Pergamon Press.

Coone, T. and I. Katz (1996) "Safe passage?", *Latin Trade*, 4(4):45–53.

Council of Logistics Management (1993) "Reconfiguring European logistics systems", Oak Brook, IL: Council of Logistics Management.

Coyle, J., E. Bardi and J. Langley (1996) *The management of business logistics*, 6th edn, pp. 29–36. St Paul, MN: West Publishing Company.

Coyle, J., E. Bardi and R. Novack (2000) *Transportation*, 5th edn, pp. 57–77. Cincinnati, OH: South-Western College Publishing.

Diaz-Matabalis, A. and M. Dresner (1996) "The modernization of ports in Latin America: A case study of Puerto Cabello", University of Maryland, working paper.

Economist Intelligence Unit (1995) *The EU 50: Corporate case studies in single market success*. London: The Economist Intelligence Unit.

Gooley, T.B. (1999) "Pan-European logistics: Fact or fiction?", Logistics Management and Distribution, Report Online (http://www.manufacturing.net/magazine/logistics/archives/1999/log0301.99/185.htm).

Grimm, C.M. and J. Plaistow (1999) "Competitive effects of railroad mergers", *Journal of the Transportation Research Forum*, 38:1–12.

Grimm, C.M. and R. Windle (1999) "The rationale for deregulation", in: J. Peoples, ed., *Regulatory reform and labor markets*. Boston: Kluwer.

Grimm, C.M. and C. Winston (2000), "Competition in the deregulated railroad industry: Sources, effects and policy issues", in S. Peltzman and C. Winston, eds., *Deregulation of network industries: The next steps*. Washington, DC: Brookings Institution.

Lambert, D., J. Stock and L. Ellram (1998) *Fundamentals of logistics management*, pp. 10–11. Boston, MA: Irwin McGraw-Hill.

Ogier, T. (1998) "A Cinderella port", *Journal of Commerce* (http://www.joc.com/Issues/980324/m1time/e57141.htm).

Wallengren, M. (1999) "Successful expansion", Mexico Connect, http://www.mexconnect.com/mex_/travel/bzm/bzmveracruzharbor.html.

Winston, C., T. Corsi, C.M. Grim and M. Evans (1990) *The economic effects of surface freight deregulation*. Washington, DC: Brookings Institution.

World Trade (1995) "So good it hurts", November, pp. 68–69.

Part 7

COMPARATIVE INDUSTRY APPROACHES

Chapter 24

RETAIL LOGISTICS

JOHN FERNIE

Heriot-Watt University, Edinburgh

1. Introduction

There was a logistical revolution in retailing in the late 1980s and throughout the 1990s. During this period, supply chain management rose in status in company boardrooms because of the impact which the application of supply chain techniques can have on a company's competitive position and profitability. The phenomenal success of British grocery retailers in the late 1980s and early 1990s could be attributed to the transformation of the grocery supply chain at that time. Similarly, in the fashion market the success of Benetton and The Limited was partly due to reducing time to market. By contrast, the slow demise of Laura Ashley is a sad reflection of a company which failed to manage its own inventory.

2. Supply chain concepts

Many of the current ideas on supply chain management have their roots in the work of Michael Porter, the Harvard Business School professor, who introduced the concept of the value chain and competitive advantage (Porter, 1985). These ideas have been further developed by academics such as Martin Christopher in the UK (Christopher, 1997). In essence, we have a supply chain model whereby at each stage of the chain value is added to the product through manufacturing, branding, packaging, display at the store, and so on. At the same time, at each stage cost is added in terms of production costs, branding costs, and overall logistics costs. The trick for companies is to manage this chain to create value for the customer at an acceptable cost. The managing of this so-called "pipeline" has been a key challenge for logistics professionals since the 1990s, especially with the realization that the reduction of time not only reduced costs but gave competitive advantage.

It was in fashion markets that the notion of "time-based competition" had most significance in view of the short time window for changing styles. In addition, the

Handbook of Logistics and Supply Chain Management, Edited by A.M. Brewer et al.
© 2001, Elsevier Science Ltd

prominent trend since the 1980s has been to source products off-shore, usually in low-cost Pacific Rim nations, which lengthened the physical supply chain pipeline. These factors combined to illustrate the trade-offs which have to be made in supply chain management and on how to develop closer working relationships with supply chain partners.

The catalyst for much of the initiatives in lead time reduction came from work undertaken by Kurt Salmon Associates (KSA) in the U.S.A. in the mid-1980s (Kurt Salmon Associates, 1993). KSA were commissioned by U.S. garment suppliers to investigate how they could compete with Far East suppliers. The results were revealing in that the supply chains were long (one and a quarter years from loom to store), badly co-ordinated, and inefficient (Christopher and Peck, 1998). The concept of quick response was therefore initiated to reduce lead times and improve co-ordination across the apparel supply chain. In Europe, quick response principles have been applied across the clothing retail sector. Supply base rationalization became a feature in the 1990s as companies dramatically reduced the number of suppliers and worked much closer with the remaining suppliers to ensure more responsiveness to the marketplace.

The importance of supply chain integration cannot be understated; however, much depends on the degree of control a company has over the design, manufacture, marketing, and distribution of their supply chains. Two of the most successful fashion retailers in Europe – Benetton and Zara – illustrate how an integrated supply chain can enhance the retail offer. Both companies draw heavily on lean production techniques developed in Japan. Their manufacturing operations are flexible, involving a network of subcontractors and, in the case of Benetton, suppliers in close proximity to the factory. Benetton was one of the first retail companies to apply the "principle of postponement" to its operations whereby semi-finished garments were dyed at the last possible moment when color trends for a season became apparent from electronic point of sale (EPOS) data at their stores. So rather than manufacture stock to sell, Benetton could manufacture stock to demand.

Zara's operation has similarities with the The Limited in the U.S.A. in that the company scouts the globe for new fashion trends prior to negotiating with suppliers to produce specific quantities of finished and semi-finished products. Only 40% of garments, those with the broadest appeal, are imported as finished goods from the Far East; the rest are produced in Zara's automated factories in Spain. The result of their supply chain initiatives is that Zara has reduced its lead time gap for more than half of the garments it sells to a level unmatched by any of its European or North American competitors.

Benetton, Zara, and The Limited, as the name of the last mentioned suggests, have narrow product assortments for specific target markets. This streamlines and simplifies the logistics network. For general fashion merchandisers, rather than specialists, the supply chain is more complex, with thousands of suppliers around the globe. Nevertheless, quick response and efficient consumer response concepts are being applied to these sectors in an effort to minimize mark downs due to out-of-season or unwanted stock in stores.

Retailers have been analyzing carefully their supply base and the cost of sourcing/distributing products to their network of stores. The difficulties experienced by Marks & Spencer in recent years led the company to review its entire supply chain. Marks & Spencer's logistics system is centered around the U.K. in that even its international distribution centers are supplied from warehouses in the south-east of England. In the case of supplying its Hong Kong stores it was taking up to 27 days to deliver goods from England by air transport. Although this has been reduced to between 7 days and 11 days, it conceals some of the problems further back up the supply chain (Jackson, 1998). Good logistics is not solely about reducing product replenishment lead times but also being able to react to changes in the market. This is a primary distribution problem, and relates to how quickly the pipeline can be turned on or off according to surges or collapses in demand. The Marks & Spencer review has meant that efforts are being made to take inventory out of the whole supply chain system through buying closer to the particular seasons and through the utilizing of air freight for much of its international sourcing.

Although grocery retailers are more oriented to their national or super-regional markets such as the EU than their clothing counterparts, the internationalization of grocery retailers and their customers has led to changing sourcing patterns. Furthermore, the increased competition in grocery markets with the resultant pressure on profit margins has acted as a spur to companies to improve supply chain performance.

3. Efficient consumer response (ECR)

ECR arrived on the scene in the early 1990s when Kurt Salmon Associates produced a supply chain report, "Efficient Consumer Response," in 1993 in response to another appeal by a U.S. industry sector to evaluate its efficiency in the face of growing competition to its traditional sector. Similar trends were discerned from their earlier work in the apparel sector: excessive inventories, long unco-ordinated supply chains (104 days from picking line to store purchase), and an estimated potential saving of U.S. \$30 billion, 10.8% of sales turnover.

The ECR initiative has stalled in the U.S.A. since 1993; indeed, inventory levels remain over 100 days in the dry grocery sector. Nevertheless, ECR has taken off in Europe since the creation of the ECR Europe executive board in 1994 with the support of Europe-wide associations representing different elements of the supply chain – AIM, the European Brands Association; CIES, the Food Business Forum; EAN International, the International Article Numbering Association;

and Eurocommerce, the European organization for the retail and wholesale trade.

It was in 1994 that initial European studies were carried out to establish the extent of supply chain inefficiencies and to formulate initiatives to improve supply chain performance (Table 1). ECR Europe defines ECR as "a global movement in the grocery industry focusing on the total supply chain – suppliers, manufacturers, wholesalers and retailers, working close together to fulfil the changing demand of the grocery consumer 'better, faster and at less cost'" (Fiddis, 1997, p. 40).

One of the early studies carried out by Coopers & Lybrand identified 14 improvement areas whereby ECR principles could be implemented. These were categorized into three broad areas of product replenishment, category management, and enabling technologies (Figure 1). Most of these improvement areas had received management action in the past, the problem was how to view the concepts as an integrated set rather than individual action areas.

As the ECR Europe movement began to gather momentum, the emphasis on much of the work conducted by the organization tended to shift from the supply-side technologies (product replenishment) to demand-driven initiatives (category

Table 1
Comparisons of scope and savings from supply chain studies

Supply chain study	Scope of study	Estimated savings
Kurt Salmon Associates (1993)	U.S. dry grocery sector	10.8% of sales turnover (2.3% financial, 8.5% cost)
		Total supply chain U.S. $30 billion, warehouse supplier dry sector U.S. $10 billion
		Supply chain cut by 41% from 104 days to 61 days
Coca-Cola Supply Chain Collaboration (1994)	127 European companies	2.3–3.4% percentage points of sales turnover (60% to retailers, 40% to manufacturer)
	Focused on cost reduction from end of manufacturers line	
	Small proportion of category management	
ECR Europe (1996, ongoing)	15 value chain analysis studies (10 European manufacturers, 5 retailers)	5.7% percentage points of sales turnover (4.8% operating costs, 0.9% inventory cost)
	15 product categories	Total supply chain saving of U.S. $21 billion
	7 distribution channels	U.K. savings £2 billion

Source: Fiddis (1997).

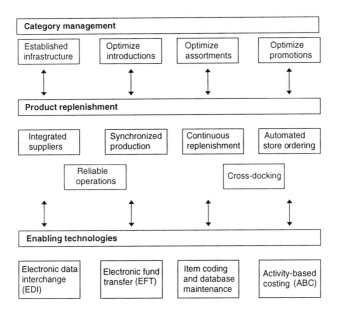

Figure 1. ECR improvement concepts (source: Coopers and Lybrand, 1996).

management). This is reflected in the early ECR project reports which dealt with efficient replenishment and efficient unit loads. While the supply side is still important, as reflected in projects currently being carried out on transport optimization and unit loads identification and tracking, the majority of recent projects have focused upon consumer value, efficient promotion tactics, efficient product introductions, and collaboration in customer-specific marketing.

Commensurate with this change in emphasis has been the topics under discussion at the annual ECR Europe conference. At its inception in Geneva in 1996, the concept was being developed and efficient replenishment initiatives were prominent on the agenda. Four subsequent conferences in Amsterdam (1997), Hamburg (1998), Paris (1999), and Turin (2000) have tended to emphasize demand-driven initiatives.

It can be argued that the early work focused upon improving *efficiencies* within the supply chain but since the late 1990s the emphasis has been on the *effectiveness* of the supply chain. Thus, the focus now is on how to achieve profitable growth, as there is little point in delivering products efficiently if they are the wrong assortment, displayed in the wrong part of the store!

The ECR Europe prime objective is to develop best practices and to disseminate these benefits to all members of the food supply chain in Europe. To date it has been highly successful in moving toward this objective. The early conferences were well attended (over 1000 delegates), but the Turin event attracted 3000 people. ECR initiatives are now formally organized in 14 European countries, and the work in these countries is formally recognized through representation on the ECR Europe executive board. The board itself is comprised of 30 senior executives from leading retailers and branded manufacturers in Europe, and established the policy agenda to initiate new pilot projects and demand and supply strategies.

It is, however, clear that ECR will not be a panacea for all companies. The improvement areas suggested in Figure 1 provide a tariff of initiatives from which companies will choose according to their own particular objectives. Each company will have a different starting point and a different agenda depending upon the current nature of supplier–retailer relationships. Nevertheless, a common theme applicable to all retailers is the limited number of relationships which are established with suppliers. The large grocery retailers deal with thousands of suppliers and have only formal partnerships or initiated pilot projects with a small number of suppliers. Winters (1998) details Somerfield's co-managed inventory trial with 12 of its suppliers and J. Sainsbury has 6-monthly supply chain forums which bring together senior supply chain staff with 19 of their counterparts (suppliers) which account for a large part of Sainsbury's volume business.

4. Differences in logistics "culture" in international markets

Clearly, ECR principles will be adopted by companies according to their own strategy with regard to supplier relationships. Table 1 shows that the Kurt Salmon report hoped for an improvement of supply chain time from picking line to consumer from 104 days to 61 days. A comparative study of European markets by GEA Consultia in 1994 shows that all of the major countries hold much less stock within the supply chain. Indeed, the U.K. figure is now around 25 days. Variations in Europe are quite marked between and within countries. Mitchell (1997) argues that few of the largest European retailers (mainly German and French companies) have proven to be ECR enthusiasts. Many of those French and German retailers are privately owned or franchise operations and they tend to be volume- and price-driven in their strategic positioning. By contrast, U.K. and Dutch firms are essentially publicly quoted, margin-driven retailers who have had a more constructive approach to supplier relations. While accepting that there are key differences in European markets, in general there are differences between the U.S.A. and Europe with regard to trading conditions. Mitchell (1997) states that:

(1) The U.S. grocery retail trade is fragmented, not concentrated as in parts of Europe.
(2) U.S. private label development is primitive compared with many European countries.

(3) The balance of power in the manufacturer–retailer relationship is very different in the U.S.A. compared with Europe.

(4) The trade structure is different in that wholesalers play a more important role in the U.S.A.

(5) Trade practices such as forward buying are more deeply rooted in the U.S.A. than in Europe.

(6) Trade promotional deals and the use of coupons in consumer promotions are unique to the U.S.A.

(7) Legislation, especially anti-trust legislation, can inhibit supply chain collaboration.

Fernie (1994, 1995), cites the following factors to explain these variations in supply chain networks:

(1) the extent of retail power,
(2) the penetration of store brands in the market,
(3) the degree of supply chain control,
(4) the types of trading format,
(5) the geographical spread of stores,
(6) the relative logistics costs,
(7) the level of information technology development, and
(8) the relative sophistication of the distribution industry.

These eight factors can be classified into those of a relationship nature (the first three) and operational factors. Clearly, there has been a significant shift in the balance of power between manufacturer and retailer during the last 20–30 years as retailers increasingly take over responsibility for aspects of the value-added chain, namely product development, branding, packaging, and marketing. As merger activity continues in Europe, retailers have grown in economic power to challenge their international branded manufacturer suppliers. While there are different levels of retail concentration at the country level, the trend is for increased concentration even in the southern European nations, which are experiencing an influx of French, German, and Dutch retailers.

By contrast, Ohbora et al. (1992) maintain that this power struggle is more evenly poised in the U.S.A., where the grocery market is more regional in character, enabling manufacturers to wield their power in the marketplace. This is changing, however, with a spate of merger activity which started in the late 1990s (Wrigley, 1998) and the roll out of Wal-Mart's supercenters, which will give the U.S. market a *national* grocery presence.

Commensurate with the growth of these powerful retailers has been the development of distributor labels. This is particularly relevant in the U.K. where supermarket chains have followed the Marks & Spencer strategy of strong value-added brands that can compete with manufacturers' brands. In the U.S.A., own-label products accounted for 15% of sales in U.S. supermarkets (Fiddis, 1997). This will change, however, with the drive by Wal-Mart to link its supercenter format and own-label strategy in addition to the expansion by European retailers such as Ahold and J. Sainsbury, which have high own-label penetration in their domestic markets.

The net result of this shift to retail power and own-label development has meant that manufacturers have been either abdicating or losing their responsibility for controlling the supply chain. In the U.K. the transition from a supplier-driven system to one of retail control is complete compared with some parts of Europe and the U.S.A.

Of the operational factors identified by Fernie (1994), the nature of trading format has been a key driver in shaping the type of logistics support to stores. For example, in the UK the predominant trading format has been the superstore in both food and specialist household products and appliances. This has led to the development of large regional distribution centers (RDCs) for the centralization of stock from suppliers. In the grocery sector, supermarket operations have introduced composite warehousing and trucking, whereby products of various temperature ranges can be stored in one warehouse and transported in one vehicle. In Holland, Albert Heijn has utilized cool and ambient warehouse complexes to deliver to their smaller-sized supermarkets, whereas the German and French retailers have numerous product category warehouses supplying their wide range of formats (with hypermarkets, depending on spread of stores, products may be delivered direct by suppliers).

The size and spread of stores will therefore determine the form of logistical support to retail outlets. Geography also is an important consideration in terms of the physical distances products have to be moved in countries such as the U.K., The Netherlands, and Belgium compared with the U.S.A. and, to a lesser extent, France and Spain. Centralization of distribution into RDCs is more appropriate to urbanized environments where stores can be replenished regularly. By contrast, in France and Spain some hypermarket operators have few widely dispersed stores, often making it more cost-effective to hold stock in store rather than at an RDC.

The question of a trade-off of costs within the logistics mix is therefore appropriate at a country level. Labor costs permeate most aspects of the logistics mix – transport, warehousing, inventory, and administration costs. Not surprisingly, dependence on automation and mechanization increases as labor costs rise (the Scandinavian countries have been in the vanguard of innovation here because of high labor costs). Similarly, it can be argued that U.K. retailers, especially grocery retailers, have been innovators in ECR principles because of high inventory costs, and high interest rates in the 1970s and 1980s. This also is true of land and property costs. In Japan, the U.S.A., and the Benelux countries the high cost of retail property acts as an incentive to maximize sales space and minimize the carrying of stock in store. In France and the U.S.A. the relatively lower land costs lead to the development of rudimentary warehousing to house forward-buy and promotional stock.

In order to achieve cost savings throughout the retail supply chain, it will be necessary for collaboration between parties to implement ECR principles. The "enabling technologies," identified by Coopers & Lybrand (1996) are available, but their implementation is patchy both within and between organizations. For example, McLaughlin et al. (1997), in their study of US retail logistics, comment that 40% of order fulfillment problems are a result of miscommunications between retail buyers and their own distribution center personnel. However, they anticipate much greater use of technologies in the early 2000s, with electronic data interchange (EDI) exhibiting most growth.

In Europe there has been some progress in achieving greater information flow between supply chain partners with the establishment of an ECR Europe EDI project team in 1995. The team has built upon its initial work on value added networks (VANs) to focus more on the role of the internet as a communications medium for EDI traffic. Nevertheless, these technologies require large user communities to achieve critical mass to make the technologies cost-efficient. The U.K. leads the rest of Europe with communities of 1000 or more, and the recent introduction of internet-based information exchanges by three of the main grocery retailers (Tesco, J. Sainsbury, and Safeway) allows them to respond immediately to fluctuations in demand.

Even in the U.K. there are technical obstacles to information flows because of the lack of acceptance of common standards. For example, the three grocers cited above have all gone their own way in developing their internet exchange systems. But even greater problems remain to be solved concerning the data, which need to be aligned in order that barcode data are consistent in databases throughout the supply chain.

One area of collaboration that is often overlooked is that between retailer and professional logistics contractors. The provision of third-party services to retailers varies markedly by country according to the regulatory environment, the competitiveness of the sector, and other distribution "culture" factors. For example, in the U.K. the deregulation of transport markets occurred in 1968, and many of the companies which provide dedicated distribution of RDCs today were the same companies which acted on behalf of suppliers when they controlled the supply chain two decades ago. Retailers contracted out because of the opportunity cost of opening stores rather than RDCs, the cost was "off-balance sheet," and there was a cluster of well-established professional companies available to offer the service.

The situation is different in other geographical markets. In the U.S.A., in particular, third-party logistics is much less developed, and warehousing is primarily run by the retailer while transportation is invariably contracted out to local hauliers. Deregulation of transport markets has been relatively late in the U.S.A., leading to more competitive pricing. Similarly the progressive deregulation of EU markets is breaking down some nationally protected markets. Nevertheless, most European retailers, like their U.S. counterparts, tend to only contract out the transport function. Compared with the U.K., the economics of outsourcing is less attractive. Indeed, in some markets a strong balance sheet and the investment in distribution assets is viewed more positively than in the U.K.

In recent years, however, the role of logistics service providers has been enhanced. This can be attributed to the internationalization of retail and transport businesses and the need for greater co-ordination of supply chain activities. The supply chain is now more complex than before. Retailers are optimizing traffic loads to minimize empty running and are backhauling from suppliers and recovering packaging waste from recycling centers. As efficient replenishment initiatives are implemented, consolidation of loads is required within the primary distribution network. Logistics service providers are better placed to manage some of these initiatives than manufacturers or retailers. Furthermore, the internationalization of retail business has stretched existing supply chains, and third-party providers can bring expertise to these new market areas. Some British companies, especially Marks & Spencer, have utilized U.K. logistics companies as they opened stores in new markets, for example the establishment of RDCs in Paris and Madrid with Exel Logistics. Similarly, the world's largest retailer has utilized the expertise of a U.K. logistics company to provide logistical support to stores acquired in Canada and Germany. Wal-Mart announced in March 1999 that SCM Europe, a subsidiary of Tibbet & Britten, would handle its logistical support to the 95 stores acquired through two acquisitions in late 1997 and 1998.

5. The internationalization of logistics practice

The gist of our discussion on differences in logistics cultures was to show that implementation of best practice principles has been applied differentially in various geographical markets. Nevertheless, the impetus for internationalization of logistics practice has been achieved through the formal and informal transfer of "know-how" between companies and countries. ECR Europe conferences, their sponsoring organizations, and national trade associations have all promoted best practice principles for application by member companies. Many of the conferences initiated by these organizations have included field visits to state-of-the-art distribution centers to illustrate the operational aspects of elements of ECR. At a more formal level, companies transfer "know-how" within subsidiaries of their own group or through formal retail alliances.

Schaumburg (1998) discusses how Ahold subsidiaries (five in the U.S.A.) benefit from voluntary synergy groups which use intranets to transfer best practices between operating companies. However, if companies have been recently acquired, the cross-cultural transition period can take time. Watson (1998), commenting upon the Shaw's operation, illustrates how it services a third fewer stores than the Sainsbury operation *but* has twice the number of items from a third fewer suppliers. She comments that best practice is not about transferring but sharing – ideas, processes, and people.

Another method of transferring "know-how" is through retail alliances. Throughout Europe, a large number of alliances exists, most of which are buying groups (Robinson and Clarke-Hill, 1995). However, some of these alliances have been promoting a cross-fertilization of logistics ideas and practices. In the case of the European Retail Alliance, Safeway in the U.K. has partnered with Ahold of The Netherlands and Casino in France. In 1994 a "composite" distribution center was a U.K. phenomenon; now, composites have been developed by Safeway's European partners. These logistics practices have not only been applied in France and The Netherlands but in the parent companies' subsidiaries in the U.S.A., Portugal, and Czechoslovakia. Harvey (1997) comments that "in the space of three years they caught up seven." Not surprisingly, the exploitation of U.K. retail logistics expertise has enabled distribution contractors to penetrate foreign retail markets, not only in support of British retail companies entry strategies but for other international retailers. Harvey (1997) argues that the success of his company (Tibbett & Britten) and other U.K. logistics specialists can be derived from the success of the fast-moving consumer goods sector but, like UK retailers, the success for the future lies with global opportunities.

6. The future

With Ahold and J. Sainsbury expanding in the U.S.A., and Tesco, Metro, Carrefour/Promodes, Kingfisher, and Wal-Mart targeting European markets, a degree of convergence of logistics practice is inevitable. The advent of the Euro and other harmonization measures in the EU make greater Europe an attractive market for these "supergroups." Further consolidation of retail markets is predicted, and opportunities to synchronize buying across chains will occur.

This discussion has centered around traditional supply chains. Clearly, much speculation exists about the nature and form of e-commerce, and the impact which non-traditional forms of retailing will have on existing supply chains. Despite the hype and the meteoric rise and fall of e-commerce shares in the stock market, internet or digital TV shopping has still to make much of an impact on overall retail sales. That is not to say that a sizeable market is unlikely in the next 10 years or so.

As a result of this uncertainty, traditional retailers are experimenting with the technologies to position themselves in the market if e-commerce does take off. To date, with the exception of accepted e-commerce retailers such as Amazon.com., Le Shop, and Value Direct, conventional retailers have tended to make incremental adjustments to their existing logistics networks. Thus, grocery retailers have picked from stores and delivered to the home or office. Some European retailers such as Albert Heijn, Delhaize, and Asda have undertaken order fulfillment from centralized facilities as opposed to from stores but offer a more restricted range.

With regard to non-food retailers, the transition to e-commerce networks would be easier because many home delivery networks already exist. In Germany a high proportion of frozen food is currently delivered to the home. As Germany also houses the world's largest mail order companies, opportunities to utilize an existing logistics infrastructure are evident. In cases where store-based retailers reposition themselves to supply this emerging market, collaboration with mail order companies is a possibility. Indeed, Arcadia has reached agreement with Littlewoods in the U.K. to access its logistics infrastructure and to collaborate in their mail order catalogues. The evidence to date, however, would appear to suggest that major changes to the existing infrastructure are a long way in the future.

References

Coopers & Lybrand (1996) "European value chain analysis study – final report", ECR Europe, Utrecht.

Christopher, M. (1997) *Marketing logistics*. Oxford: Butterworth-Heineman.

Christopher, M. and H. Peck (1998) "Fashion logistics", in: J. Fernie and L. Sparks, eds., *Logistics and retail management*, chap. 6. London: Kogan Page.

Fernie, J. (1994) "Quick response: an international perspective". *International Journal of Physical Distribution and Logistics Management*, 24(6):38–46.

Fernie, J. (1995) "International comparisons of supply chain management in grocery retailing", *Service Industries Journal*, 15(4):134–147.

Fiddis, C. (1997) *Manufacturer–retailer relationships in the food and drink industry: strategies and tactics in the battle for power*. London: FT Retail & Consumer Publishing, Pearson Professional.

GEA Consultia (1994) *Supplier–retailer collaboration in supply chain management*. London: Coca Cola Retailing Research Group Europe.

Harvey, J. (1997) "International contract logistics", *Logistics Focus*, April, 2–6.

Jackson, P. (1998). "Taking coals to Newcastle: the movement of clothing to Hong Kong and the Far East", in: J. Fernie and L. Sparks, eds., *Logistics and retail management*, chap. 4. London: Kogan Page.

Kurt Salmon Associates (1993) *Efficient consumer response: enhancing consumer value in the supply chain*. Washington, DC: Kurt Salmon Associates.

McLaughlin, E.W., D.J. Perosio and J.L. Park (1997) *Retail logistics & merchandising: requirements in the year 2000*. Ithaca: Cornell University.

Mitchell, A. (1997) *Efficient consumer response: a new paradigm for the European FMCG sector*. London: FT Retail & Consumer Publishing, Pearson Professional.

Ohbora, T., A. Parsons and H. Riesenbeck (1992) "Alternative routes to global marketing", *The McKinsey Quarterly*, 3:52–74.

Porter, M.E. (1985) *Competitive advantage*. New York: Free Press.

Robinson, T. and C.M. Clarke-Hill (1995) "International retailing trade and strategies", in: P.J. McGoldrick and G. Davies, eds., *International alliances in European retailing*. London: Pitman.

Schaumburg, T. (1998) "Best practice in logistics", Conference paper at CIES/FMI Global Logistics Conference, Boston.

Watson, J. (1998) "The successes and challenges of transferring best supply chain practices between Europe and North America", Conference paper at CIES/FMI Global Logistics Conference, Boston.

Winters, J. (1998) "Effective implementation of co-managed inventing (CMI)", in: J. Fernie and L. Sparks, eds., *Logistics and retail management*, chap. 7. London: Kogan Page.

Wrigley, N. (1998) "European retail giants and the post-LBO reconfiguration of US food retailing", *International Review of Retail, Distribution and Consumer Research*, 8(2):127–146.

Chapter 25

CITY LOGISTICS AND FREIGHT TRANSPORT

RUSSELL G. THOMPSON
University of Melbourne

EIICHI TANIGUCHI
Kyoto University

1. Introduction

The movement of goods within urban areas is vital since cities are the center of economic and social life in modern civilization. However, freight movement often puts a considerable strain on urban transport infrastructure and imposes high social costs in terms of crashes and environmental intrusion. Congestion levels in urban areas are rising as a result of increased urbanization and car usage. Cities are now facing global competition for investment, and trade with an efficient transport system is necessary for sustained economic prosperity.

The holistic goals of an urban freight system are to enhance economic performance while minimizing the adverse social and environmental effects. These generally involve decreasing costs and increasing efficiency while containing environmental damage and social intrusion as well as ensuring high levels of safety (Hicks, 1977). This chapter initially identifies the key stakeholders involved in urban freight and then describes city logistics – an integrated approach to urban freight problems. A summary of major issues for each key stakeholder involved in urban goods movement is presented as well as a number of case studies describing several effective city logistics initiatives that have already been implemented.

2. Key stakeholders

There are a number of participants and key stakeholders involved in urban freight transport, including:

(1) shippers and receivers (customers),
(2) freight carriers,

Handbook of Logistics and Supply Chain Management, Edited by A.M. Brewer et al.
© *2001, Elsevier Science Ltd*

(3) residents, and
(4) administrators (government).

Each group has different roles, problems, and objectives. With such a range of disparate views and conflicting interests it is difficult to find solutions that are acceptable to each group. The consumers' paradox, where customers are demanding more convenient retail facilities and access to markets as well as becoming increasingly concerned about the presence of trucks in local areas, epitomizes the increasing pressures being placed on urban freight systems.

3. City logistics

Urban freight transport continues to face many challenges. Freight carriers are expected to provide higher levels of service within the framework of just-in-time (JIT) transportation systems with lower costs. Congestion levels on urban roads have been constantly rising due to the increasing levels of traffic demand. The environmental problems caused by traffic have become major issues in many cities. Large trucks produce a substantial amount of air pollution in urban areas by emitting NO_x, suspended particle material (SPM), and other gaseous or airborne pollutants. Energy conservation is also an important issue, not only because of the limited amount of natural resources available but also for reducing CO_2 emissions to limit global warming. Traffic accidents involving large trucks often produce serious impacts for the community. This section introduces the concept of city logistics and outlines some common initiatives.

The concept of "city logistics" (e.g., Ruske, 1994; Kohler, 1997; Taniguchi and van der Heijden, 2000) has the potential for solving many of these difficult and complex problems. City logistics can been defined as "the process for totally optimizing the logistics and transport activities by private companies in urban areas while considering the traffic environment, the traffic congestion and energy consumption within the framework of a market economy" (Taniguchi and Thompson, 1999). The aim of city logistics is to globally optimize logistics systems within an urban area by considering the costs and benefits of schemes to the public as well as the private sector. Private shippers and freight carriers aim to reduce their freight costs while the public sector tries to alleviate traffic congestion and environmental problems. City logistics usually includes one or more of the following initiatives:

(1) advanced information systems,
(2) co-operative freight transport systems,
(3) public logistics terminals,
(4) load factor controls, and
(5) underground freight transport systems.

Advanced information systems have become more important in rationalizing existing logistics operations. Several researchers have investigated co-operative freight transportation systems (Ruske, 1994; Taniguchi et al., 1995; Kohler 1999; Nemoto, 1997; van Duin and Jagtman, 1999) that allow a reduced number of trucks to be used for collecting or delivering the same amount of goods. Public logistics terminals located in areas surrounding a city can be helpful in promoting co-operative freight transportation systems (Janssen et al., 1991; Taniguchi et al., 1999). Load factor controls are relatively new initiatives, and two European cities (Copenhagen and Amsterdam) introduced a certificate system for freight carriers who deliver or collect goods with the central city areas in 1998. Underground freight transport systems are innovative solutions for urban freight transport problems (Koshi et al., 1992; De Boer, 1999; Ooishi and Taniguchi, 1999). Private companies with varying degrees of support provided by the public sector usually operate city logistics initiatives. To realize the full potential of city logistics initiatives it is therefore crucial that an effective partnership between both the private and public sector is established and maintained.

Quantification of the consequences of city logistics initiatives is necessary for their evaluation and planning in urban areas. Predicting the impacts of city logistics initiatives for evaluation purposes requires modeling to be undertaken. Models should describe the behavior of each key stakeholder. They should also incorporate the activities of freight carriers including transporting and loading/unloading goods at depots or customers. Models must also describe the traffic flow on urban roads of freight vehicles as well as passenger cars. Models are also required to quantify the changes in costs of logistics activities, traffic congestion, emissions of hazardous gases, noise levels, etc., after implementing city logistics initiatives.

Modeling city logistics is a challenging exercise, because there are many complicated logistics activities for each of the stakeholders as well as many different evaluation criteria available for assessing the impacts of city logistics initiatives (Taylor and Button, 1999; Visser and Binsbergen, 1999). Therefore, the modeler must be very careful about what sorts of activities of stakeholders should be considered and what sort of evaluation criteria should be predicted. In addition, modeling transport on road networks is an important component of city logistics models. Freight vehicles represent only a part of total traffic on the road network. City logistics models need to consider both freight and passenger vehicles and focus on the impacts produced by freight vehicles. This requires a separate treatment for both freight and passenger vehicles in the formation of origin and destination matrices and traffic assignment procedures.

The evaluation of the city logistics initiatives is essential. There are many evaluation criteria, including the financial costs, hazardous gas emissions, congestion levels on urban roads, energy consumption, level of service for customers, and the labor force. Evaluation criteria for each stakeholder must be determined. As city logistics aims to totally optimize the urban freight systems, we need to incorporate the multiple evaluation criteria for examining the city logistics initiatives. To achieve this, multi-objective optimization models and multi-agent models will be useful. For example, Yamada et al. (1999) applied multi-objective programming method to identify the optimal location of logistics terminals. van Duin et al. (1998) studied multi-stakeholder models for decision support. These attempts help consider multiple evaluation criteria from multiple stakeholders viewpoints of city logistics.

3.1. Residents

Residents who live, work, and shop in the city do not welcome large trucks using local roads, nevertheless these vehicles carry commodities that are necessary for them. They would like to minimize traffic congestion, noise, air pollution, and traffic crashes near their residential and retail areas. Within the commercial zones of urban areas, retailers want to receive their commodities at a convenient time. However, this sometimes conflicts with residents who desire quiet and safe conditions on local roads. Large trucks using local streets can substantially reduce the amenity of residential areas.

Some cities have implemented regulations for limiting the maximum size of trucks (for example, 7.5 tons in total weight) that can enter inner city areas. As a result, more small trucks entered the cities instead of large trucks, and traffic congestion increased. To solve this dilemma, some measures have been proposed and implemented.

The idea of time windows has been widely applied for trucks entering central areas of cities for picking up/delivering goods at customers' premises. Setting time windows requires co-ordination among residents, retailers, and freight carriers.

Another initiative is controlling the load factor of trucks. This measure has been tested in Copenhagen and Amsterdam since 1998. Trucks are permitted the enter city center only if their load factor exceeds a certain level, for example 80%. In Copenhagen this type of truck is labeled with a green tag, implying environmental friendliness to residents.

A new co-operative system of electric vans started in Osaka, Japan, in 1999. With this system an organization provides electric vans at various parking places to be used co-operatively by many companies for carrying goods to customers. Users can pick up an electric van at a parking place and leave the van at the nearest parking place to their destination, returning to the office by subway or bus. This system can reduce the number of trucks traveling without carrying any goods. Electric vans themselves are better for the environment than the normally used gasoline or diesel pick-up trucks.

To achieve a good environment for residents, the public sector is becoming more actively involved in co-ordinating the conflicts among residents, retailers, and freight carriers. Kohler (1999) reported that a round table meeting of these stakeholders was helpful to discuss problems and implement city logistics initiatives in Kassel. Establishing such a platform for discussing urban freight and environmental problems is crucial before implementation. Moreover, public financial support permits testing and trials of innovative ideas, such as the experiment involving electric vans in Osaka.

3.2. Shippers and receivers

The demand for goods has become increasingly complex. JIT manufacturing requires minimum inventory levels, resulting in more frequent but smaller deliveries. Often inventory levels are kept low to help minimize storage costs (see Chapter 12). Contemporary retailing often involves niche marketing with high stock turnovers requiring regular deliveries. Narrow time windows specified by shippers and receivers has led to increased numbers of trucks being used to transport smaller consignments of goods. Economies of scale are lost, resulting in higher marginal transport costs.

There is increasing emphasis on quality issues as defined by shippers and receivers when goods are moved within urban areas. Reliability, security, and flexibility are key features of contemporary logistics requirements. This often leads to higher frequencies of deliveries, requiring sophisticated management and control techniques.

Another major issue is ease of access to retail outlets. There are often inadequate loading and unloading facilities due to poor building and planning regulations. In response, many cities have recently developed design standards to facilitate access to new retail and commercial premises. Loading zones used by cars often present accessibility problems for delivery trucks in retail and commercial areas.

The demand for urban freight is now seen as interdependent, with such considerations as warehousing, inventory control, information systems, production, and retail location. This adds more complexity to the distribution system and increases the challenges for the optimization of transportation costs. Supply chain management involves an integrated approach to logistics where the key elements of production and consumption are co-ordinated. Rapid response systems involves the application of integrated information technology. Recent developments in vehicle location systems (e.g., global position systems) and in-vehicle navigation systems allow real time demands to be satisfied. Advanced technology can also improve the security of goods and vehicles as well as increasing the reliability of deliveries (see Chapter 34). Loading bays or docks can be booked in advance using internet-based systems to help reduce delays at large terminals such as ports.

3.3. Carriers

Improved network conditions for freight vehicles has the potential to substantially reduce the cost of urban goods movement (Ogden, 1992). Designated truck routes that are designed and managed to allow trucks to travel efficiently can reduce the costs of goods distribution. In many cities the road geometry presents many challenges to drivers of large vehicles. Parked vehicles add to the delays. Often the operation of signal systems does not include consideration of the performance of trucks in their timing procedures. For example, trucks with their lower acceleration often experience larger delays compared with cars on arterial roads with co-ordinated signal systems.

The condition of road pavements can increase distribution costs, adding to delay and safety problems. Often, poor pavement conditions on curb lanes, vertical obstructions, narrow lanes, poles, signs, shop awnings, etc., can cause damage to vehicles and the goods being carried. This can result in increased operating costs as well reduce the quality of stock, particularly in the agricultural sector.

Inner-city retail and commercial areas often suffer from a lack of adequate loading and unloading facilities, either off-street or on-street. Dedicated loading/unloading facilities for trucks servicing large commercial, industrial, and retail developments are desirable. There are a number of benefits of adequate off-street loading facilities, including reduced costs of deliveries (i.e., drop time) and improved security for goods and vehicles. Inadequate off-street loading facilities often results in curbside loading and unloading that can lead to considerable road safety and efficiency problems for carriers and other road users.

How to increase the utilization of fleets is a major issue for operators. Often, there is an under-utilization of capacity, with most vehicles not carrying full loads. New technology (e.g., the internet) allowing booking systems for co-ordinating back-loads can help alleviate this problem. Transporting freight outside normal working hours, especially during the evening, offers some potential for increasing the utilization of vehicle fleets. Reduced travel times and fuel consumption would be experienced by freight operators. However, increased staff costs and security concerns for operators as well as noise impacts for residents considerably reduce the feasibility of night operations in many situations.

Third-party logistics operators are becoming more prevalent in urban areas. Specialized transport companies are typically being contracted to transport goods for large manufacturers and retailers. With this type of operation, manufacturers or retailers do not have to own and maintain a fleet of vehicles. This allows more sophisticated logistics systems and expertise to be developed.

Operators and drivers are often encouraged to enhance their safety and efficiency by self-regulation and education programs. The trucking industry often lacks the expertise to enable recent developments in computer-based vehicle routing and scheduling systems and automated monitoring systems to be implemented. However, this technology has the potential to substantially reduce operating costs. Desktop software can be used to increase the efficiency of fleets by incorporating optimization procedures that utilize geographical information systems to manage customer demands (e.g., location and time windows), vehicle fleets, and urban traffic networks. Due to the recent developments in meta-heuristic techniques, microcomputer software can now be used to provide decision support for logistics managers for many complex practical transport problems.

Drivers usually have to determine what traffic links to use when travelling through urban areas. Drivers when selecting routes within urban traffic networks have been found to be most sensitive to congestion, distance, and the number of turns. Owner drivers seem to be most sensitive to toll roads. Drivers of articulated vehicles generally place a greater emphasis on road width, road vertical alignment, turns, signals, and congestion than drivers of rigid vehicles.

Recent developments in vehicle technology have provided significant economic and safety benefits. More efficient engine performance and design (e.g., increased power-to-weight ratio and electronic fuel injection), braking systems (e.g., anti-lock), tires, and suspension systems have substantially increased the productivity of vehicle fleets.

3.4. Government

Public administrators within a city attempt to enhance the economic development of the city and increase employment opportunities. They also aim to alleviate traffic congestion, improve the environment, and reduce crashes. They play a major role in resolving any conflicts among the key stakeholders who are involved in urban freight transport, as they are not biased toward any of the parties. Therefore, the administrators have a major role to play in co-ordinating and facilitating city logistics initiatives.

The public sector has historically been involved in the provision of transport infrastructure and regulation of the freight industry. While these roles still exist to some degree, there are a number of new roles that are emerging and which are changing the relationship between governments and other key stakeholders.

Governments are becoming more actively involved in facilitating and promoting economic development. The integration of land use and transport planning is becoming accepted as a responsible approach by government. Goods movement is now being considered when determining land use zoning, densities, and the location of terminals.

The public sector also has a major role to play in the standardization and harmonization of regulations to reduce the costs of goods distribution. This extends to determining environmental standards using vehicle design rules to ensure that noise and emissions are acceptable and equitable. Road safety is also a responsibility of governments. This involves periodically reviewing vehicle design standards, licensing, and road geometry to reduce the number of crashes involving trucks. The major challenge in many countries is to develop policies and regulations that can reduce the environmental and social costs of road freight while not significantly increasing the costs for shippers and carriers.

The efficient management of the road network is vital to the economic well-being of an urban area. Governments are moving away from policies designed to meet unrestrained demands for urban road space to managing transport networks and constraining growing traffic volumes. Demand management techniques and intelligent transport systems offer high potential for increasing the capacity of existing road systems without having to construct additional infrastructure.

Contemporary urban transport planning requires a well-defined network of designated freight routes to be established. This involves a functional hierarchy be developed with primary and support routes that are designed and managed for freight vehicles. Traffic management techniques can be implemented to limit property access and provide priority for trucks using parking and parking controls. Geometric design standards for truck routes, incorporating adequate lane widths, turning circles, lateral clearance, and sight distance, can help facilitate the movement of large vehicles on these routes. However, remedial treatments are expensive and thus are often not implemented on a uniform scale.

Demand management measures can be implemented to reduce the effects of congestion, especially during peak periods. Trucks often avoid these peak periods at considerable costs. Road-pricing policies and ride-sharing schemes can be developed. The environmental capacity of roads determines the maximum volume of traffic that can be accommodated within a predefined upper limit of impacts. However, determining acceptable levels of annoyance levels for communities and developing cost-effective solutions is often a challenging exercise.

Many governments are establishing and co-ordinating forums for facilitating dialog between various stakeholders. This aids the identification of problems and issues for groups. Often, conflicts can be resolved and effective relationships developed, since governments are frequently limited in their ability to respond to needs of the carriers and shippers due to their lack of understanding of industry and community issues.

There is a need for the public sector to extend the "traditional economic analysis" undertaken as part of an evaluation program of major road proposals, since it involves only estimates of total benefits at consumers' own valuation of the services (travel) affected, as revealed by their demand behavior (road usage). This methodology involves comparing direct costs with expected direct benefits, which are savings in user costs and road maintenance costs, at a system level. This type of analysis is often called a "partial equilibrium" method as it takes out a small section of the economy and analyses it as if it were independent of other sections of the economy. It is "partial" because interactions with other traffic networks and regional economic and production growth are assumed to be constant rather than variables subject to change.

A broader level of economic analysis involves a system-wide or macro-economic analysis that includes the estimation of direct user benefits and benefits from wider economic growth (including off-road user benefits) and also traffic interactions across a whole road network to be taken into account. This level of analysis is rarely undertaken because of the complexity involved. However, significant regional and national benefits can often be gained by constructing a new road.

A range of secondary benefits are also likely to be realized once a new road is constructed. Secondary benefits are those benefits to road users or their customers not measured in standard road user impact relationships. These are an effect of a cost reduction in the product distribution chain. They would include the reduced damage to goods while in transit as well as the effects of fleet management and logistics management costs, not incorporated in the standard road user cost algorithms. Also, economies of scale through the use of more efficient vehicles and more efficient utilization of land for production purposes would be likely. Flow-on effects to consumers and industry in the form of employment growth and lower consumer prices of commodities are likely after the construction of a new road.

There is currently a lack of information relating to the existing freight movement patterns within many cities. However, this type of information is important to ensure that appropriate infrastructure is provided and that existing facilities are managed properly. Due to the complex nature of urban goods movement patterns, surveys are often expensive, and the data quickly become dated. Determining the frequency and nature of freight movement surveys that provides adequate information for planning, monitoring, modeling, and assisting vehicle operations is an important issue.

Governments also have a major role to play in encouraging experimental programs and trials, especially with city logistics schemes. This often involves establishing working groups and promoting involvement from industry as well as publicizing the outcomes.

3.5. Case studies

There have recently been a number of city logistics schemes successfully implemented in various cities. Since each city has its unique characteristics, often city logistics schemes have to be tailored for specific problems and the local environment. This section presents several case studies that highlight the potential of city logistics to address a number of complex issues for each key stakeholder involved in urban freight distribution.

Advanced information systems

Advanced information systems have become important in rationalizing existing logistics operations. In general, advanced information systems for pick-up/delivery truck operations have three important functions:

(1) to allow drivers and the control center to communicate with each other,
(2) to provide real-time information on traffic conditions, and
(3) to store detailed historical pick-up/delivery truck operation data.

The third function has not been fully discussed in the literature, but it is very important for rationalizing logistics operations. A Japanese milk-producing company experienced one successful application of historical operations data. After introducing a satellite-based information system for 1 year, the company was able to reduce the number of pick-up/delivery trucks by 13.5% (from 37 to 32 vehicles) and increase its average load factor by 10% (from 60% to 70%). A computer-based system was used to store detailed historical data of the pick-up/delivery truck operations, including start/arrival times at the depot and customer information, as well as waiting times, truck speeds, and routes traveled. The company was able to analyze these data and change its routes and schedules to substantially increase the efficiency of the vehicle fleet. This type of system can reduce both freight transport and environmental costs within a city.

Co-operative freight transport systems

Based on a survey by Kohler (1999), the co-operation between competitive freight carriers in delivering goods to the inner city of Kassel in Germany was investigated. In this system, a neutral freight carrier collects goods from five freight carriers and delivers them to shops in the inner city. After introducing this system the total time traveled by trucks was reduced, and queues of trucks for waiting to deliver goods on streets were also reduced. Originally this system started with 10 freight carriers; five companies now remain in this co-operative system.

Another outstanding case is the co-operative delivery system among 11 department stores in Osaka, Japan. Here, two department stores with depots adjacent to each other exchange their goods to be delivered in the neighborhood of the depot. This has led to a considerable reduction in travel time for trucks, personnel work hours, and total costs.

As observed in these two cases, co-operative freight systems can substantially reduce transport costs as well as environmental impact.

A new urban freight transport system co-operatively using electric vans has been tested in Osaka, Japan (Taniguchi et al., 2000). The main idea behind the system is that one organization provides electric vans at various public parking places to be used co-operatively by many companies. Tests have been conducted in the central area of Osaka City using 28 electric vans equipped with advanced information systems with the participation of 79 voluntary companies. The test system has operated well, without any serious problems. About 24% of users return the electric vans to a different parking place from the initial one after delivering goods to the customer, and used public transport (subway or bus) for their return journey. This behavior can contribute to reduced truck traffic without loads in urban areas and hence a better environment.

Public logistics terminals

A good example of a platform for city distribution can be seen in Monaco. This platform is provided by the government and operated by a private freight carrier for delivering goods to city areas. The company is subsidized by the government to provide a delivery service with lower prices than normal. This system helps reduce the required number of trucks used for deliveries. In Japan the first multi-functional logistics terminal is to be built in Seki near Nagoya. This logistics terminal is referred to as a "logistics town," and has various functions such as the trans-shipment of goods, assembling products during distribution, and provision of warehouses and wholesale markets. This project is being planned and executed by a group of companies from different industries with the support of the national, prefecture, and municipal governments.

Controlling load factors

Controlling the loadings of pick-up/delivery trucks is a relatively new initiative compared with conventional regulations such as vehicle weight limits, designated times for trucks to enter city centers, and the control of vehicle emissions. Two European cities (Copenhagen and Amsterdam) introduced a certificate system for freight carriers who deliver or collect goods within the central city areas in 1998.

In Copenhagen, only vehicles with a certificate (green sticker) are allowed to use public loading/unloading terminals in the inner city. This certificate can only be issued to vehicles satisfying the following two conditions:

(1) the load factor is over 60%;
(2) the vehicle is less than 8 years old.

Companies owning vehicles are required to produce a report on the load factors of their vehicles every month. To maintain certification, a company must have an average load factor during the previous month above 60%.

In Amsterdam, vehicles weighing over 7.5 tons are not permitted to use streets other than main roads unless a certificate is obtained. To obtain this certificate, vehicles must satisfy the following three conditions:

(1) the load factor is over 80%;
(2) the length of the vehicle is less than 9 m;
(3) the engine conforms to Euro II emission standards.

The police have the authority to inspect the load factor of vehicles. This initiative assumes that higher load factors produce a lower environmental impact.

Underground freight transport systems

Underground freight transport systems are innovative solutions for urban freight transport problems. Koshi et al.(1992) estimated the impact of building an underground freight transport system in the central area of Tokyo, Japan. The results of this study indicate that NO_x and CO_2 emissions would be reduced by 10% and 18%, respectively, energy consumption would be reduced by 18%, and the average travel speed would be increased by 24%. Ooishi and Taniguchi (1999) studied the economic feasibility of an underground freight transport system in Tokyo and concluded that this project would have an internal income rate of 10% if the infrastructure were to be constructed by the public sector.

The dual-mode truck (DMT) was developed and tested by the Public Works Research Institute of the Ministry of Construction in Japan. This new type of automated electric truck can travel both on a dedicated lane in an underground tunnel, using self-guidance and an external electricity supply, and on normal streets, driven manually using internal batteries. A similar idea was proposed in The Netherlands (De Boer, 1999), and the feasibility of underground freight transport system between Aalsmeer and Schiphol airport for carrying flowers was investigated. An automated self-guided truck named the "Combi-road" system was also developed and tested by a group of private companies for carrying large containers (12 meters) in The Netherlands.

References

De Boer, E. (1999) "Designing the structure and alignment of an underground logistic system; the case of the Amsterdam airport ULS", in: *Selected Proceedings of the 8th World Conference on Transport Research*, pp. 601–612. Oxford: WCTR.
Hicks, S. (1977) "Urban freight", in: D.A. Hensher, ed., *Urban transport economics*, Cambridge: Cambridge University Press.

Janssen, B.J.P. and A.H. Oldenburger (1991) "Product channel logistics and city distribution centers; the case of the Netherlands", in: *OECD Seminar on Future Road Transport Systems and Infrastructures in Urban Areas*, pp. 289–302. Paris: OECD

Kohler, U. (1997) "An innovating concept for city logistics", *Proceedings of the 4th World Congress on Intelligent Transport Systems (ITS)*, CD-ROM. Berlin: Vertico.

Kohler, U. (1999) "City logistics in Kassel", in: Cairns, E. Taniguchi and R.G. Thompson, eds., *City logistics I. 1st International Conference on City Logistics*, pp. 261–272. Kyoto: Institute of Systems Science Research.

Koshi, M., H. Yamada and E. Taniguchi (1992) "New urban freight transport systems", in: *Selected Proceedings of the 6th World Conference on Transport Research*, pp. 2117–2128. Lyon: WCTR.

Nemoto, T. (1997) "Area-wide inter-carrier consolidation of freight in urban areas", *Transport Logistics*, 1(2):87–103.

Ogden, K.W. (1992) *Urban goods movement*. Aldershot: Ashgate.

Ooishi, R. and E. Taniguchi (1999) "Effects and profitability of constructing the new underground fright transport system", in: E. Taniguchi and R.G. Thompson, eds., *City logistics I. 1st International Conference on City Logistics*, pp. 303–316. Kyoto: Institute of Systems Science Research.

Ruske, W. (1994) "City logistics – solutions for urban commercial transport by cooperative operation management", OECD Seminar on Advanced Road Transport Technologies, Omiya.

Taniguchi, E. and R.G. Thompson (1999) *Report on the 1st International Conference on City Logistics*. Kyoto: Institute for City Logistics.

Taniguchi, E. and R.E.C.M. van der Heijden (2000) "An evaluation methodology for city logistics", *Transport Reviews*, 20(1):65–90.

Taniguchi, E., T. Yamada and T. Yanagisawa (1995) "Issues and views on co-operative freight transportation systems", in: 7th World Conference on Transport Research, Sydney.

Taniguchi, E., M. Noritake, T. Yamada and T. Izumitani (1999) "Optimal size and location planning of public logistics terminals", *Transportation Research*, 35E(3):207–222.

Taniguchi, E., S. Kawakatsu and H. Tsuji (2000) "New co-operative system using electric vans for urban freight transport", *Urban Transport and the Environment for the 21st Century VI*. WIT Press.

van Duin, J.H.R., P.W.G. Bots and M.J.W. van Twist (1998) "Decision support for multi-stakeholder logistics", in: Proceedings of the TRAIL Conference, Scheveningen.

van Duin, J.H.R. and E. Jagtman (1999) "Best of both? Combining the modelling insights of Japanese and Dutch practices", in: E. Taniguchi and R.G. Thompson, eds., *City logistics I. 1st International Conference on City Logistics*, pp. 117–132. Kyoto: Institute of Systems Science Research.

Taylor, S. and K. Button (1999) "Modelling urban freight: what works, what doesn't work?", in: Cairns, E. Taniguchi and R.G. Thompson, eds., *City logistics I. 1st International Conference on City Logistics*, pp. 203–218. Institute of Systems Science Research.

Visser, J.G.S.N. and A. van Binsbergen (1999) "Urban freight transport policy and the role of modelling", in: E. Taniguchi and R.G. Thompson, eds., *City logistics I. 1st International Conference on City Logistics*, pp. 187–202. Kyoto: Institute of Systems Science Research.

Yamada, T. and E. Taniguchi and M. Noritake (1999) "Optimal location planning of logistics terminals based on multiobjective programming method", in: *Urban transport and the environment for the 21st Century V*, pp. 449–458. Kyoto: WIT Press.

Chapter 26

MANUFACTURING LOGISTICS

TONY WHITEING
University of Huddersfield

1. Introduction

It is often claimed that the essential principles of logistics and supply chain management can be applied to many different business processes across a wide range of industries, including service industries. However, most mainstream textbooks widely adopted for logistics and supply chain management teaching have focused primarily on manufactured goods. Coyle et al. (1996) and Bowersox and Closs (1996) are the classic examples of their genre. Through successive editions, these texts have evolved from their roots essentially in physical distribution management towards a more comprehensive and integrated coverage of supply chain activities. Recent editions contain far more material on inbound logistics, once the province solely of purchasing and supply texts. They also incorporate modern thinking on supply chain relationships (e.g., on partnership sourcing), and they set out the basic just-in-time business philosophy. What they fail to do is to provide thorough and rigorous treatment of the principles and methodology of modern manufacturing in the context of the supply chain.

As a consequence, students wishing to gain a more complete understanding of manufacturing principles and systems must make recourse to operations management texts such as those by Martinich (1997) and Dilworth (2000). In a similar vein, it is only comparatively recently that such books have started to explain production and operations management against the background of the overall supply chain (see, e.g., Waller, 1999). It remains the case that even recent texts with titles such as *Integral Logistics Management* (Schonsleben, 2000), which have their roots in the operations management camp, provide remarkably little information on physical distribution activities, notably transport and warehousing. A rare common thread running through texts, regardless of whether their origins lie in physical distribution, procurement, or operations management, is the extensive coverage of inventory management principles, although even here the perspective can vary significantly.

Handbook of Logistics and Supply Chain Management, Edited by A.M. Brewer et al.
© *2001, Elsevier Science Ltd*

Hence, while academics, practitioners, and consultants now preach the virtues of fully integrated supply chains, it is not easy to obtain a rounded view of how to achieve these from textbooks. It is important that this issue be addressed, because there is a growing need for the effective integration of manufacturing with other supply chain activities. Great strides have been made over the last 20 years in both physical distribution management and in procurement. There are a number of reasons for this advance, including the widespread adoption of information technology (IT), the pressure to reduce inventory levels, and the adoption of the total quality approach, particularly on the procurement side. Further progress in supply chain development and integration is therefore coming to depend on the more complete integration of manufacturing with both procurement and distribution. In a world characterized by increasingly global procurement, with major benefits from large-scale production (yet with a need to differentiate products in the eyes of the customer), and growing levels of traffic congestion, which make distribution service levels harder to maintain, the tensions are plain to see. The development of more flexible manufacturing systems, allowing the economic scheduling of smaller production batch sizes and increasing the opportunity for inventory reduction, adds a further impetus to the need for more integrated supply chain management. It is therefore timely to explain the principal elements of the manufacturing logistics toolkit, to review the key trends in manufacturing management, and to suggest the way forward for supply chain management in manufacturing.

2. Manufacturing in the demand responsive supply chain

While there is no doubt that many businesses have tried to take on board the need to be more responsive to the needs of the final customer, this is not to suggest that all such businesses have abandoned supply-push manufacturing and logistics systems in favor of demand-pull systems. Not all businesses are able to manufacture and distribute in line with known customer orders. An obvious example is where provision of production capacity adequate for the peak seasonal demand would be an unrealistic proposition.

Technological developments have in many cases increased the flexibility of manufacturing facilities, notably in batch production systems where rapid product changeover capability has allowed the scheduling of smaller batches. Production in smaller batches, closer in line with customer demand, now incurs relatively small cost and time penalties. In many industries, however, it remains the case that economies of scale require a large total throughput if unit costs are to be contained. Hence flexibility is a particularly important issue for "the manufacture of products which have been forced to the middle of the continuum between variety and volume" (Brown, 1996). In recent times, considerable attention has been paid to the scope for "mass customization" (Schonberger, 1996) and postponement techniques (Harrison et al., 1999). The aim here is to keep products or product families as standardized as possible during manufacturing, to be adapted to meet more precise customer preferences as late in the overall process as possible.

Even if production runs are scheduled in accordance with known customer orders, it is not necessarily the case that those manufacturers will pull in materials and components from their suppliers on a purely "as required" basis. Supplies may suffer from seasonal availability (although this is less of a problem in a world of global sourcing), or from seasonal price fluctuations. Suppliers may be unable to supply sufficient materials at short notice, or they may be reluctant to supply very small quantities.

The principle of pulling products and materials along the supply chain precisely as required for the next stage of the operation is a sound one that can provide major benefits, not least in terms of inventory reduction. The reality is that, in most cases, supply becomes uncoupled from demand somewhere upstream. The various possible locations for such uncoupling are usefully explored using decision point analysis (Hines et al., 2000). The realistic challenge is to push the point of uncoupling as far upstream as possible. Given the great efforts put into supplier development in recent years, it is probably fair to say that if uncoupling can be pushed upstream beyond assembly and manufacturing operations, then in many situations it should be possible to push it through major first-tier suppliers as well.

2.1. Just-in-time manufacturing

The scope for just-in-time (JIT) management has been discussed widely in the context of supply chain management in recent years (see Chapter 13). The origins of JIT are quite clearly in manufacturing management, notably in the Toyota manufacturing system introduced in the late 1960s, which incorporated, among other things, a pull scheduling technique based on kanban cards (Voss, 1987). Voss is quick to point out that JIT has evolved from these manufacturing origins, and in the context of modern business is an overall philosophy or approach to managing the business, rather than any one technique or even a set of techniques for manufacturing. His broad definition of JIT is "an approach that ensures that the right quantities are purchased and made at the right time and quality, and that there is no waste."

Major changes to procurement and distribution almost inevitably follow in the wake of JIT implementation, but it is manufacturing which is usually transformed most fundamentally. Voss highlights the need to facilitate and streamline the flow of production through the manufacturing system. Cellular manufacturing and group technology frequently replace traditional production-line working practices. Techniques for set-up and changeover time reduction make production in small batches easier and more economic. Scheduling is based on pull principles, ideally according to demand, rather than in order to produce for stock. Reduction in inventory levels, notably of work in process, but also of inbound materials and components, then leads to the desire to change the system of procurement. More precise quantities are ordered at shorter lead times. In certain situations the intention is to organize delivery direct to production or assembly locations, rather than into store. The removal of the inventory safety net inevitably leads to a sharp focus on the quality of materials supplied. Preferred supplier relationships or single sourcing systems often follow.

Successful implementation of JIT can give a wide range of benefits throughout the business, notably inventory reduction (particularly in work in process), set-up time reduction, lead-time reduction, space reduction, increased productivity, and improved quality. Despite these benefits, the problems associated with the introduction of JIT should not be underestimated. Implementation will require fundamental change in philosophy and outlook on the part of all concerned, as well as significant capital investment.

3. The tools of the trade

3.1. Materials requirements planning

Materials requirements planning (MRP) is a well-established technique that has been used since the 1960s to assist in the management of materials inventories and the procurement of supplies (see Chapter 13). MRP systems take the master production schedule (MPS) for a given period of time and identify materials requirements on the basis of the "bill of materials" (i.e., materials and parts required) for each product involved. If inspection of inventory records shows that materials stock on hand for any scheduled production is inadequate, orders are generated, bearing in mind the ordering and delivery lead times involved so that delivery can be achieved prior to commencement of production.

3.2. Manufacturing resource planning

By the late 1970s, MRP had been expanded and developed into manufacturing resource planning (MRP II), which has been described by Tincher and Sheldon (1995) as "one of the fundamental tools that many high performance businesses use today." They further describe MRP II as "a structured approach, a process way of thinking, and a formal way to manage a manufacturing company" (Tincher and Sheldon, 1995). In other words, in a similar vein to JIT, discussed above, MRP

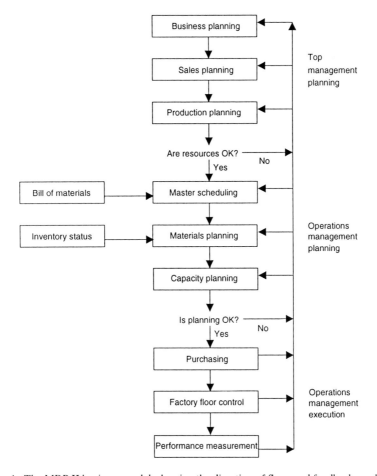

Figure 1. The MRP II business model, showing the direction of flows and feedback mechanisms (adapted from: Tincher and Sheldon, 1995).

II is to be regarded as a structured approach to the running of a manufacturing business, much more than a manufacturing technique per se.

Tincher and Sheldon set out MRP II as a "business model" (Figure 1). The core of the system is the operations management planning section. Here the master production scheduling (MPS) and materials planning, utilizing the bills of material and information on inventory already on hand, are essentially the same as in the MRP approach described above. Procurement is also likely to proceed as in the MRP approach. The MPS is usually a weekly plan detailing quantities of each item to be produced and intended batch sizes.

The MRP II approach, however, has a number of significant enhancements over and above the earlier MRP approach. The first of these is the explicit incorporation of top management planning. This stage starts with a statement of overall objectives in the business plan. Once this is in place, a sales plan is generated. This is a statement of intended sales by product line or family, probably on a yearly basis. The next step is the generation of a production plan, which is a broad statement of production resource requirements, again probably for the coming year. A second major enhancement is the explicit inclusion of capacity planning and capacity checks, to ensure that planned production can be achieved without undue bottlenecks. The operations management execution stage includes detailed factory planning, involving allocation of work to specific machines and labor, in order to provide at least a daily schedule to the shopfloor. Monitoring and reporting systems facilitate formalized performance measurement. Wallace (1990) identified two enhancements worthy of particular mention: the ability of MRP II to translate physical units into financial terms, and the ability to undertake simulation ("what if") analysis. Another enhancement is the extensive use of feedback mechanisms to ensure that plans are being achieved, that capacity is adequate but not excessive, and that the overall business targets are being met. Tincher and Sheldon (1995) make great play of the need to "close the loop." They classify formalized closed-loop implementations, which are used by top management to run the business and achieve 95% reliable performance in all elements, as "class A" MRP II.

MRP II is clearly a powerful tool for manufacturing companies but, as highlighted by Wallace (1990), implementation can be difficult. "Implementing MRP II properly requires a great deal of time and effort on the part of many people throughout the company. Data must be made more accurate, people must be educated and trained, new software must be acquired and installed, new policies and procedures must be developed and made operational, and on and on" (Wallace, 1990). Wallace also emphasized the need for implementation to be carried out by the company's own management, who must maintain a high level of commitment to the project. Attempts to use consultants and software suppliers for this purpose have frequently ended in failure.

3.3. The relationship between JIT and MRP/MRP II

MRP in particular predates JIT, and many companies have implemented this (and indeed MRP II) long before any consideration of JIT. In principle, MRP and MRP II can be applied in both supply-push and demand-pull environments, and hence are broadly compatible with the use of JIT. Voss (1987), for example, states that "MRP and JIT can mutually support each other," although he warns that "major adaptations must be made."

One likely change is in the nature of procurement. Traditionally, companies would use MRP to identify procurement requirements. Order sizes would, however, be based on economic order quantities, rather than on precise requirements. In other words, materials would be deliberately over-ordered, and would be received into store rather than sent directly to production. Hence the need to check stock on hand prior to ordering, to use up previously stockpiled materials. Under the JIT logic, the method of identifying requirements, based on intended production and the bill of materials, is essentially the same, but orders are more likely to be for precise requirements, subject to shorter lead times and for delivery to the point of use. In such circumstances, material quality assumes much greater significance.

3.4. Distribution requirements planning

Distribution requirements planning (DRP) employs logic essentially similar to that in MRP. In this case, the challenge is to minimize the distribution time between release from the factory and receipt by the customer. Success in this respect means that the customer's order is fulfilled, while at the same time production can be scheduled as late as possible on the factory floor. DRP techniques come into their own particularly in hierarchical distribution channels, where successive movement, handling, and storage operations can consume considerable amounts of time (Dilworth, 2000). The requirement to satisfy the demands of major multiple retailers in terms of "quick response" and efficient consumer response (ECR) is leading to increased interest in such methods from the manufacturing industry.

3.5. Enterprise resource planning

A more recent development, building on earlier systems such as MRP II and DRP, is the emergence of enterprise resource planning (ERP). Dilworth (2000) describes ERP as "an integrated information system that can record transactions throughout a company to enter an order, produce the order, ship the order and account for the payment" (Dilworth, 2000). In terms of supply chain management, therefore, ERP provides the potential to schedule all distribution, production, management, and procurement activities. The step forward with ERP is essentially one of width of coverage of activities, rather than any fundamental change in logic. In many ERP systems MPS, MRP, MRP II, and DRP are retained as the essential logic underlying the scheduling of production and the management of the procurement process.

The development of ERP has depended on the explosion of computing and IT, particularly in terms of information processing capability and data storage capacity. The information requirements for ERP can be immense, and rely on state of the art developments such as fourth-generation computer programming languages and massive relational databases or "data warehouses." The essential principle underlying ERP is that various IT applications are linked together, allowing company-wide communication and co-ordination. This is facilitated by the use of client-server rather than mainframe computer systems (Ptak and Schragenheim, 2000). Such systems also allow the sharing of data with key suppliers or major customers.

While ERP has been implemented successfully in manufacturing companies, a number of shortcomings can be identified. The first is that, despite the inclusion of scheduling tools such as MPS and MRP, ERP is essentially accounting and finance oriented. Enterprise-wide resource requirements calculated using ERP can be readily converted into costs. In other words, its primary role is in running the accounting side of the business. Another criticism is that computational limitations restrict the number of constraints that can be considered. Infeasible plans may be produced as a result. The sequential nature of the ERP process, through the various stages such as MPS and MRP, means that re-planning is a time-consuming exercise. Computing and IT limitations also reduce the scope for "what if" analysis. As a result of these shortcomings, ERP is more effective as an execution system than as a planning tool.

The scale of the implementation task and the complexities of the software required dictate that most ERP users purchase software developed by major suppliers, such as SAP/R3. As was the case with MRP II, many organizations find that this route to implementation can be fraught with difficulty and delay. One important issue relates to the extent of modification and customization that needs to be made to the "standard" ERP package. Education and training of those involved in the implementation and use of ERP systems are essential for success (Ptak and Schragenheim, 2000).

3.6. Advanced planning and scheduling systems

Further increases in the power of computing and IT are now allowing the problems associated with ERP to be addressed. In some recent releases of ERP, the conventional sequential suite of master production scheduling, requirements planning, and capacity planning modules has been replaced by advanced planning and scheduling (APS) systems (Langenwalter, 2000). APS systems take advantage of very powerful computer technology to incorporate optimization techniques for the scheduling and planning tasks. They are designed "to help manufacturers determine, then actually execute, optimal decisions" (Langenwalter, 2000).

Optimization at strategic, tactical, and operational planning levels is possible through the use of appropriate methods of linear or mixed-integer programming. Such optimization techniques, together with the ability to consider far more constraints, greatly increases the suitability of ERP as a planning as well as an execution tool.

APS systems can also be integrated with forecasting and demand management systems, customer order systems, production control systems, and transport, warehouse and inventory management systems, thus allowing ERP to be incorporated in powerful supply chain management systems. Langenwalter (2000) also makes reference to "supply chain execution" systems, which are the latest and most advanced APS systems. These allow optimization across several plants, depots, and warehouses, for example to determine which plant can manufacture the required output most profitably or which depot should be used in order to maximize distribution service levels. He goes on to suggest that "even more advanced APS systems integrate not only the manufacturer's plants, but also integrate suppliers and customers directly" (Langenwalter, 2000).

4. The way ahead

As we move into the 21st century, manufacturing industry faces an exciting and challenging time. Manufacturers face increasing customer service demands, especially from powerful multiple retailers. Time compression and the requirement for "quick response" and ECR are set to continue. There seems likely to be continued pressure for inventory reduction, not least in the grocery sector, where remaining shelf life is a major issue. Demand-pull logistics systems are likely to continue to increase in popularity. The need for effective management of manufacturing is therefore becoming critical.

As authors cited earlier in this chapter are at pains to point out, techniques such as MRP II and ERP are not simply ways to manage a manufacturing system. They are intended as supporting mechanisms for the wider management of a manufacturing company. Clearly, they are able to fulfil this role, although to a greater or lesser extent depending on a wide range of factors, not least the ability, enthusiasm, and commitment of all relevant stakeholders. What will no doubt become critical in the near future is the ability of ERP (or its successor) to break out from the boundaries of the individual organization, both upstream and downstream along the integrated supply chain. As suggested by Langenwalter (2000), the latest ERP systems incorporating APS are already starting to address this issue, although at very considerable cost in terms of computing capability. Further developments in computing and IT will allow more progress to be made.

One way ahead is the use of such tools across the "virtual" or "extended" enterprise. A feature of developments such as the "smart car" by MCC is that principal first-tier suppliers operate on the assembly site and undertake a very significant proportion of the overall engineering and assembly processes involved (Christopher, 1998). Systems that can integrate manufacturing and supply chain activities across the various members of the virtual enterprise are likely to be of very major benefit to all concerned.

What is less clear is how ERP can effectively be extended upstream and downstream in a more general sense. The stumbling block here is that large customers and suppliers will be involved in many (potentially competing) supply chains. A major supplier of, for example, electrical components may well be involved in the supply chains of most of the major car producers. Is it realistic to expect such suppliers to be linked directly to the ERP systems of all their major customers? What are the IT systems implications, for example?

Another concern is that the complexity and high cost of ERP may restrict its use to larger and wealthier companies. If this is the case, widespread use of ERP along integrated supply chains would be likely to disenfranchise smaller companies, particularly smaller suppliers and third-party logistics providers. The dominant position of major supply chain players would be reinforced as a result. Very significant reductions in the cost of computing and IT could offset this development, but reductions on a sufficient scale may be some time away.

The extent of change and upheaval likely to face manufacturers and their major supply chain partners as they struggle to implement ever more sophisticated systems should not be underestimated. Continual training and education is becoming a vital need. In addition, there is the issue of impersonalization – the organization seemingly controlled by the computer. Hence the organization of the future will need to pay much greater attention to staff retention and human resource management (Langenwalter, 2000).

On a more speculative note, the use of ERP-type systems to manage the integrated supply chain may start to affect the balance of supply chain power. The trend, at least in the U.K. in recent years, has been for multiple retailers to gain the upper hand at the expense of manufacturers. Sophisticated forecasting, purchasing, and IT systems employed by the major retailers are partly responsible for this development. More integrated supply chain planning, by definition, would suggest a more equal distribution of supply chain power, although this may be a somewhat naïve view. If manufacturers were more able and willing to become the driving force behind the specification of supply chain management systems (as for example they appear to be in the automotive sector), then the focus of supply chain planning and operation may return to manufacturing.

In conclusion, whatever the precise way ahead, it is clear that logistics and supply chain management in the manufacturing sector is becoming increasingly challenging. While the planning and control systems continue to grow in power, their complexity already calls for high levels of skill on the part of those responsible for their implementation and operation.

As a final point, perhaps the time is approaching when the textbooks will need to be rewritten. Such is the complexity of the modern integrated supply chain that any attempt to produce a text that comprehensively explains all supply chain activities will result in a book of encyclopedic proportions. Maybe a better way ahead is through succinct overviews of supply chain strategic management, leaving the detail to more specialized texts on manufacturing management, physical distribution, or procurement. It is notable that there have been few attempts to write the latter in recent years. Publishers and authors have chased the wider market generated by the growing interest in integrated supply chain management, and the less fashionable parts of the market, including manufacturing logistics, have received much less attention.

References

Bowersox, D.J. and D.J. Closs (1996) *Logistical management: The integrated supply chain process*. New York: McGraw-Hill.

Brown, S. (1996) *Strategic manufacturing for competitive advantage*. Hemel Hempstead: Prentice Hall.

Christopher, M. (1998) *Logistics and supply chain management: Strategies for reducing cost and improving service*, 2nd edn. London: Pitman.

Coyle, J.J., C.J. Bardi and C.J. Langley (1996) *The management of business logistics*, 6th edn. St Paul, MN: West Publishing.

Dilworth, J.B. (2000) *Operations management: Providing value in goods and services*, 3rd edn. Fort Worth: Dryden Press.

Harrison, A., M. Christopher and R. van Hoek (1999) *Creating the agile supply chain*. Corby: Institute of Logistics and Transport.

Hines, P., R. Lamming, D. Jones, P. Cousins and N. Rich (2000) *Value stream management: Strategy and excellence in the supply chain*. Harlow: Financial Times/Prentice Hall.

Langenwalter, G.A. (2000) *Enterprise resources planning and beyond: Integrating your entire organisation*. New York: St Lucie.

Martinich, J.S. (1997). *Production and operations management: An applied modern approach*. New York: Wiley.

Ptak, C.A. and E. Schragenheim (2000) *ERP: Tools, techniques and applications for integrating the supply chain*. New York: St Lucie.

Schonberger, R.J. (1996) *World class manufacturing: The next decade*. New York: Free Press.

Schonsleben, P. (2000). *Integral logistics management: Planning and control of comprehensive business processes*. New York: St Lucie.

Tincher, M.G. and D.H. Sheldon (1995) *The road to class a manufacturing resource planning (MRP II)*. Chicago: Buker.

Voss, C.A., ed. (1987). *Just-in-time manufacture*. Berlin: Springer-Verlag.

Wallace, T.F. (1990) *MRP II: Making it happen*. Essex Junction, VT: Oliver Wight.

Waller, D.L. (1999) *Operations management: A supply chain approach*. London: International Thomson Business Press.

Chapter 27

MARITIME LOGISTICS

MARY R. BROOKS
Dalhousie University, Halifax

GRAHAM FRASER
Kent Line International, Saint John

1. Introduction

In previous chapters of this book, logistics and logistics management have been discussed. General customer requirements drive carrier offerings, setting both the pace and the direction. The management of the supply chain has been well established by those outside the maritime industry. Logistics in the maritime field has not developed as might be expected from the progress seen in other modes of transport. Ocean carriers, in general, have lagged in adoption of supply chain activities. While ocean carriers now encounter demands quite different from those faced 10 years ago, they have not played a leading role in supply chain management (SCM) to the extent that many hoped.

This is not to say that ocean carriers have failed to focus on providing better services to their customers. For many years, the integration of logistics activities has occurred in some sectors of bulk shipping (e.g., tankers, ore, coal trades) as well as in trades using specialized carriers (e.g., gas carriers, car carriers, some refrigerated ships). Such integration has resulted largely from cargo owners and ship owners being one and the same, or because shipper–carrier alliances have promoted their development. Moreover, in the last few years, better integration has occurred in trades where control of the chain was seen as paramount; for example, grain is now transported by container in cases where the buyer wishes to ensure product integrity from source to end market (e.g., seed grains and organic grains (without genetic modification)).

This chapter focuses on the maritime logistics associated with container shipping. This is the primary means of transporting parts and finished goods, and thus that part of the transport market where SCM should be expected to be best developed. It does not focus on the logistics of moving goods by sea, or on those SCM issues dealt with in previous chapters. It contemplates strategic management

Handbook of Logistics and Supply Chain Management, Edited by A.M. Brewer et al.
© 2001, Elsevier Science Ltd

decisions that marine carriers must make when they choose the roles they will play (or not play) in the provision of maritime logistics services.

The chapter begins by identifying the roles carriers may choose to play in supply change management and the activities these roles entail. The choice of role may result in the carrier committing to active SCM participation. The drivers of change in carrier management strategy are then discussed, and the reasons why these activities may have been chosen by a carrier are elaborated. The changed logistics environment is illustrated by an example of what may happen in the trade between a developed country and an emerging market. The chapter closes with a discussion of the risks to be managed and other thoughts on the future of maritime logistics.

2. Possible carrier roles in supply chain management

As a carrier grows its business, many opportunities for new business development or acquisition abound. In assessing these opportunities, carriers may find themselves asking: Should the company expand its operation beyond the core business of providing ocean carriage into investment in terminals, inland carriage services, or other value-added activities? Should the company establish an operation to provide logistics services, which may or may not be at arm's length from ocean carrier operations? Should the company do both of these? Should the company invest in information technology (IT) to provide solutions to customer problems, eventually growing the IT part of the business as a core activity? Finally, should the company, if it establishes a logistics service business, take on the greater role of managing the supply chain on behalf of its key customers, or many of its smaller customers? If so, should the logistics business be operated at arm's length from the carriage business? None of the answers will necessarily be right, but in asking and answering these questions and in making choices about how to grow or shape the business the carrier must examine its competitive position in order to determine an appropriate strategy for its business.

There are three broad roles an ocean carrier can play in SCM:

(1) a *supplier of ocean carriage* (including contracting of terminal handling arrangements) to a supply chain managed by another company, usually a manufacturer;

(2) a *supplier of more than one link in the distribution network* (ocean carriage plus inland carriage; or ocean carriage plus value-added services) used by a supply chain manager; or

(3) a supplier of any of the above services and *manager of the supply chain* on behalf of client companies.

It is less likely today that the ocean carrier will choose a strategic direction based on the first of these roles, given the opportunities the others afford in terms

Box 1
Possible supply chain management activities

Product design
• Production planning and scheduling (agile production as possible output)
• Materials requirements planning (MRP)

Supplier relations
• Demand management (order acquisition and processing, and including sales forecasting)
• Inventory systems for each link in the chain
• Supplier and vendor choice (including transport supplier choice)
• Distribution requirements planning (including choice of channel functions to acquire effective customer response and to enhance reverse logistics if necessary)

Customer relations
• Transport documentation services
• Logistics information systems (design, management, integration; may include electronic data interchange (EDI))

---

of improving the carrier's operating margin. Drewry Shipping Consultants (1996) highlighted the poor financial performance of the industry, noting that operating margins in the 90s were prevalent for the largest of the container carriers. (Operating margins in the U.S. rail business, by comparison, were in the low 80s at the time.) Clearly, carriers could not continue the strategies of the past. More recently, some have bet the company on investments in new post-Panamax tonnage, hoping that economies of scale would grant needed operating margin improvements. Others, in the early 1990s, established logistics subsidiaries to broaden their service offerings. In the beginning, most of these subsidiaries were tied to the carrier's brand name, to build on existing brand equity. By the end of the decade, this was not necessarily the case, as some carriers sought to isolate the two types of business. The desire to separate activities arose, in part, because carriers discovered that buyers were unwilling to pay carriers a premium for logistics services. By isolating the activities, however, returns on logistics services could be maintained at higher levels (e.g., 7–9%) and not eroded to the 2–3% return being seen for purely ocean carriage services.

Once the carrier contemplates moving beyond simple ocean carriage, its analysis of its situation will identify opportunities for economies of scale and the firm-specific advantages it has to exploit in the market. To build a successful strategy, the carrier needs to exploit its advantages, be they access to superior IT solutions, owned terminal or warehouse facilities, or excellently managed land-side transport operations. The carrier determines the range of activities to offer in the context of its own expertise and competencies. The geographic scope over which they should be offered will also be identified. Box 1 identifies a sample of SCM activities that may be added to the carrier's offering of ocean services. Through the addition of any one or a combination of these activities, the carrier hopes to customize its competitive strategy in a way that will give it a unique competitive position judged by the market as worthy of a premium price. The key question is: What role should it play in the network and therefore what activities should it offer? Should this go so far as to include SCM?

Only some carriers will have the expertise needed to become supply chain managers. What is that expertise? It includes

(1) an understanding of rates and services,
(2) modeling capability to optimize the network to meet benchmark performance criteria,
(3) analytical skills for the extraction of management information generated by the network established, and
(4) production and consumer knowledge to ensure that all parties in the supply chain experience a successful outcome.

While the first three skills are competitive necessities in today's maritime industry, the last one makes it unlikely that a carrier, even one owning warehouses, inland transport companies, and terminals, can take on the role of supply chain manager.

Furthermore, even if a carrier wants to become the supply chain manager, the historical relationship between cargo interest (or its representative) and carrier may frustrate the effort and limit the role to distribution requirements planning, transport documentation services, and logistics service provision. Shippers traditionally viewed their relationships with carriers as based on transactions rather than as partnerships. While the relationship between carriers and cargo owners has migrated towards partnerships, for many cargo owners a residual distrust limits scope for change. The provision of transport services is an agency relationship; the carrier never owns the goods but provides a service necessary to change the location utility of the goods. The traditionalist would feel that, without ownership, the commitment necessary for successful SCM would not exist. However, many third party logistics services (called 3PLs in North America) are taking on the SCM role without taking ownership of the goods. Therefore, the role of ocean carrier as supply chain manager is not entirely implausible.

3. Drivers of change

According to Brooks (2000), an ocean carrier's choice of competitive position in the marketplace is usually a function of four intersecting drivers of change:

(1) alteration of the regulatory environment in which the carrier operates,
(2) changing trade or competitor dynamics,
(3) the carrier's internal assessment of the effectiveness of its existing strategy, and

(4) the preferences of its management in their implementation of strategy, including their assessment of the risks associated with the chosen strategy.

Ignoring the regulatory environment until later (see Section 4), there is no doubt that one of the most important drivers of an ocean carrier's strategy is the changing nature of customer demands. While the traditional buyer of ocean carriage was quite satisfied to engage transport services on a transactional basis, container by container, that customer must now compete with other manufacturers that have embarked on time-driven just-in-time (JIT) production and delivery systems. Also, many North American manufacturers spun off their in-house logistics operations in the late 1980s and early 1990s to focus on their core business operations. These divestitures enlarged the market for 3PLs and, at the same time, encouraged the customer to expect more from transport services, in terms of both operational performance and responsiveness. This led some carriers to explore questions about the extent of their participation in logistics activities beyond their core ocean business.

The new customer wants tight delivery schedules to be met, damage to be minimized, and performance to be monitored. Quality is the watchword. Efficient consumer response is driving the transport buyer's choice of carrier, routing, warehousing, and so on, and this may or may not be through a 3PL. The business climate for ocean carriers serving this part of the market has changed. As a result, carriers must improve their performance in an era of declining margins, increasing consolidation, and fleet reinvestment. New expectations strain old ways of doing business, but also afford new opportunities. The trend to out-sourcing all non-core activities on the part of manufacturers opened the door to global specialization. While the globalization of the trading environment has prompted some carriers to think global in their competitive positioning, it has also encouraged a very few to enter the logistics services market.

Inbound Logistics' annual survey of "third party" suppliers (Anon., 1999a) noted that the top 100 U.S. 3PLs in 1999 were predominantly non-asset based (56 of the 100) and offered electronic data interchange (EDI) (93 of the 100), SCM services (86 of the 100), JIT services (73 of the 100), and tracking and tracing (72 of the 100). A closer look at the 36 asset-based 3PLs offering intermodal services revealed the importance of those services: EDI (31 of the 36), SCM services (29 of the 36), JIT services (22 of the 36), and tracking and tracing (26 of the 36). Of these 36, however, only three (APL Logistics, Crowley Logistics, and SeaLand Logistics) were businesses related to marine carriers. All three indicated they offer SCM services.

By 2000, the number of asset-owning 3PLs in the top 100 offering intermodal services had grown to 40 (Anon., 2000). The 2000 listing looked at a changed set of activities and the term "supply chain management" was not used, although many of the SCM functions (as suggested by Box 1) were itemized. Of the 40, 11 3PLs offered internet-based SCM services. However, for almost all, the assets owned were tractors and trailers. Both APL Logistics and Crowley Logistics now categorized their services as non-asset based. Only one marine-related business considered itself asset based: Maersk Logistics, the entity resulting from the merger of SeaLand Logistics and Mercantile. Maersk Logistics offers the full set of logistics services surveyed by *Inbound Logistics* as well as internet-based SCM services.

Clearly, with the exception of Maersk, large ocean carriers serving the U.S.A. are not choosing to provide SCM services as part of their strategy. The first question is: Why not? The second question is: What are they doing instead? A large part of the answer to the first question lies in the risks perceived.

4. Risks to be managed

Part of the strategy formulation process a carrier undertakes will be driven by the perception of risk held by the carrier's management team and by those who own the cargo (and therefore are responsible for dealing with cargo damage in transit). Risk management is therefore a key element to be addressed in the design of the maritime logistics system, and the perception of risk by all parties to the SCM system must be acknowledged. There are several types of risks to be considered.

First, there is the risk arising from network complexity. As the number of players and service nodes increase, the potential for service failures also rises, possibly exponentially. According to Turnbull et al. (1992), JIT processes work best with a specific type of supplier relationship – the *kieretsu* of Japan, where suppliers and contractors are vertically aligned and the single supplier–customer relationship is prevalent. Then the supplier is not faced with juggling production to meet the needs of several buyers simultaneously. If the carrier becomes the SCM, there are numerous contact points to be secured, and the share of attention available for each is diminished. Reactive exception reporting may replace proactive vigilance, increasing the risk for the cargo owner.

Second, is the risk associated with demand volatility. Out-sourced logistics works best for the cargo owner when demand is relatively stable; volatility increases the pressure within the system for building buffer stocks. SCM is needed to address such volatility so as to optimize the system for all players. Without SCM, buffer inventories would of course be larger and less optimal. The optimal solution for the cargo owner may be out-sourcing or it may be in-house management of the supply chain.

As a result of demand volatility, the risk of rate instability is also ever-present. Carriers have reduced the risk of rate fluctuations through participation in conferences. Since the early 1900s, conferences have been granted antitrust immunity in most jurisdictions so that they may stabilize the supply of shipping on the trade routes serving the country granting the immunity. Conference members consider that such anti-trust immunity serves to stabilize rates and enable those rates to be raised with tightening demand. The regulation of container carriage, particularly that applying to conferences and consortia, has changed in the last decade and the changes have, for the most part, been subtle. The relevance here is that antitrust regulation in Europe has limited the ability of conference members to integrate inland and terminal activities under one umbrella without jeopardizing their antitrust immunity. Likewise, consortia regulation has frustrated integration. To date, for the majority of carriers, the protection afforded by their antitrust immunity status has left them less inclined to contemplate taking on an SCM role.

Another risk arises with the choice of supply chain manager. The strongest link rather than the weakest link should be the manager; it is not optimal to put the weakest link in the position of monitoring performance. Ocean shipping has certainly had difficulty managing its on-time delivery performance from a weather point of view on some trade routes, particularly the North Atlantic in winter. Furthermore, as the speed of ships increases, fewer ships are required to service the same demand. The result is that the margins for scheduling and schedule integrity become narrower; delay due to weather or other factors (such as labor strife) has a greater impact as it becomes obvious to the ship operator that time cannot be extracted from the system without port call omission. The knock-on effect in the worst case becomes the loss of a voyage with the consequential loss of many millions of dollars of contribution. (Although current multimodal rules for most routes have no "cost of delay," consequential business loss may be litigated successfully, impacting a carrier's insurance premiums for protection and indemnity insurance (Kindred and Brooks, 1997).)

There are also questions about the choice of supply chain manager from a conflict of interest point of view. While there are only a few ocean carriers engaged in SCM activities, and most have isolated their 3PL operations from their ocean shipping operations, many shippers and other 3PLs are concerned about the potential for conflict of interest between the two related businesses. Will Maersk Logistics specify its own related business over an alternate carrier that may provide superior services or better prices? This concern is a very real one to cargo interests.

Many of those involved in the supply chain are reducing the risks they see as inherent in either transport or supply chain contracting by tracking the performance of those providing the services. It is no longer uncommon to find that report cards are developed to monitor performance, be it performance of a transport supplier or a 3PL. The reduction of risk is particularly critical in a JIT network, be it manufacturing based or retail based.

Finally, there is the business risk as seen by investors in transportation enterprises – the impact of logistics on shareholder value and hence stock prices.

It has been pointed out recently (Anon., 1999b) that, on a shareholder value basis, non-asset-based logistics suppliers outperform asset-based logistics suppliers, reflecting the higher returns from managing information flows and the lower returns from pure transportation services.

With the exception of very few ocean logistics suppliers, SCM has remained the purview of large cargo interests, be they manufacturers with retail operations or time-driven suppliers. While the trade press has focused on those with a story to tell, the role of ocean carriers never appears to extend beyond the provision of third-party distribution planning, software solutions, or asset ownership land-side. Because carriers never acquire ownership of the goods, their strategic role is bounded. The risks to be managed and the skills base to undertake the activities inherent in SCM explain why it has not proven an attractive path for growth to many carriers.

In summary, risk management considerations have left manufacturers inclined to control the supply chain or contract its management to 3PLs; few carriers appear willing to offer the services, particularly given the skills required to deliver many of the activities noted in Box 1.

5. The implementation of the chosen strategy

The carrier's service strategy may or may not be executed in a standardized way across geographic markets. While the benefits of a standardized global strategy have been argued by many, the reality is that many businesses engage in local adaptation of global implementation plans. In addition, different strategies may be developed for different geographic markets (a multidomestic approach). For example, APL Logistics focuses its activities on dynamic IT applications that assist its customers to manage their supply chains. APL Logistics boasts of clients, such as DuPont, The Gap, GM, and Toyota, with global supply chain needs. However, through various related companies, such as APL Business Logistics Services and ACS Logistics, APL Logistics offers a varied menu of activities, but one that is not uniformly available across the globe (Davison, 1999).

The current global marketing literature (e.g., Johansson, 2000) cites very few examples of global execution without some local adaptation. Differences in execution result in part from the nature of the markets being served: mature, growth or emerging. The markets where SCM is being implemented include all three, yet much of what has been written about SCM takes a developed country perspective. While containerization and SCM have revolutionized the execution of trading activities, the technology and systems have not been implemented uniformly. In most developed countries, logistics solutions can include a full range of activities of the type noted in Box 1; this will not likely be true for emerging markets. Differing market conditions do not encourage similar execution, and logistics solutions developed at home may not be readily transferable to a new market. Local labor costs and practices may make highly productive, developed country approaches less relevant. The carrier needs to identify the resources needed, their availability, and the ability of local organizations to adapt to new ways of operating. Of course, the capital to invest in new approaches and the return on investment from them are critical considerations to the implementation of the carrier's strategy.

While the role of SCM in a mature market may not be feasible, a carrier may be in a good position to take on the SCM role in trades involving emerging markets given its position in the trade, at the middle. In developing countries, fragmentation of distribution networks is a reality not always understood by developed country manufacturers, paving the way for possible greater participation than experienced in a developed country to developed country trade. At the very least, the carrier may be in a position to introduce improvements to the existing transportation network to improve overall supply chain efficiency.

6. Carrier logistics strategy in a developed country to emerging country trade

In late 1999, Contship Containerlines Limited, a carrier established in the North European/Indian sub-continent trade since 1977, identified an opportunity to extend its presence by developing a separate service into the U.S. east coast. While the container trade from Europe was well established, with competing lines and consortia operating on both direct and indirect bases, the India/U.S. trade was only covered by a few relay operators. Through trade and customer research, Contship confirmed that there was a high degree of interest, particularly from the Indian sub-continent, for a direct service with superior transit times to U.S. east coast ports. It was determined that an effective port range could be maintained by operating four fast container ships on a biweekly schedule.

One of the key elements in the strategic decision about the opportunity was the traditional problem of suitable ship type. The U.S. trades generally operate with ungeared vessels, given the well-developed shore-side container-handling facilities. On the other hand, few Indian ports are as well developed, and feeder operations generally require geared vessels (having their own shipboard cranes). The use of geared vessels often incur double-handling and slower turnaround. Contship was anxious to take advantage of India's growing share of world trade and compete against slower operations using geared vessels with a higher speed service direct to a U.S. east coast port. The use of gearless ships was integral to the success of the opportunity.

Having investigated and established the viability of a service product, one that did not previously exist, Contship opened discussions with selected other shipping lines in the Indian trade that had either current or historical involvement with the

U.S.A. After a number of meetings with these prospective partners, Contship agreed with both The Shipping Corporation of India (SCI) and CMA CGM commenced this service; each party either directly provided gearless tonnage or guaranteed to slot charter from the others.

This first stage was important as it confirmed the involvement of SCI, India's national carrier, in the initiative and also provided an economically sound platform for the other two established carriers, who wished to grow geographically. Timing was a critical issue. It had become known in the trade that the leading relay operator was itself intending to establish a direct service, albeit likely not within the first half of 2000. This created, for Contship Containerlines and the other lines, additional incentive to act decisively and immediately. The initial participants were able to fix tonnage prior to a strengthening of rates in the charter market created by a reduced supply of suitable, fast, medium-sized ships.

A critical factor was the capability of both Mumbai and Colombo to handle gearless ships, thereby enabling Contship and its partners to offer a fast, direct service from the Indian sub-continent to the U.S. east coast. This meant that the Indian sub-continent port selection was straightforward, and reflected the limited choice available to berth gearless tonnage. Both Mumbai and Colombo were able to offer 24 hour a day operations, with handling rates of approximately 25 moves per crane-hour. Consider that as little as 5 years previously, operations for a similar number of container exchanges could have taken up to 10 days without dedicated terminal facilities.

While direct ports of call on the Indian sub-continent were never in question, in the U.S.A. there were many choices for the carriers, in terms of both geographical port calls and also terminals within those ports. The decision process created considerable debate, with selection criteria driven by differing logistical requirements, and operations undertaken by each of the lines. In the southern U.S.A., the choice was between the benefits of calling Charleston or Savannah. Although only 2 hours apart by road, Charleston was preferred for the westbound import trade, whereas the eastbound export trade favored Savannah. The final decision was to call Charleston because of the greater volume of Indian sub-continent export trade requiring this port of entry, as determined by both the location of many consignees and their greater degree of comfort when dealing with local customs authorities.

Among the partners, the choice of northern U.S. east coast ports of call generated the most debate. In New York, where a significant number of terminal operators are located, the decision had to be made between individual stevedoring and terminal operators and also their physical location within the Port of New York/New Jersey. Here some of the lines' distinctly different logistics operations polarized the consortium. Inasmuch as some were using New York only for the local market and hinterland, they were prepared to use more centrally located terminals in the New York/Brooklyn area. Another carrier was using New York for both the local hinterland and intermodal routing to the U.S. Midwest, and therefore required terminals to be located on the New Jersey side of the port adjacent to rail facilities. The carrier using New York as a local destination was, for U.S. Midwest traffic, using on-dock rail facilities at the service's second U.S. port of call, Norfolk, and could therefore accept both the location and cost advantage of a Brooklyn terminal. Eventually the lines agreed to differ in terms of intermodal operations with the service calling at both New York (Jersey side) and Norfolk, where separate intermodal operations were conducted.

This example illustrates how any particular trade or developments in liner services may differ on an import/export basis for both trade and carrier reasons. Different lines within the same consortium, as has been seen, often develop divergent strategies. The example is for a new container trade having only modest volumes and less demanding requirements than, for instance, a major east–west trade, such as the transpacific, with its larger number of carriers, significant shipper supply chain sophistication, and intense cost focus. Such trade dynamics will drive differing and competing carrier approaches.

7. The future of maritime logistics

The rise of 3PLs is a natural outgrowth of an increasingly specialized commercial environment where smaller shippers are able to convert in-house fixed costs of logistics management services into variable cost out-sourced activities, thereby reducing expenses and improving profitability. 3PL services themselves have both scale economies and processes available to partially pass on to smaller manufacturers who would otherwise not benefit from consolidation. For a very few ocean carriers, the decision to diversify into the related business of 3PL services has been a logical avenue for growth; others have chosen to grow differently. The question is, what are they doing instead?

What many are doing instead is making improvements in service provision so that the transport process is more seamless or more efficient, and engaging in alliances to better service their customers. This does not mean that ocean carriers will be forever only transport companies. The out-sourcing of activities en bloc affords the astute carrier seeking better profitability the opportunity to grow the business through acquisition of better returns for IT services. Management of information can be a competitive weapon that can be spun into value-added services for other carriers, as American Airlines has done with its Sabre technology, creating a now independent company. American Airlines' investment in Sabre (a computer reservation and revenue management system) helped American Airlines build customer solutions and entrap customers who liked the results. Perhaps the growth of logistics providers with roots in the ocean shipping business is a repetition of the Sabre experience. Logistics IT can be both a competitive weapon and a service differentiator, but this window of opportunity is closing fast as companies such as Nortel Networks, through its global logistics business unit, provides IT solutions and then out-sources the logistics operations, thereby taking a significant position in the business of SCM as a fourth party.

References

Anon. (1999a) "The top 100 3PLs", *Inbound Logistics*, July, 61–71.
Anon. (1999b) "The markets take notice", *Canadian Transportation and Logistics*, November/ December, 34–37.
Anon. (2000) "The top 100 3PLs", *Inbound Logistics*, July, 47–63.
Brooks, M.R. (2000) *Sea change in liner shipping: Regulation and managerial decision-making in a global industry*. Oxford: Pergamon Press.
Davison, D. (1999) "Logistics leader", *Containerisation International*, August, 49–51.
Drewry Shipping Consultants (1996) *Global container markets: Prospects and profitability in a high growth era*. London: Drewry Shipping Consultants.
Johansson, J.K. (2000) *Global marketing: Foreign entry, local marketing and global management*. Boston: Irwin McGraw-Hill.
Kindred, H.M., and M.R. Brooks (1997) *Multimodal transport rules*. The Hague: Kluwer Law International.
Turnbull, P., N. Oliver and B. Wilkinson (1992) "Buyer–supplier relations in the UK automotive industry: Strategic implications of the Japanese manufacturing model", *Strategic Management Journal*, 13:159–168.

Chapter 28

AIR-FREIGHT LOGISTICS

AISLING J. REYNOLDS-FEIGHAN
University College Dublin

1. Introduction

The air transport industry has gradually increased its share of global passenger and freight traffic, largely at the expense of rail transport (Button et al., 1998). This trend has accelerated in the last 30 years. For the past decade, air-freight traffic growth has outpaced air passenger traffic growth by 1–2% points each year. The air-freight sector has been transformed since the late-1970s. Prior to deregulation of air cargo in the U.S.A. in 1977, the industry offered limited products, with heavy reliance on several intermediaries and a significant dependence on air passenger operations. The industry can now be characterized as a sophisticated, innovative sector relying heavily on new electronic technologies and offering a wide range of transport and logistical products through dedicated specialist cargo operators. With increasing emphasis on the globalization of trade and economic activity, air-freight growth is expected to continue to outpace air passenger traffic growth and be greatest in Asian markets (i.e., intra-Asia, North America–Asia, Europe–Asia, Australasia), despite the recent economic crises in the region. The correlation between world gross domestic product (GDP) and world air-freight traffic forms the basis for traffic forecasts. Because of the cyclical nature of GDP growth, air-freight traffic growth is also subject to cyclical effects.

The process of physical distribution of freight has become a highly sophisticated operation, with increasingly greater reliance being placed on the use of new technology to assist in the movement, storage, and tracking of consignments. Transport is but one component in this logistics chain. In this chapter, the air-freight sector is examined in terms of its structure, organization, and role in the supply chains of shippers. Section 2 looks at the organization and structure of the air-freight sector, identifying the main groups of players and their distinguishing characteristics. Section 3 examines the main issues influencing pricing in the sector. This is followed in Section 4. by an overview of the main trends in the industry in the recent period, which are heavily influenced by U.S. domestic market trends. The final section looks at constraints facing the sector and the

Handbook of Logistics and Supply Chain Management, Edited by A.M. Brewer et al.
© 2001, Elsevier Science Ltd

future prospects for the industry, particularly in light of the reshaping of consumption patterns facilitated by electronic commerce.

2. Air-freight industry organization

Air-freight markets are difficult to delimit and analyze, for a number of reasons. The air-freight providers are a heterogeneous group of operators, offering different types of services and different levels of logistical expertise. There are three main categories of air-freight operators:

(1) line-haul operators,
(2) Integrated/courier/express operators, and
(3) niche operators.

Line-haul operators move cargo from airport to airport, and rely on freight forwarders or consolidators to deal directly with customers. Line-haul operators can be:

(1) *All-cargo operators* (scheduled and non-scheduled), moving only freight in dedicated freighter or cargo aircraft (e.g., Cargolux (European Union(EU)), Arrow Air (U.S.A.)). All-cargo operators offer relatively high reliability and have the capability to move large volumes over long distances.
(2) *Combination passenger and cargo operators*, which use both dedicated freighter aircraft and the belly holds in passenger aircraft to move freight (e.g., Lufthansa (EU), United Airlines (U.S.A.)). For the combination carriers, the cargo operations are mainly long-haul, with a large amount of freight being interlined on to shorter haul feeder services. The high utilization of long-haul aircraft justifies the purchase of new aircraft for these services.
(3) *Passenger operators*, which use the belly holds in passenger aircraft. Passenger carriers have tended to view cargo as a by-product of passenger operations. They are seen to offer the lowest prices and the least reliable service (GECAS, 1994). Passenger carriers move cargo in the belly holds of passenger aircraft, where it has traditionally taken second place to passenger services. Unlike passenger services, shippers do not have access to price information analogous to passenger computer reservation systems (CRSs). Freight forwarders play an important role' in consolidating shipments for line-hauliers.

Integrated/courier/express operators move consignments from door to door, with time-definite delivery services (e.g., UPS, Federal Express, TNT, DHL). These integrated carriers operate multimodal networks, combining air services with extensive surface transport to meet customer demands. The integrators offer a variety of products to shippers, and supplement air services with extensive ground transport to provide time-definite delivery with continuous shipment tracking and, if necessary, logistical expertise to support just-in-time (JIT) inventory control strategies. In order for integrators to be able to offer door-to-door next-day deliveries, they require night-time operations. In terms of aircraft requirements then, they need to operate quiet, reliable aircraft with low utilization levels (as few as 2 hours/day flying time in some cases). Integrators seek to purchase a combination of new aircraft, with high capital costs and better utilization on long-haul segments, with less expensive renovated second-hand aircraft for the medium-haul operations with lower utilizations. The integrated carriers initially began offering services in the small parcel/document sector, but now typically offer a broad range of services in terms of maximum weight and dimension restrictions. The association of integrators with purely express freight is no longer valid. The integrators have focused their attention on the premium high-yield traffic. Legislative changes in the U.S.A. have permitted *integrated freight forwarders* (e.g., Emery Worldwide, Airborne Express (U.S.A.)) to line-haul their consignments themselves, and since 1994 interstate ground operations for all carriers have been deregulated.

Niche operators operate or leverage specialized equipment, or indeed expertise, in order to fill extraordinary requirements (e.g., Heavylift (The Netherlands) and Challenge Air Cargo (U.S.A.)). These operators attract business through their capabilities for handling outside freight or special consignments, including line-haul to locations with poor infrastructure facilities. For chartered freight and niche operators, the discontinuous use of aircraft makes it financially preferable to acquire freighter aircraft on a second-hand basis.

The air-freight industry was dominated until the mid-1980s by the line-haul carriers. The integrated carriers rapidly increased their market share in the U.S. domestic market (following deregulation in 1977), and more recently in international air-freight markets (Carron, 1981). There are several important distinctions between passenger demand and shipper demands for air transport services. These distinctions place a different set of constraints and operating conditions on carriers depending on whether they are carrying cargo, passengers, or both. Freight comes in a large variety of shapes, densities, and sizes, and must be loaded onto and off aircraft by equipment and handlers. Large units may have to be carried in freighter-only aircraft. The routing of cargo (including the number of stops or transfers) is unimportant to the shipper. What is important is the lapsed time between pick-up and delivery. For passengers, however, their preference is typically for daytime, non-stop flights. Shippers' preferences are for night-time carriage of goods, with early morning delivery.

One of the most significant differences between passenger and freight air transport (a factor which significantly affects the economic viability of cargo operations) lies in the fact that passengers typically travel on round-trip journeys,

Table 1
Air-freight tonnage for the top 20 freight airports (1998)

Cargo rank	Airport	Region	Airport code	Freight (tonne)	% change in freight 1997–1998	Airport rank for passenger volumes
1	Memphis	U.S.A.	MEM	2 368 975	6 1	87
2	Los Angeles	U.S.A.	LAX	1 861 050	–0.7	3
3	Miami	U.S.A.	MIA	1 793 009	1.5	12
4	Hong Kong	Asia	HKG	1 654 356	–8.8	23
5	Tokyo	Asia	NRT	1 637 521	–5.8	32
6	New York	U.S.A.	JFK	1 604 422	–3.7	16
7	Frankfurt/Main	Europe	FRA	1 464 955	–3.3	7
8	Chicago	U.S.A.	ORD	1 441 829	2.5	2
9	Seoul	Asia	SEL	1 425 009	–9.1	20
10	Louisville	U.S.A.	SDF	1 394 999	3.7	171
11	Singapore	Asia	SIN	1 305 592	–3.9	34
12	London	Europe	LHR	1 301 251	3.3	4
13	Anchorage	U.S.A.	ANC	1 289 266	2.3	144
14	Amsterdam	Europe	AMS	1 218 746	0.9	11
15	Newark	U.S.A.	EWR	1 094 383	4.5	13
16	Paris	Europe	CDG	1 067 255	–0.5	9
17	Taipei	Asia	TPE	916 881	0.4	58
18	Atlanta	U.S.A.	ATL	907 208	4.9	1
19	Dayton	U.S.A.	DAY	893 239	9.8	240
20	Indianapolis	U.S.A.	IND	812 664	22.6	115

Source: Airports Council International (1999).

while cargo travels from a point of production to a point of consumption. Matching demand with inbound and outbound capacity is a difficult task and can lead to different network organizations for freight services compared with passenger services. For combination carriers, this can pose difficulties, since freight demand and passenger demand for principal destinations may not coincide. Carriers will take account of inbound and outbound requirements in considering whether or not to provide service on a route, and in deciding on the segments of the route and capacity available on each of the segments. Table 1 lists the top 20 air-freight airports in the world in 1998. The table includes the rank of each airport in terms of air passenger traffic, and highlights the distinctions in network organization of combination carriers and integrated carriers. The line-haul combination carriers tend to focus their cargo operations on international gateway airports, allowing consolidation or break-out loads to be transferred between long-haul and short-haul services. The integrated carriers focus their operations at cargo hubs that do not necessarily have very high volumes of passenger traffic. Memphis and Indianapolis airports in the U.S.A. are principal and secondary hubs, respectively, for Federal Express; Louisville, also in the U.S.A. is the primary hub for UPS; and Dayton is the primary hub for the U.S.A. integrated forwarder Emery Worldwide.

3. Air-freight pricing

Air-freight services are sold and marketed in a number of different ways. The line-haul operators sell a relatively small proportion of their cargo space directly to their customers. The greater proportion of their space is sold through general sales agents (GSAs) or freight forwarders, who negotiate with the airlines for fixed amounts of space. The agents or forwarders then sell on the freight space to customers.

The line-haul airlines publish their cargo tariffs as agreed at International Air Transport Association (IATA) tariff conferences. In practice, only a small percentage of customers pay these published tariffs, which can be considered as an upper-band on air cargo rates. As with passenger fares, discounting is widely applied, and in the case of cargo the rates will be determined on the basis of a number of characteristics and circumstances, including the following:

(1) volume, density, and weight;
(2) commodity type;
(3) routing;
(4) season;
(5) regularity of shipments;
(6) imports or exports; and
(7) priority or speed of delivery

Consolidated shipments, aggregated by forwarders and carried by the line-haul operators, typically travel under a single air waybill (AWB). The freight forwarders bundle a variety of services and expertise and offer shippers a wide range of logistical and transport options. These include collection and delivery of shipments door to door, complete paperwork and documentation for customs purposes, customs clearance, tracking of shipments, and inventory management and control. The freight forwarders act as wholesalers and earn their profit by maximizing the difference between what they pay airlines and other carriers and what they can charge shippers. The integrated operators offer a variety of products or services depending on (1) the weight of the consignment and (2) the speed of delivery required by the customer. Discounting is applied to these services on the basis of the volume and regularity of custom. However, because each consignment is treated as a separate piece of freight, with an individual AWB and customs declaration, the integrated carriers provide and practice electronic tracking of individual shipments, and levy charges individually. Customs services in many jurisdictions now operate electronically, so that consignments receive clearance en route to their destination airport. The customs authority can notify the operator of consignments that will need to be cleared on the ground, and this information can be forwarded to the customer via the tracking system.

4. Recent trends in air freight

In global terms, the dominant air cargo flows are in three main markets: Asia–North America, the North Atlantic (i.e., North America–Europe), and Europe–Far East. The Europe–Asia market is expected to have one of the top growth rates over the period 1997–2017. Boeing (1998) estimates that air freight on this sector will grow by 7.3% per annum. Indeed, any markets involving Asia are expected to experience the highest growth rates in the next 10 years. Intra-Europe freight has the lowest forecast growth rate of 4.5%. The international air express market is expected to grow at a tremendous rate over this period, and this market is served primarily by the integrators. Boeing forecasts an annual growth rate of 18%, which they claim will result in express services accounting for approximately 40% of the total international cargo business by 2015. It currently accounts for 5% of the total market. This mirrors the U.S. experience, where express services accounted for 4% of the U.S. market in 1977, and with an average annual growth rate of 25% express operators or integrators claimed in excess of 60% of the U.S. domestic market in 1998. It is believed that this experience in the U.S. raised customer expectations for air-freight services worldwide.

Air-freight markets are shifting as the economic growth pattern of developing countries accelerates past that of already industrialized economies. The main influences or drivers behind these trends are (Reynolds-Feighan and Durkan, 1997):

(1) the primary influence of world economic activity (world GDP is the best single measure of global economic activity, with a high correlation between changes in world GDP and changes in world air cargo revenue tonne-kilometers (RTKs));
(2) the impact of the range of services in the express and small package market;
(3) inventory management techniques;
(4) deregulation and liberalization;
(5) national development programs; and
(6) the stream of new air-eligible commodities and the growth of e-commerce.

Air-freight is a significantly more expensive mode of carriage of goods than other modes, and will be used when the value per unit weight of shipments is relatively high and the speed of delivery is an important factor. Under these circumstances, the transport costs can comprise a small proportion of the revenue associated with the products. The advantages which movement by air offers shippers are the speed, particularly over long distances, the lower risk of damage, security, flexibility, accessibility for customers, and good frequency for regular destinations (Simmons, 1994). For integrated operators, the guaranteed delivery and the facility to track consignments gives customers additional advantages over standard air-freight carriage. These superior qualitative differences give rise to higher rates for integrated services. Over shorter distances, air transport faces stiff competition from surface modes and from combined road and sea services. Air-freight demand varies by season, and this is taken into account by carriers supplying airlift capacity.

The passage of the Domestic All-Cargo Deregulation Statute of 1977 in the U.S.A. eliminated the Civil Aeronautics Board's (CAB) control over entry into and exit from the all-cargo market (Carron, 1981). Control over air-freight rates was curtailed and eventually eliminated. Carriers were permitted the right to refuse specific types of freight, and they became liable for the full value of the freight carried. The deregulation of air freight raised cargo rates as expected (Taneja, 1979), but gave shippers greater choice among carriers with respect to rates, consequential damage, and excess value charges (Taneja, 1979). Under CAB regulation of air freight, all-cargo operators were unable to generate reasonable profits with the result that the quantity and quality of service were deteriorating. Furthermore, it was generally felt that freight carried by air traveled longer distances than was necessary because surface modes could not be used to support the carrier's operation (Taneja, 1979). Integrated carriers now offer multimodal services that take advantage of the distance, cost, and time trade-offs offered by the different modes. In the European and Asian markets the integrated carriers have recently increased the size of their international operations. Indeed, within Europe, it is estimated that the integrated carriers now perform most of the total intra-European RTKs (Reynolds-Feighan, 1994).

Within Europe, competition from surface modes has had, and will continue to have, a downward impact on air-freight growth rates (4–5% per annum for 1997–2017). This factor, along with a relatively low overall economic growth rate, explains the below average long-term growth rate for air freight. "Air trucking," which involves the movement of air cargo by road under AWB, has been expanding at a rate of 15% per annum since 1975, according to Boeing (1998), with an estimated 7340 trips per week in Europe in 1997. Boeing suggests that the number of routes served within Europe has expanded from 38 in 1975 to 386 in 1995. In 1971, international airlines through IATA introduced and adopted IATA Resolution 507b, which clearly defined the circumstances under which trucking could be undertaken. The main circumstances involved:

(1) a lack of available space on aircraft;

Table 2
Comparison of integrated and non-integrated services

Integrated carrier/freight forwarder	Non-integrated operation with air trucking
Shipper	
Integrator: Picks up consignment Tags and electronically traces consignment until delivery Line-hauls package from airport to airport Clears customs Delivers to destination	*Agent*: Consolidates multiple shipments under a single AWB Delivers to airport bond *Air trucker*: Picks up consignment and delivers to another airport bond *Airline*: Line-hauls consignment from airport to airport *Agent*: Arranges customs clearance, collection and delivery
Consignee	

(2) consignments that could not be handled on aircraft operated by an airline due to the size, weight, or nature of the consignments (certain commodities may only be shipped in freighter or all-cargo aircraft), or because the carrier refused carriage on some other grounds;

(3) carriage by air would have resulted in delayed transit times or in carriage not being accomplished within 12 hours of acceptance; and

(4) carriage by air would have resulted in missed connections.

Today the practice of air trucking is predominantly oriented towards moving intercontinental freight traffic to gateway airports. This process is shown diagrammatically in Table 2, which helps to illustrate the distinguishing characteristics of non-integrated operations compared with integrated operations.

5. Constraints and future prospects

Several factors can be identified as significant constraints on the growth of air freight. These include the significant growth of air trucking and the reduction in freight carrying capacity of the passenger airlines. In the longer term, the integrated operators and all-cargo airlines can be expected to increase their share of the air-freight market, as passenger carriers are forced to charge more realistic cargo rates that are in line with the costs of producing the services. Passenger carriers have been facing declining passenger and freight yields (revenue per seat-kilometer or tonne-kilometer) as competition has forced efficiencies on many aspects of their operations.

Environmental regulations have impacted on the air freight sector by forcing a reduction in the number of older, noisier aircraft available, and have delayed or altered the infrastructure planning process and contributed to the capacity constraints at many airports, particularly in Europe. The noise and pollution requirements now in place at many of the large airports raise operating costs for many carriers. The congestion of air transport infrastructure has been identified in several studies as a major bottleneck in the development of competitive air passenger and freight transport markets in domestic and international markets (Reynolds-Feighan and Button, 1999). Finally, security problems are a significant factor constraining the growth and development of both express operations and air trucking.

The emphasis on multimodal transport operations and on greater integration of transport with other logistical services will dominate freight developments in the next two decades. While e-commerce eliminates the need for the physical distribution of some products and services, it is dramatically altering the pattern of consumption and generating new sources of business for the air-freight industry.

References

Boeing (1998) *1998/99 world air cargo forecast.* Seattle, WA: Boeing Commercial Airplane Group.

Button, K.J., K. Haynes and R. Stough (1998) *Flying into the future: Air transport policy in the European Union.* Cheltenham: Edward Elgar.

Carron, A.S. (1981) *Transition to a free market: Deregulation of the air cargo industry.* Washington, DC: Brookings Institution.

GECAS (1994) *Air cargo: An industry study.* Shannon: GECAS.

Reynolds-Feighan, A.J. (1994) "EC and US air-freight markets: network organisation in a deregulated environment" *Transport Reviews,* 14(3):193–217.

Reynolds-Feighan, A.J., and K.J. Button (1999) "An assessment of the capacity and congestion levels at European airports", *Journal of Air Transport Management,* 5(3):113–134.

Reynolds-Feighan, A. J. and J. Durkan (1997) *The impact of air transport on Ireland's export performance.* Dublin: Institute of International Trade of Ireland.

Simmons, J. (1994) "Benefits of different transport modes", in: *ECMT Economic Research Centre, Round Table 1993.* Paris: ECMT.

Taneja, N. (1979) *The U.S. air freight industry.* Lexington, MA: Lexington Books.

Chapter 29

BULK COMMODITY LOGISTICS

KEITH TRACE
Monash University, Clayton

1. Introduction

This chapter begins with a description of a typical bulk commodity logistical system, setting the scene for a discussion of the general principles underlying the movement of bulk commodities. The focus then shifts to the economics of land and sea transport. The section on land transport explores the logic underlying the choice of mode, while the discussion of sea transport focuses on economies of vessel size and the institutional structure of freight markets in the dry bulk and tanker trades. The chapter concludes with a case study of Australia's Pilbara iron ore exporters, focusing especially on Hamersley's logistical chain.

The term "bulk commodity" refers to any granular, lumpy, loose, or powdery substance (e.g., mineral ore, coal, sand, gravel, grain) that is handled in large volumes. A distinction may be made between dry bulk commodities (e.g., iron ore, coal) and liquid bulks (e.g., crude and refined oil). Generalizing, a bulk commodity is one that has some or all of the following attributes:

(1) it may be moved by pipeline or conveyor belt;
(2) it is handled in sufficiently large quantities to fully utilize a road or rail truck or bulk carrier;
(3) it has a relatively low value/weight (or volume) ratio; and
(4) it has been subject to limited processing.

A low value/weight (or volume) ratio implies that logistics costs are more likely to be a critical determinant of competitiveness for bulk commodities than manufactured goods. Insofar as the trend is towards smaller, lighter, and more valuable manufactured goods, elaborately transformed manufactures are able to better withstand the impact of logistics costs. However, the cost of transporting and handling bulk commodities may be reduced if value adding takes place further upstream in the logistics chain (e.g., the trend towards earlier beneficiation of iron ore).

Handbook of Logistics and Supply Chain Management, Edited by A.M. Brewer et al.
© 2001, Elsevier Science Ltd

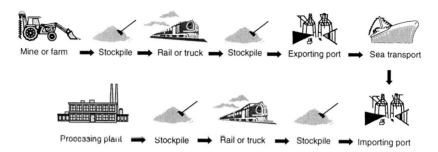

Figure 1. Schematic representation of a bulk commodity logistical system.

The handling characteristics of bulk commodities, which are influenced by a range of factors, including grain size and weight, water content, surface adhesion, ease of flow, and extent of compaction, differ significantly. Each commodity has its distinctive handling characteristics. While bulk handling systems have many common features, they are usually custom designed for specific products and functions.

The logistical issues posed by bulk commodities may be explored most effectively using a systems approach. A typical bulk commodity logistical system is shown in Figure 1. The system begins with the production of the raw commodity, which might be crude oil, coal, mineral, or farm produce. We assume the commodity in question is transported by road, rail, pipeline, or conveyor belt to the nearest port, where it may undergo some processing. It is then exported by sea to another country or region, unloaded, and transported (by road or rail) to a factory or processing plant, where it will be transformed into the finished product.

The logistical system comprises two land journeys and one sea journey, several intermodal transfer points, and four stockpiles, located at the point of production, the loading port, the discharging port, and the destination. The commodity moves through the system as a series of discrete shipments, with stockpiles acting as buffer stocks. Stockpiles are required not only because production and consumption rates may differ, but also because the capacity and service frequency of the individual transport modes vary. For example, while a bulk carrier typically carries a much larger load than a truck or train, ocean transport usually operates less frequent services than rail or road transport.

It is important to recognize that the logistical system includes eight or more handling operations as the bulk commodity is transferred from one form of transport to another and as it is moved into and out of the four stockpiles. Intermodal transfer points and buffer stocks impose time and cost penalties on bulk transport systems, while operational variations due to weather and human failures introduce uncertainty and risk. Uncertainty and risk are among the most expensive and least controllable elements of the supply chain. Effective supply chain management controls and reduces such risk.

As Frankel (1999) notes, the direct and indirect costs incurred along the supply chain are chargeable to various participants, including the shipper, transport provider, intermodal facility operator, and the consignee. There is a tendency for the control of supply chains to become the responsibility of a single agency. In such cases, a single firm, for a consideration, takes responsibility for organizing the entire supply chain, including the assumption of risk and responsibility for cost outcomes. This tendency is less common in raw material than in manufacturing supply chains.

Whatever the organizational form adopted, the aim is to move the commodity as cheaply and efficiently as possible. An ideal logistical system for bulk commodities would be designed so that the individual components of the system link together into an efficient whole. When planning such a system, certain general principles must be kept in mind.

2. General principles of bulk commodity logistics

Our description of a bulk commodity logistical system suggests that cost may be reduced in one of four ways:

(1) by taking advantage of economies of scale in transport and cargo handling;
(2) by reducing the number of times commodities are handled;
(3) by integrating the various elements of the transport chain; and
(4) by reducing the size of stocks held.

Note, however, that most methods of cost reduction require capital expenditure and some cost-reducing "solutions" to work in opposition to each other. For example, exploitation of economies of scale in sea transport may require larger rather than smaller stockpiles. In such situations, trade-offs are inevitable (Stopford, 1997).

2.1. Economies of scale

Unit costs can be reduced by taking advantage of economies of scale in both land and sea transport. For example, employing trucks with higher mass limits, especially B-Doubles, B-triples, or road trains, reduces the cost of road transport, while the cost of rail transport may be lowered by the employment of more powerful and fuel-efficient locomotives, the use of high-capacity wagons, and the operation of unit (permanently coupled) trains. Similarly, the cost of sea transport can be reduced by the use of larger vessels – the larger the vessel, the lower the unit cost of carriage. Reductions in unit costs through economies of scale usually require substantial capital expenditure. The general rule applying in bulk commodity logistics is to move the largest possible unit or quantity as far as possible toward the next stage in the production sequence before breaking the lot or load into smaller units (Sims, 1991).

2.2. Handling costs

The cost of cargo handling represents a major outlay. The use of high productivity cargo handling equipment can minimize the handling cost per tonne and cut delays in loading and unloading trucks, wagons, and bulk carriers. More radical solutions are possible at the cost of higher levels of capital expenditure. Processing plants can be relocated to reduce the number of transport legs and hence reduce the need for cargo handling. For example, land transport costs may be minimized if steel mills or alumina refineries are built at or relocated to coastal sites.

2.3. Integration of the stages in a supply chain

Cargo handling can be made more efficient by integrating the various stages in the logistics system. For example, bulk cargoes may be shipped in a form that minimizes the handling costs incurred throughout the relevant logistics chain. Improved information systems and close co-operation between the various participants may minimize the need for stockpiles (e.g., by loading a vessel directly from a unit train rather than drawing from a stockpile). An ideal logistical system would integrate all stages in the supply chain (land transport, storage, terminals, ships) into a single balanced system. Examples may be found in the oil and steel industries.

2.4. Reducing stocks

Reducing the volume of a bulk commodity held in storage reduces cost. However, as noted above, a logistical system must handle parcels of cargo of a size acceptable to both producers and their customers. While it might be cheaper to transport manganese ore in a 200 000 dwt bulk carrier, the steel industry actually ships it in much smaller vessels. Even if steel plants had the capacity to accommodate larger stockpiles of manganese ore, inventory costs might exceed the savings in freight costs. The size of the parcel in which a bulk commodity is shipped represents a trade-off between the economies of vessel size and the cost of holding inventory.

3. Land transport

Bulk commodities may be transported by rail, road, conveyor belt, or pipeline. The technological and cost characteristics of the individual transport modes differ significantly; the optimal mode of transport varying according to a range of factors, including distance, volume, and the specific characteristics of the commodity in question. Thus, while conveyor belts are normally used for short-distance transport of coal and minerals, rail is suited to the transport of large volumes of coal and ores over relatively long distances and the cost characteristics of road favor it for commodities requiring multiple loading and/or discharge points. In the absence of regulation, forcing shippers to use a particular mode of transport, modal choice will depend on the nature of the commodity, the scale on which it is shipped, and the number of customers or destinations.

Many bulk commodities are transported over relatively long distances from mine or farm to processing plant or port. As a result, freight charges typically account for a significant proportion of free-on-board (FOB) costs. The Bureau of Transport and Communication Economics (1992) estimates that transport charges represent 15–45% of the FOB cost of Australian minerals, depending on the commodity in question and its location.

Rail transport offers the lowest cost method of transportation where large volumes of a dry bulk commodity are transported over long distances and where the commodity in question is consigned to a single destination, usually a port or processing plant. Rail operates most efficiently where unit trains shuttle between terminals designed to load and unload the commodity rapidly and economically. Coal, iron ore, other minerals, and grain are normally carried by rail.

Rail transport enjoys economies of scale and scope. A rail line may cost Australian $1 million or more per track kilometer. The cost of loading and unloading facilities, depending on scale and sophistication, is in the range Australian $10–100 million. The denser the traffic flow along a given rail line, the further the fixed or overhead costs are spread and the lower the unit cost of rail transport. Under normal operating conditions, the high level of fixed costs implies that the marginal cost[*] of carriage lies below average cost. Freebairn and Trace (1988) have estimated that, under Australian conditions, the long-run marginal cost[†] of rail carriage is 33% lower than average cost for a 4 million tonne annual rail haul, and 16% lower for a 10 million tonne haul.

---

[*]Marginal cost is the incremental increase in total cost that result from a one-unit increase of output (e.g., running an additional train along a given rail line).

[†]The long-run marginal cost (LRMC) measures the cost of increasing output when all inputs can be varied. For example, LRMC measures the cost of operating an additional train when the capacity of the relevant rail line can be increased or decreased.

Economies of scope arise where the cost of producing two or more outputs by means of one set of facilities is less than the cost of producing the same outputs in separate facilities. In cases in which a rail line is used by several types of traffic (e.g., coal and grain) or by several mines producing the same commodity, economies of scope stem either from the employment of previously underutilized physical plant (e.g., track and signaling systems) or from the use of shared inputs, such as marketing expertise.

Road has a competitive advantage over rail when the commodity in question has a distribution system serving multiple destinations. Petroleum products are typically handled by road rather than rail transport because the distribution system adopted by the oil companies requires multiple discharge points. Road is also competitive when rail transport is unable to offer through transport from origin to destination and when the rail haul is short (typically less than 80 km). However, rail can be competitive over short distances, provided that rail facilities exist at both the origin and the destination, and providing also that the volume to be transported is relatively large. For example, rail is competitive for the transport of coal over a distance of less than 30 km in the Hunter Valley of New South Wales. The competitive balance between road and rail may be determined by constraints or bottlenecks. For example, rail has a competitive edge when the capacity of an export wheat terminal to accept road deliveries is limited.

Today's shippers of bulk commodities are usually free to choose their preferred mode of transport. However, some commodities remain subject to regulations restricting shipper choice. Australia provides an example, as historically rail transport was provided by protected state-owned enterprise. Competitive threats to state rail systems, such as that by the automobile, led to regulation designed to protect rail's share of freight and passenger traffic. Until recently, Queensland, New South Wales, and Victoria required that coal, minerals, petroleum, and grain be transported by rail. While many restrictions have been removed in the past 30 years, some remain in force. Queensland Rail (QR) still benefits from a legislative monopoly of coal and mineral haulage. The Queensland Government's strategy has been to raise the finance necessary for capital expenditure (including the building of new lines or the upgrading of the existing infrastructure, as well as the purchase of new locomotives and wagons) by way of an up-front payment from coal or mineral producers. QR has charged mineral producers a freight rate that has been shown to be substantially above the marginal cost of carriage (Freebairn and Trace, 1992). Under the Competition Policy Agreement (1995), QR's monopoly of coal transport will cease in the near future.

Deregulation may lead to substantial efficiency gains. Prior to 1988 the Australian grain marketing and transport system was subject to extensive regulation. Grower complaints led to an investigation of the Australian grain-handling system by the Royal Commission into Grain Storage, Handling and Transport (1988). The Commission's report recommended:

(1) the removal of discriminatory restrictions preventing grain growers from taking advantage of road transport;

(2) that rail systems should be able to act commercially in their pricing and investment decisions and should not be constrained by non-commercial objectives; and

(3) the costs incurred by the alternative modes of transport should accurately reflect the real resource cost of providing the service.

Following the report, reforms were implemented aimed at achieving an integrated and efficient rail and road transport system. In general, deregulation has led to lower costs for the transport of export wheat. Rail has retained most long-distance export grain transport as a result of its greater cost efficiency, although export wheat sourced from silos situated relatively close to a port is normally carried by road. Wheat destined for domestic markets is usually transported to local mills by road transport (Bureau of Transport and Communication Economics, 1992).

The shippers of bulk commodities either chose the mode of transport that minimizes cost or the mode that offers the preferred mix of cost and service quality. However, economists argue that the minimization of private cost may not provide an acceptable solution where externalities[*] are present (i.e., where private and social costs diverge). For example, certain mines in the Southern Coalfield of New South Wales are not connected to the rail system. Coal from these mines is transported to Port Kembla by road, giving rise to negative externalities, notably congestion and pollution caused by coal dust (Bureau of Transport and Communication Economics, 1992). From an economist's perspective, comparisons between two or more transport systems should focus on the extent of the divergence between private and social costs. As far as is practicable, pricing should aim to internalize the externalities generated by transport.[†]

4. Sea transport

The technological and cost characteristics of sea transportation make it most efficient for the high-volume, long-distance movement of relatively low-value commodities. Four principal groups of commodities are transported in bulk by sea: dry cargoes (coal, iron ore); petroleum (crude oil, refined products); liquefied

---

[*]An externality is an unpriced cost or benefit imposed on one agent by the actions of a second; costs are referred to as negative externalities, benefits as positive externalities.

[†]For example, an efficient-road-user charge should take account of the external costs generated by road vehicles, including pavement damage, congestion, noise, accidents, and environmental damage. If road-user charges reflect these costs imperfectly, distortions in resource allocation will occur.

gases (LNG and LPG); and liquefied chemicals (ammonia). The fundamental division is between the dry bulk and the liquid bulk (tanker) trades. Both tankers and (dry) bulk carriers usually carry a single (homogeneous) cargo. Shipping markets in both sectors are characterized by a large number of ship owners, many of whom own only one or two vessels. Historically, this has meant that there has been limited scope for concerted action by owners to rationalize the bulk trades. Bulk shipping is often characterized as a "boom or bust" business, with intense competition for cargoes at times of surplus capacity leading to very low freight rates.

4.1. Dry cargo trades

Commodities handled in the dry bulk trades not only have a granular or lumpy composition, enabling them to be handled by automated equipment such as conveyors and grabs, but are typically shipped in relatively large volumes. The dry bulk trades consist of five "major bulks" (iron ore, coal, grains, bauxite and alumina, phosphate rock) and numerous "minor bulks." The major bulks, each of which has shown a distinctive growth path over the past 20 years,[*] are the driving forces behind the dry cargo market (Stopford, 1997).

The size of vessels employed in the dry bulk trades represents a trade-off between several factors: economies of vessel size; the parcel size in which cargo is available; the depth of water alongside wharves; and the capacity of cargo-handling facilities to accommodate large vessels. Substantial cost savings are achievable by using large vessels.[†] However, such vessels face two important constraints. First, there may be a physical limit to the size of cargoes that an industry is able or willing to accept at any one time. For example, stockpile size may be limited to 20 000 tonne. Second, the depth of water in approach channels or alongside wharves may constrain ship size, as may cargo-handling rates in terminals. Such trade-offs lead to the employment of vessels of varying sizes across trades.

At the smaller end of the range, "handy" bulk carriers (10 000–30 000 dwt) are the "maids of all work" in trades where parcel size and/or draft restrictions rule out the employment of larger vessels. Improvements in port facilities and/or

---

[*]Shipments of iron ore grew rapidly until the mid-1970s. Thereafter the growth rate slowed appreciably. The shipment of grain grew fairly steadily until the mid-1980s, then stagnated, albeit with substantial year-to-year fluctuations.

[†]Economies of vessel size stem from three sources: capital costs per tonne of capacity fall substantially as vessel size increases; the ratio of crew members to carrying capacity declines as vessel size increases; and the cost of fuel per tonne of cargo carried also tends to fall as vessel size increases.

investment in improved terminals have enabled some commodities to employ Handymax (30 000–50 000 dwt) bulk carriers. Panamax (55 000–70 000 dwt) bulk carriers service the coal, grain, bauxite, and the largest of the minor bulk trades, while Capesize bulk carriers (75 000–300 000 dwt), which are heavily dependent on the carriage of coal and iron ore, serve the upper end of the market. While large bulk carriers are purpose built for particular cargoes, especially iron ore, smaller (Panamax-size carriers and below) vessels may, over time, carry a variety of commodities.

While bulk shipping may be thought of as dividing into numerous submarkets, each with its own supply and demand characteristics, the fact that vessels of various classes compete at the margin implies a level of interdependence between market segments.

Four types of contractual arrangement are commonly used for the shipment of bulk cargoes: voyage charter, contract of affreightment, time charter, and bareboat charter.[*]

A *voyage charter* provides transport for a specific cargo from port A to port B for a fixed price per tonne. Arrangements for the charter are laid down in a charter party and, if all goes well, the specific ship that has been chartered arrives at A on the due date, loads the cargo, and transports it to B. Once the cargo is discharged at B, the transaction is complete.

Under a *contract of affreightment* a ship owner agrees to carry a series of cargoes or part cargoes between A and B, at intervals that are defined in the contract, for a fixed price per tonne. The ship owner, who is responsible for providing a suitable ship on each occasion, is able to plan the use of his ships in the most efficient manner.

A *time charter* gives the charterer operational control of the vessel carrying his cargo, while leaving its ownership and management in the hands of the ship owner. The ship owner pays the operating costs of the vessel (crew, maintenance, repair costs), while the charterer pays all voyage expenses (bunker, canal, port, cargo-handling costs).

If a company wishes to have complete operational control of a vessel, but does not wish to own it, a *bare boat charter* may be negotiated. Under this arrangement an investor, who may be a financier rather than a professional ship owner, purchases the vessel and hands it over to the charterer for a specific period, usually 10–20 years. The charterer has operating control of the vessel and is responsible for operating and voyage costs.

Futures markets, such as the Baltic International Futures Exchange (BIFFEX), enable shippers and ship owners to hedge against fluctuations in charter rates.

---

[*]The first formal freight market, the Baltic Exchange, opened in London in 1883, replacing an informal market that had developed in certain City coffee houses (Stopford, 1997).

The freight futures market is based on the Baltic Freight Index (BFI), an index covering freight rates on 11 different trade routes. The hedging operation involves trading contract units on the freight futures exchange.[*] The price at which individuals are prepared to buy and sell forward depends on their expectations of the future. If the market expects freight rates to rise during the next 3 months, the price of contract units for settlement at the end of the 3-month period will be higher than the current BIFFEX index.

4.2. Tanker trades

Liquid cargoes shipped by sea comprise three main groups: crude oil and petroleum products, liquefied gas (primarily LNG and LPG), and liquid chemicals (e.g., ammonia, phosphoric acid). Together these commodities account for half of world sea-borne trade, with crude oil and petroleum products generating most of the volume. We concentrate here on the shipment of crude oil and petroleum products.

Crude oil is usually transported from the oilfield to the coast by pipeline. Assuming it is not refined locally, the crude oil is then loaded into large crude carriers and shipped to its destination, usually an oil refinery, where it is discharged at a bulk terminal. Since crude oil is carried in very large tankers, loading and discharging facilities are generally located in deep-water locations. The refined product is distributed by road, rail, or shipping.

The major oil producers operate sophisticated and cost-minimizing logistical networks. For example, sea-borne oil is transported in massive oil tankers, enabling the industry to capture economies of scale, operating to schedules that minimize round voyage times. Tankers sail with full cargoes, loading and unloading is accomplished speedily, and waiting time is minimal. Until the mid-1970s, major oil companies operated tanker fleets including owned and chartered tonnage. The typical major owned the shipping required for its base load shipments, supplementing its own fleet by chartering independently owned tankers (mostly from Greek or Norwegian owners) to meet peak load requirements. Following the OPEC I oil price increase (1973–1974) and the subsequent fall in demand for oil, the major oil companies decided that oil transport was no longer a core business. They now rely on the charter market to a much greater extent than in the pre-1973 era.

The tanker fleet may be divided into several segments: Handy (10 000–50 000 dwt), Panamax (50 000–70 000 dwt), Aframax (70 000–100 000 dwt), Suezmax (100 000–200 000 dwt), VLCC (200 000–300 000 dwt), and ULCC (over

---

[*]Contract units can be bought and sold at a rate of Australian $10 per BIFFEX index point, with units traded ahead for settlement at 3-month intervals.

300 000 dwt). Since tankers in the various size classes are not interchangeable, each segment operates as a distinct submarket. Whereas oil companies employ as large crude carriers as possible, constrained only by the size of the crude oil flow and the depth of water available, most oil products are carried in Handy size tankers.

5. Case study: Pilbara iron ore

The case study of bulk commodity logistics focuses on Hamersley Iron Ore (Hamersley) operations in western Australia. The case study illustrates the logistical principles discussed earlier in this chapter. Note in particular the importance of economies of scale, the desirability of integrating the various stages in the production process, and the role of the logistics system in creating the export product.

The post-World War II discovery of large-scale iron ore deposits in the Pilbara region of western Australia led to the growth of a major export industry.* The development of Hamersley's Mount Tom Price mine in the mid-1960s was followed by the exploitation of other ore bodies, including Goldsworthy and Mount Newman by Broken Hill Proprietary Ltd (BHP) and Pannawonica by Robe River Iron Ore Associates (Robe River). The Pilbara iron ore deposits are located some 200–400 km inland. The ore is transported via privately owned rail lines to ports at Dampier, Cape Lambert, and Port Hedland. It is then loaded into large bulk carriers for export to Asia (especially Japan and Korea) and Europe.

As a result of the development of the Pilbara mines, Australia is a major iron ore exporter. World sea-borne iron-ore trade is estimated to have been 423 million tonne in 1997. Australia (with 1997 exports of 147 million tonne) and Brazil (136 million tonne) account together account for two-thirds of world iron-ore shipments.

Hamersley opened its Tom Price mine, with its rail link to Dampier, in 1966. Somewhat later, the company opened mines at Brockman, Paraburdoo, Channar, Marandoo, and Yandicoogina. Production of Hamersley ore involves:

(1) drilling and blasting the ore and removing waste;
(2) crushing, blending, and stockpiling the ore at the mine;
(3) reclaiming the ore from the stockpile and loading the trains;
(4) rail transport of the ore;
(5) stockpiling and blending of the ore at Dampier to achieve export grade product;

---

*Iron ore is also mined in Tasmania and South Australia.

(6) reclaiming the ore from the stockpile and loading it on to bulk carriers; and
(7) transport by bulk carrier to overseas steel mills.

Hamersley exports two products: "lump," which is suitable for direct charging to a blast furnace, and "fines," which undergo a sintering process to produce artificially constructed lump. Hamersley exports a specified grade of ore.[*]

The geology of the Pilbara area is complex and the quality of iron ore variable. Each of Hamersley's mines produces ore of a different average grade. In general, the quality of Tom Price ore exceeds Hamersley's export specification, while the average grade produced by the other mines is below that specified in export contracts. It is therefore necessary to blend ore from the various mines to achieve Hamersley's export grade. Hamersley's logistical system plays a critical role in creating the export product.

Hamersley ore is crushed, blended, and stockpiled at each mine. The process of blending ore in batches ensures that the ore in any given batch is of a consistent grade. Once blended the ore is stockpiled to await rail shipment. The need for a stockpile at the individual mines arises because the production process is continuous, whereas trains are dispatched from time to time. Stockpiles also ensure the railway can continue to operate, even if there is a temporary disruption within the mine.

The approximately 400 km rail line runs from the Hamersley Ranges via the Fortescue River Valley, the Chichester Range, and the coastal plain to Dampier. All mines except Channar transport their ore from mine to port via spur lines connecting into the main (Tom Price–Dampier) line. Ore from Channar is carried via a 20 km conveyor belt to Paraburdoo, for loading on to trains that also carry Paraburdoo ore. The total length of rail line operated by Hamersley is 638 km. The rail system handles some 70 million tonne of ore annually.

The 2.2 km long trains consist of two locomotives and 200–210 ore wagons. Each wagon has a theoretical payload of 105 tonne. On average, trains carry 22 000 tonne of ore (gross train weight averages 27 000 tonne). Typically, trains complete a round trip every 24 hours.[†] At the current production level (approximately 70 million tonne/year), the Hamersley rail system handles an average of nine laden trains a day (and a similar number of return empty consists) over a 360-day working year. However, the rail system must be capable of handling higher numbers of trains a day to allow for peak export demand and for periods

[*]The discussion of the quality of Hamersley ore and the blending process is drawn from the report of the recent Court case between Hamersley and the National Competition Council (Federal Court of Australia, 1999).
[†]The Mount Tom Price–Dampier journey takes approximately 7 hours. Unloading at Dampier takes about 5 hours; the Dampier–Mount Tom Price return journey also takes about 7 hours; and loading at Mount Tom Price takes about 4 hours. The total cycle time is approximately 24 hours.

when production has been disrupted by inclement weather (the Pilbara area is cyclone prone), industrial disputation, or other factors. Rail Management Services, the consultant employed by the National Competition Council in a recent Federal Court Case, estimated the theoretical maximum capacity of the existing infrastructure to be in excess of 200 million tonne/year (Rail Management Services, 1999)

The rail track is predominantly single line, with passing loops at approximately 20 km intervals. An approximately 28 km double-track section of the line provides operational flexibility. The maximum opposing grade to loaded trains on the main line (i.e., Mount Tom Price–Dampier) is 0.33% (approximately 1 in 300), while empty trains returning to Tom Price face a maximum grade of 2% (1 in 50). (Australasian Railway Association, 1998)

Ore is unloaded by rotary dumping at Parker Point and East Intercourse Island (Dampier) and conveyed to stockpiles. Ore from the various mines is blended in these stockpiles, train schedules from the various mines being adjusted according to the blending requirements of each batch of iron ore. Each batch of export iron ore has a different recipe and requires a different blend of ores from the various mines. In other words, the order of train arrivals is controlled in such a way that the port stockpile meets the specifications of Hamersley's export product: "The essence of the batch system ... is the operation of Hamersley's mines as one single unit ... activity within each mine is coordinated with the activity in all other mines to provide ore which is fed into conveying systems (including the railway) for blending at the port to create Hamersley's export product" (affidavit of S.M. Walsh, as quoted in Federal Court of Australia (1999)).

The ore is reclaimed from the stockpiles by means of a grab and belt conveyor. The ore is weighed and samples are taken to ensure that the ore meet Hamersley's export specifications. The conveyor feeds shiploaders at the Parker Point and East Intercourse Island wharves. The shiploaders, in turn, funnel the ore into the holds of the ship in accordance with a prearranged loading plan.

The East Intercourse Island wharf (341 m long, depth of water alongside 21.5 m) has ore loading equipment capable of loading vessels at the rate of 7500 tonne/h. It can accommodate vessels of up to 250 000 dwt. In contrast, the Parker Point wharf (268 m long, depth of water alongside 17.2 m) has ore loading facilities capable of loading at 6000 tonne/h.

As noted above, the size of vessels employed in the bulk trades is determined by trade-offs between several factors: economies of vessel size; the parcel size in which cargo is available; the depth of water alongside wharves; and the capacity of the existing cargo handling facilities. In the iron-ore trades, large-scale producers of iron ore such as Hamersley sell directly to major iron and steel producers with plants located in Asia and Europe. Loading and unloading facilities allow the producers and their customers to benefit from the economies of scale residing within the logistics chain. For example, the Capesize vessels (75 000–300 000 dwt) employed carrying iron ore are among the largest vessels employed in the dry bulk trades.

The carrying capacity of these Capesize vessels makes it necessary to establish sizeable stockpiles at Dampier. Whereas Hamersley trains carry 22 000 tonne of ore, the bulk carriers carrying the export cargoes vary in the range 100 000–250 000 dwt capacities. If we assume that the average vessel carries 150 000 tonne of iron ore, Hamersley's production of approximately 70 million tonne/year would require about 460 sailings (i.e., roughly 1.3 sailings per day).

References

Australasian Railway Association (1999) *Yearbook and industry directory.* Melbourne: Australasian Railway Association Inc.

Bureau of Transport and Communication Economics (1992) "Relative efficiencies in the transportation of commodities", AGPS, Canberra, Report No. 76.

Federal Court of Australia (1998) *Hamersley Iron Ore Pty Ltd v. National Competition Council,* [1999] FCA 867 (28 June 1999).

Frankel, E.G. (1999) "The economics of total trans-ocean supply chain management", *International Journal of Maritime Economics,* 1(1):61–9.

Freebairn, J.W. and K. Trace (1988) "Principles and practice of rail freight pricing of bulk minerals", Centre of Policy Studies, Monash University, Clayton.

Freebairn, J.W., and K. Trace (1992) "Efficient railway freight rates: Australian coal", *Economic Analysis and Policy,* 22(1):23–38.

Rail Management Services (1999) "Application of Robe River Iron Associates for declaration of the Hamersley Iron Railroad: Consultant's report on rail costs", Rail Management Services Pty. Ltd, Sydney.

Sims, Jr., E.R. (1991) *Planning and managing industrial logistics systems. Advances in Industrial Engineering,* vol. 12. Amsterdam: Elsevier.

Stopford, M. (1997) *Maritime economics.* London: Routledge.

United Nations Conference on Trade and Development (1998) *Review of maritime transport 1998.* New York: United Nations.

Chapter 30

EXPRESS DELIVERY

DIETER SAGE
Hamburg

1. Introduction

The concept of "express delivery" has been with us through history, servicing our need for communication and exchange of information. These early delivery services were mainly provided by private couriers, particularly for the transportation of urgent messages. These services existed relatively unchanged until the last few centuries, when emerging national postal monopolies began to constrict traditional courier deliveries (Manner-Romberg, 1998).

The first signs of the modern express delivery concept as a specific sector of freight transportation were seen in the U.S.A. toward the end of the 1960s. It was in response to the requirements of shippers, which conventional freight transportation was unable to fulfil – the need for fast and reliable delivery of time-sensitive or important documents and packages. These requirements evolved due to fundamental changes in industrial production and distribution. During the last 30 years the general framework of economies has undergone numerous changes: changes in customers' demands, the development of new technologies, shorter life-time cycles of products, new competition from developing countries, and the substitution of standard mass-produced articles by more individualized products of higher value.

At the same time, in the production sector there was increasing focus on a few key products, the out-sourcing of parts of the production process, and the procuring of the necessary components, with the aim of increasing efficiency gains by economies of scale. Furthermore, being forced to reduce costs in order to meet fierce competition, the production sector began to decentralize production, and plants were relocated to regions that were more suited to production (e.g., reduced transportation costs and better logistics). The need to reduce stock costs has led to the just-in-time (JIT) delivery principle, which is characterized by more frequent delivery of materials at the right time and at the right place in the production process (see Chapter 13).

Handbook of Logistics and Supply Chain Management, Edited by A. Brewer et al.
© *2001, Elsevier Science Ltd*

The need to reduce stock costs (especially in central urban areas, where rents are high), the increasing importance of individualized consumer goods, as well as more stringent consumer demands, characterize developments in many areas of the retail trade, which has led to requirements for more frequent and faster deliveries with reduced consignment sizes (see Chapter 24).

All these developments have necessitated improvements in logistics and transportation to achieve reliable and fast deliveries, not only of important documents, spare parts, and patterns or models for products, but also of "ordinary" industrial and consumer goods. These requirements, accompanied by improvements in communication and information technology, logistics, and means of transport have enabled the development of express delivery services as an important part of the freight transportation sector and, furthermore, have stimulated a dramatic growth in the provision of these services during the past 20 years, which is ongoing and expected to continue.

In Germany, the most important market for express delivery in Europe, the turnover of the express delivery services amounted to about Euro 9 billion in 1998, and growth between 1990 and 1995 was 52%, and about 6–8% anually in the following years (Manner-Romberg, 1995; Belohradsky, 1997). Generally, the prospects for future growth are more promising in Europe than in the U.S.A., because development of the express delivery market is not as advanced in Europe. Furthermore, in contrast to the U.S.A., in Europe business-to-consumer delivery plays only a minor role. Business-to-business express delivery predominates in Europe, but most express delivery services have started to explore business-to-consumer delivery to increase their turnover (Belohradsky, 1997). This is to be seen against the background of the development of e-commerce, which is expected to lead to increasing demand for transportation and delivery.

2. Definition of express delivery

It is difficult to find a uniformly acceptable definition for the term "express delivery." As commonly understood, this term applies only the rapid delivery of goods and documents using fast modes of transport. However, in addition to these basic characteristics, the concept of express delivery varies from country to country and from one operator to another. No strict definition of express delivery exists. According to Brax (1996), express delivery services can generally be characterized as follows:

- door-to-door delivery from the point of collection to the point of delivery;
- fastest possible delivery; and
- provision of identification systems to ensure that consignments can be traced at any point of the delivery chain.

"Express delivery" can also be considered as a collective term comprising three concepts:

(1) courier services;
(2) express delivery services (in its narrowest sense); and
(3) parcel delivery services.

To avoid misunderstandings, the concept of "express delivery services" (in the narrow sense) will here be referred to as "express delivery services," while the more general concept of express delivery will be referred to as "CEP services" (from the abbreviation of the three main forms of express delivery: courier services, express delivery services (in the narrow sense), and parcel delivery services).

A *courier service* is a professional delivery service in which goods (usually documents, small samples, patterns, or important spare parts up to 5 kg in weight) are accompanied personally during all stages of transportation from sender to addressee, without re-routing. Courier services are mostly found in inner-city areas and use cars, motorcycles, or bicycles. However, intercity deliveries between major national economic and administrative centers are also of importance. Courier services play only a minor role in international deliveries, mainly due to high transportation costs. The most important feature of courier services is the personal accompaniment of the transported goods. Courier services are the fastest possible form of express delivery, and generally the most expensive. The cost factor is of less significance for inner-city deliveries because several documents and parcels may be collected during a single journey, but cost considerations become more important for longer distances.

Express delivery (in the narrow sense) is a professional delivery service which carries goods from the sender to the addressee by grouping together large numbers of units and distributing them internally in accordance with a flexible transportation program with guaranteed delivery times. In Germany the guaranteed delivery times (e.g., same-day delivery, next-day delivery (24 hours)), usually apply to all distances (Bohnhoff et al., 1998). It is also possible to negotiate specific delivery times. The main feature of express delivery services is the guaranteed delivery time. Examples of these services are DHL, Federal Express, and TNT.

Specific forms of express delivery services are the express freight systems. These services specialize in the express delivery of large amounts of goods for industry. Generally, long-term contracts exist, and deliveries are undertaken for a few major clients. This form of express delivery service is to some extent integrated into the customer's logistics system. These services are often tailored for specific areas of industry (e.g., the pharmaceutical sector) or are the result of the outsourcing of transportation and logistics from industry to logistic service providers, and often carried out as just-in-time deliveries (see Chapter 16).

Parcel services are professional delivery services which primarily convey parcels (including documents) in accordance to a fixed, defined transportation program, using logistic networks with fixed running times for their specific goods (Bohnoff et al., 1998). Parcel services are the most standardized and automated of the express delivery services, coupling fast delivery times with low prices. To meet the requirements of the standardized transportation program of a parcel service, the transported goods have to be standardized to allow automated transportation and re-routing. Parcels in this context must conform to specific requirements (e.g., maximum weight 31.5 kg, maximum length 175 cm, maximum circumference 300 cm) (Bohnhoff et al., 1998). The requirements will differ between companies and countries, and there is a current trend for their abandonment for heavier goods (Brax, 1996). Automated transportation programs allow delivery times that come close to those of the express delivery services or even meet them, but exclude guaranteed delivery times and agreements on specific delivery times. Although no guarantees are given, the standardized, defined transportation process with fixed running times allows fairly reliable "expected delivery times" for parcels to a given destination – in practice, quasi-fixed delivery times. Parcel services represent the majority of all express delivery services, particularly for domestic deliveries. Examples of parcel services are UPS and Direct Parcel Distribution.

Generally, courier services and express delivery services are more expensive than parcel services. While courier services are predominantly limited to local, regional, or national areas, express delivery services and parcel services operate worldwide. This distinction is to some extent theoretical and academic. Furthermore, the terms "parcel services", "express delivery services," and "courier services" are often used with different meanings and interchangeably, so it is not always clear what is meant. Although for each of the three concepts specialized companies exist, in reality many delivery companies provide services that are a mix of the three concepts.

3. Organization of the CEP services system

In the following sections the main features of CEP services are described.

3.1. Transportation modes

CEP services, in particular parcel services and express delivery services, are intermodal and multimodal transportation services. They are multimodal because they use both roads, as the main mode, and also air and rail – and even waterborne transport was investigated in a field test on the river Rhine (Bohnhoff et al., 1997). Air and rail are used much more in the U.S.A. than in Europe, where transportation is almost exclusively by road. However, parcel services may use rail when this mode of transport is able to fulfil the express delivery requirements in terms of times and quality. Air transportation is used in large domestic markets, such as Canada and the U.S.A., and for intercontinental transportation.

Multimodal transportation is a characteristic of express delivery services and parcel services, and in many cases more than two modes of transport is used during a delivery. This is particularly evident in transportation by air. In contrast, courier services in European markets generally use only one mode, usually road, because direct delivery without re-routing and the shorter distances involved do not usually require changes in the mode of transport. Multimodal transportation is not uncommon in the U.S.A. and Canada.

3.2. Transported goods

All goods could be transported by CEP services. However, the features of express delivery services mean that certain types of goods are more suited to transportation by such services. Generally, goods and documents are transported that are time-sensitive and meet the requirements (i.e., dimensions and weight) of CEP services transportation programs. Additional features that may influence the use of express delivery services are the importance and value of the goods. Also, security is very important, and courier services are well suited to goods and documents that need high security, such as important documents (e.g., contracts, documents of tender, concepts, plans, photos), patterns, display items, and vital spare parts.

From the logistics point of view, frequent delivery of small amounts of goods is important for some customers (e.g., the retail trade, in particular small independent shops). This reduces stock costs and enables retailers to react quickly to customers' requirements because, if an item is not held in stock, it is possible to obtain it within a few days. Parcel services in particular are used for deliveries to small retail traders in cities, and also for deliveries of ordinary consumer goods.

Although large industrial companies have their own logistic networks or use specialized companies for transportation of materials, CEP services are also used where the amounts of products requiring transportation are small or where they have to be transported to a diversity of different destinations.

Last, CEP services are used in countries where the national postal services are unreliable or insecure for the transportation of goods or documents.

3.3. Networks – direct delivery versus re-routing

Generally, CEP services offer door-to-door delivery. The main difference is between courier services, on the one hand, and express delivery services and parcel services on the other. As already mentioned, courier services are characterized by the personal accompaniment of the parcels or documents being

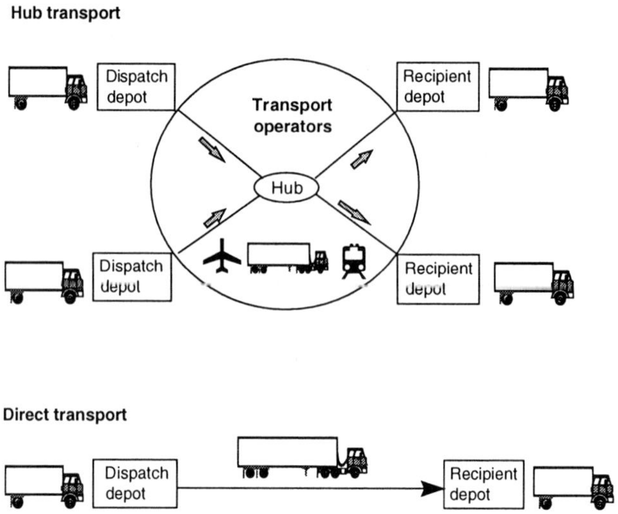

Hub transport

Direct transport

Figure 1. Hub transport/direct transport (source: Bohnhoff et al., 1998)

transported. This means that there is no re-routing or consideration of consignments. The goods are either transported from the shipper to the consignee directly as a single item, or individual items are collected and delivered. In the latter case the courier service collects the goods during a collecting tour and then delivers them on a delivery tour (e.g., within a city). This calls for a careful organization of the route (i.e., the order in which to deliver the items). This relatively simple transportation organization enables small firms or new entrants to run courier services. Generally, courier services offer the shortest transportation times, this often means less than an hour for inner-city deliveries.

Totally different and more sophisticated are the networks that express delivery and parcel services have to operate to fulfil their customers' requirements (i.e., a worldwide service with relatively low costs but of high quality).

The larger CEP operators tend to operate in a similar way. The organization for each country or geographical area is basically the same, consisting of one or more major sorting depots and several smaller satellite depots. Small satellite depots are located throughout the country, each serving a definite region. Within this region are lightweight vehicles that collect the goods during the day. The goods are then transported to the satellite depot (dispatch depot) and registered. In larger geographical markets, air transportation often fulfils longer distance transportation needs.

Depending on the number of items for a particular destination, one of two options is chosen for further transportation (Figure 1). If a there is a certain number of items for one specific destination (typically for deliveries between major economic centers), all the items are trans-shipped to heavy goods vehicles and transported directly to the destination satellite depot. This direct transportation has the advantage of shorter transportation times. Furthermore, the costly trans-shipment and sorting process in the hub is avoided. However, this direct transportation is only possible for large operators who need to transport a large number of items to a single destination and for those few routes where the number of items allows the utilization of the full capacity of at least one heavy goods vehicle.

Generally, most items are transported in bulk each evening by heavy goods vehicles or aircraft from the satellite depots to the major sorting depots. These central depots are normally located near motorways and/or airports in or near the geographical center of the country. The hubs are usually huge depots with very sophisticated sorting facilities. The efficient use of the automatic sorting facilities and the resulting reduction in costs depends on there being a significant proportion of standardized items passing through the facilities. Therefore, these automatic facilities tend to be used predominantly by parcel services that involve the transport of a large number of standardized parcels. The use of sorting facilities enable efficient and quick automatic sorting of parcels according to their destination. Huge automatic conveyer machines and belts are used, incorporating large scanners that automatically read and process the destination information, which is carried in a barcode attached to each parcel. As well as sorting the parcels according to their destination, the barcode allows the progress of parcels to be tracked and traced. Parcels that do not meet the standardization requirements or that carry a barcode that cannot be read by the scanners are diverted to conveyer belts where registration and sorting is carried out manually.

At the end of this sorting process the goods are loaded on heavy goods vehicles or aircraft according to their destination, and transported during the night or in the very early morning to satellite depots (recipient depots). At the satellite depots all the parcels are registered and trans-shipped to light vehicles for delivery the next day. The difference between a "dispatch depot" and a "recipient depot" is a theoretical distinction and depends on the time of day. During the afternoon, when all the collected goods arrive at the satellite depot, it is a "dispatch depot"; in the early morning, when the goods arrive from the sorting depot for delivery, the satellite depot acts as a "recipient depot". The movement between the depot and the hub is often referred to as line-haul.

During the sorting process in the hub, goods for foreign destinations are grouped together and transported via road or air to the destination countries. Goods for foreign destinations must often meet requirements specific to the country of destination, such as customs regulations or transportation documentation; also, national transportation monopolies may have to be taken into account. Thus a significant amount of paperwork as well as skilled staff are required, and processing of the goods is costly. To reduce costs, processing for foreign destinations is generally centralized at one or only a very few depots.

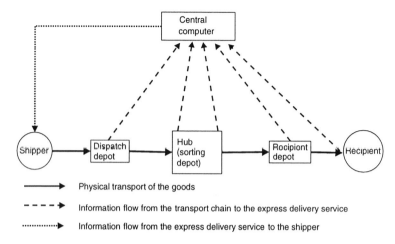

Figure 2. Tracking and tracing

3.4. Communications and information technology

Besides costs and delivery time, quality of service is a major requirement of customers of CEP services. In particular, for time-sensitive goods the shipper and the consignee need to be informed of when the item will be delivered or at which point in the transportation chain the goods are at any given moment.

All leading operators in the CEP market have made major investments in information technology systems, commonly referred to as "tracking and tracing systems" (see Chapter 34). These systems use barcode technology and laser equipment to monitor the progress of a shipment from the point of collection to the point of delivery. A barcode is affixed to each item, enabling the identification of the item. In addition, the collection and recipient depot can be given in the barcode. At every point of trans-shipment, registration of the shipment is possible at the collection depot (or even when collected from the shipper), when entering the hub, when leaving the hub, at the recipient's depot, and at the point of delivery (Figure 2).

The scanned information, together with the date, time, and location, is included and stored in the operator's computer system. The progress of a shipment can thus be given immediately to an inquiring customer. Furthermore, some larger operators allow their customers to trace the progress of their shipments directly via the internet (by using a password or identification number).

These tracking and tracing systems are a direct consequence of the transportation of large amounts of goods and the re-routing process in the hubs. This technology is the only means by which single items can be identified during the transportation process and customers informed of progress. These systems are usually not necessary for smaller courier services that generally have a better overview of their transported goods, since they move smaller amounts over shorter time spans and distances.

4. The market for CEP services

The availability of data in the field of CEP services is relatively poor. The reasons for this are the imprecise definitions, the lack of official statistics, and the keen competition, which encourages companies to be secretive (see Chapter 33; Savy, 1996; Bohnhoff et al., 1998).

Until the 1970s, the responsibility for the express delivery of parcels and letters was often the task of national postal services, particularly in Europe and North America. Since then, many monopolies have been abolished, and private companies have been given opportunities to create services in these fields. Today, in most countries the transportation of parcels and express delivery is wholly or largely deregulated. Only transportation of ordinary letters remains the task of national postal services in some countries, but even this is expected to be deregulated soon. The liberalization of express delivery has led to a more than proportional growth in the private CEP sector, partly to the disadvantage of national postal services.

Today, while there are few regulatory barriers to market entry, other constraints exist. These barriers to market entry are technical and economic, in particular for express delivery services and parcel services. It takes a lot of know-how, time, and finance to set up an efficient, high-quality network, especially in larger countries (Savy, 1996), and to provide a worldwide service. As a result, the emergence of new parcel or express delivery services is rare. Changes in the market structure mainly arise due to mergers and takeovers, or through changes in the fields of activity of companies already involved in similar areas of transportation.

A little different is the situation regarding the courier sector. The characteristics of these services, which are mainly provided locally and regionally or specialize in the transportation of specific products, result in lower barriers to entry in terms of finance and organization. Examples are inner-city bicycle courier services and courier services that concentrating on specific market areas (e.g., the transportation of documents between law courts and lawyers).

5. The future of CEP services

The further development of the transportation sector has to be seen in the context of the general development of the economy, since transportation depends to some extent on the general economic situation. However, some specific developments and changes within the economy are expected to influence considerably the further growth of the CEP market. These developments and changes include the continuation of the trends that have led to the emergence of CEP services in the past. In particular, the demand for faster transportation, the pressure for transportation services as a result of increasing trade, and the trend toward smaller products of higher value are expected to stimulate further growth in the express delivery sector.

Just-in-time delivery in some industrial sectors (automobile, information technology, pharmaceutical industries) is anticipated by many to be systematically extended (Brax, 1996), and thus will contribute to further growth in CEP services.

The trend toward increasingly compact products in several sectors of industry is expected to improve the cost–benefit ratio of express delivery by decreasing the transportation cost share. Smaller products will enlarge the market for CEP services, because more items will be of a size suitable for carriage by express delivery. Also, the increasing value of products requires rapid transportation, because companies want to reduce the interest costs bound up in stock and inventories.

Globalization and increased worldwide trade are expected to continue to grow, accompanied by further reductions in trade barriers. Out-sourcing and the focus on a few key products to increase efficiency gains by economies of scale (specialization advantages) are expected to increase in the production sector (Teufers, 1996). Furthermore, decentralization of production is expected to increase, by firms moving plants to or creating plants in regions where production of components or finished products is the most favorable (Bohnhoff et al., 1998).

The ongoing integration of the European Union (EU), creating a common economy with increased trade relations, is increasing demand for transportation, including express delivery. A reduction in customs regulations, existing monopolies of postal services (letters), and differing transport regulations is also taking place with this economic integration. The EU to include central and eastern European countries will facilitate development of these relatively new markets for the express delivery sector. The impications of the North American Free Trade Area are likely to take longer to be realized.

At a more micro-level, the requirements of the retail trade in central urban areas are, to a large degree, met by express (in particular parcel) services (Schaefer and Sage, 1997). The ability of CEP services to deliver small units and small amounts of goods very frequently so that shops with low stock can ensure that customers' orders are satisfied quickly is expected to contribute to the further growth of the express delivery sector.

Although most parts of the express delivery sector are now totally deregulated and liberalized, some parts still remain as national monopolies (in particular the delivery of letters). The liberalization and deregulation process seems likely to continue, providing new opportunities for CEP services. In Germany, for example, the market for mass delivery of letters was opened up in 1998. Free competition for the whole of the letter market is expected to be achieved by 2002 (Anon., 1996).

Some sectors of the economy already use express services extensively and on a regular basis. The high-technology sector (information technology, aerospace industries) is one in which the savings in stockholding costs generally more than offset the higher costs of express delivery. Other sectors, such as mail order and the car distribution business, use express delivery both for financial reasons and to improve their image by ensuring that stock and spares are rapidly available to customers (Brax, 1996).

5.1. Impediments to CEP services

Although the market for CEP services today is widely deregulated, there are still some impediments to the efficient development and growth of the market due in general to the increasing demand for transportation services (Bohnhoff et al., 1998).

Whereas there are only a very few impediments to transportation within a country, mainly in connection with the remaining national postal monopolies, impediments to international and cross-border transportation affect the efficiency of express delivery. These impediments result mainly from different national regulations concerning the transportation of freight, different customs regulations, and different organizational structures of existing national CEP services. These impediments lead to inefficiency, in particular at borders, because additional documents for customs or taxation have to be produced and carried along the transported chain. Furthermore, different attitudes toward express delivery exist (e.g., the definition of a parcel differs between countries). Historical differences between countries may also hinder the efficiency of the transportation chain. For example, different trailer connections for heavy goods vehicles can prohibit the use and changeover of vehicles; and some countries use numerical zip codes, whereas others use alphanumeric systems, which affects the production, scanning, and processing of labels.

6. Outlook

Despite the remaining impediments to its growth, the express delivery sector is very efficient, and with further deregulation, liberalization, and privatization in national markets this sector is expected to expand. In Europe, in particular, the economic convergence of the EU, and the ensuing harmonization of legislation and regulations, is expected to reduce most of the insitutional impediments to cross-border transportation. As a result, the express delivery sector can be seen as a very efficient part of the transportation sector with generally good future prospects in terms of further development and growth.

Where problems may remain, or even grow, is within urban areas. The "last-mile problem" is particularly acute for express services where speed and reliability are at a premium. Increased levels of urban traffic congestion pose serious problems to suppliers trying to offer these service attributes, problems that are unlikely to diminish in the future. Conversely, the rise in urban sprawl in countries such as the U.S.A. and Canada makes it difficult to develop, at low cost, a full distribution network.

References

Anon. (1996) "Bonn: Wettbewerb für Massensendungen", *DVZ*, 30 May.

Belohradsky, E. (1997) "Schnelldienste verdienen am Globalisierungs-Fieber", *Die Welt*, 16 September.

Bohnhoff, A., O. Tegg, H. Schäfer, D. Sage, R. Jorna and J. Soergel (1997) "European parcel transport – state-of-the-art and trends in the parcel sector, Aachen". Report 1 of the Case Study Parcel Services in the MINIMISE project, funded by the European Commission under the Transport RTD Programme of the 4th Framework Programme.

Bohnhoff A., O. Tegg, H. Schäfer, D. Sage, R. Jorna and J. Soergel (1998) "European parcel transport – situation, needs and contribution for a better transport system organisation, Aachen". Deliverable 4 of the MINIMISE project, funded by the European Commission under the Transport RTD Programme of the 4th Framework Programme.

Brax, B. (1996) "France", Express delivery services – report of the hundred and first round table on transport economics, Paris, pp. 5–23.

Manner-Romberg, H. (1995 et seq.) "Der KEP-Markt – Marktbeobachtung" (http://www.kurier.com/bdkep/)

Manner-Romberg, H. (1998) "KEP-Markt – Wegweiser und Marktübersicht für den schnellen Versand, Hamburg" (http://www.kurier.com/bdkep/quovadis.htm).

Sage, D. and H. Schaefer (1997) "City-Logistik in Siegen (project-concept)", Aachen, unpublished report.

Savy, M. (1996) "France", Express delivery services – report of the hundred and first round table on transport economics, Paris, pp. 5–23, 27–55.

Teufers, H.-P. (1996) "Der Paketmarkt im Wandel", *Aktuelle Themen der Güterverkehrswirtschaft*.

Part 8

RESEARCH AND DEVELOPMENT

Chapter 31

HAZARDOUS GOODS

KATHLEEN L. HANCOCK

The University of Massachusetts at Amherst

1. Introduction

The effective and efficient transport of goods is essential to global economic viability. Many of these goods can be hazardous if not handled properly. Officials within the transportation industry define hazardous materials as substances or materials that may pose an unreasonable risk to health, safety, and property when transported, and which have been so designated. These include hazardous substances, hazardous wastes, marine pollutants, and elevated-temperature materials, and are categorized as belonging to one of the nine hazard classes shown in Table 1 (Keller and Associates, 2000).

Although information on transportation of hazardous materials is incomplete and sometimes inconsistent, the U.S. Departments of Transportation and Commerce (1999) estimated that approximately 1.6 billion tons were shipped in the U.S.A. in 1997, accounting for 264 billion ton-miles. As a percentage of total goods shipped, this represents 14.1% of tons and 9.9% of ton-miles. Table 2 provides shipment characteristics of this information by mode of transportation. With this level of exposure, the potential for harm is high and warrants consideration by shippers and transporters of how to minimize potential risks when determining how and where goods move from origin to destination.

2. Regulations

Recognition of these risks has prompted the introduction of regulations whose fundamental objectives are to (1) provide safety for workers and the public, (2) protect property and the environment, (3) minimize cost to the public, industry, and government, and (4) minimize economic and social disruption.

The basis for many international, national, regional, and modal regulations originated with the *United Nations recommendations on the transport of dangerous goods*, first published in 1957, now available in its eleventh revised edition (United

Handbook of Logistics and Supply Chain Management, Edited by A.M. Brewer et al.
© *2001, Elsevier Science Ltd*

Nations, 1999). This document includes a comprehensive criteria-based classification system for substances, standards for packaging hazardous materials, a system of communicating the hazards of substances in transport through hazard communication requirements that cover labeling and marking of packages, placarding of tanks and freight units, and documentation and emergency response information that is required to accompany each shipment. Currently, these

Table 1
Categorization of hazardous materials

Class	Hazard
1	**Explosives**
1.1	Explosives with a mass explosion hazard
1.2	Explosives with a projection hazard
1.3	Explosives with predominantly a fire hazard
1.4	Explosives with no significant blast hazard
1.5	Very insensitive explosives; blasting agents
1.6	Extremely insensitive detonating articles
2	**Gases**
2.1	Flammable gases
2.2	Non-flammable, non-toxic compressed gases
2.3	Gases toxic by inhalation
2.4	Corrosive gases
3	**Flammable liquids**
4	**Flammable solids**
4.1	Flammable solids
4.2	Spontaneously combustible materials
4.3	Dangerous when wet materials
5	**Oxidizers and organic peroxides**
5.1	Oxidizers
5.2	Organic peroxides
6	**Toxic materials and infectious substances**
6.1	Toxic materials
6.2	Infectious substances
7	**Radioactive materials**
8	**Corrosive materials**
9	**Miscellaneous dangerous goods**
9.1	Miscellaneous dangerous goods
9.2	Environmentally hazardous substances
9.3	Dangerous wastes

Table 2
Transportation of hazardous materials in the U.S.A. in 1997

Mode of transportation	Tons		Ton-miles		Average miles per shipment
	Number (thousands)	%	Number (millions)	%	
For-hire truck	336 363	21.5	45 234	17.1	260
Private truck	522 666	33.4	28 847	10.9	35
Rail	96 626	6.2	74 711	28.3	853
Water	143 152	9.1	68 212	25.9	S
Air (includes truck and air)	–		95	–	1 462
Pipeline	432 075	27.6	S	S	S
Multiple modes	6 022	0.4	3 061	1.2	645
Other/unknown	17 459	1.1	1 837	0.7	38
Total	1 565 196	100.0	263 809	100.0	113

Source: U.S. Department of Transportation and U.S. Department of Commerce (1999).
Key: –, data cell equal to zero or less than 1 unit of measure; S, data do not meet publication standards because of high sampling variability or other reasons.

recommendations are being reformatted as a "Model Regulation." This will provide structural harmony of transport regulations and easier adoption within national legislation of countries throughout the world, eliminating the need for countries to reissue the regulation in their own legislative formats. In the U.S.A., current regulations originated with the Transportation Safety Act of 1974, followed by the hazardous materials regulations consolidation of 1976, and most recently with The Hazardous Materials Transportation Uniform Safety Act of 1990 (HMTUSA) and its numerous amendments that are codified in 49 CFR Parts 100–199.

3. Logistics for hazardous materials transport

In response to these regulations and good business practice, shippers and carriers have implemented techniques for assessing risks associated with transporting hazardous materials and for routing and scheduling vehicles to minimize these risks (Shaver and Kaiser, 1998). As a result, significant advances in methods for performing these assessments have occurred over the past few decades.

Decisions about the movement of hazardous materials involve many interested parties, or stakeholders. Therefore, tools and models for providing input to these decisions range from those that minimize (or maximize) a single objective to those that find the "best" solution between several, often competing, objectives. Data used for these decisions range from deterministic, where values are known or measurable, to stochastic, where uncertainty of values is included directly in the decision-making process.

3.1. Risk assessment

Risk assessment can be defined as a formal process for understanding risk. The first step in managing hazardous materials transport is, thus, the understanding of risks associated with it. This process has taken various forms based on need and available resources, from the use of simple measures as proxies for risk, to direct measures of consequences, to constructed indices that are then used as comparative measures.

At a basic sketch planning level, government and hazardous materials industry officials have used an estimate of shipments to characterize hazardous materials traffic in the U.S.A., implicitly using number of shipments as a measure of the level of risk to the transport community and general public. However, as noted in that document, the number of shipments provides only partial information about the risks inherent in hazardous materials transportation. Many additional factors should be considered for a comprehensive risk assessment to be performed, including type and quantity of material; type and amount of packaging, handling, and transport; and type of infrastructure over which the hazardous material is being shipped.

As more detailed assessments are required, many analytic techniques have been extended from those developed and used by the nuclear power industry. Fault tree analysis methods and the more commonly recognized three-stage framework shown in equation (1) have been implemented:

$$R = P_1 \times P_2 \times P_3, \tag{1}$$

where R is the risk measurement along a transport segment, P_1 is the probability of an undesirable event (crash), P_2 is the probability of a release given an event has occurred, and P_3 is the estimation of the magnitude of consequences given that a release has occurred.

Each component of equation (1) produces one or more probability distributions, with two of the three being conditional distributions. The resulting distribution provides an estimation of risk for a given type and amount of material (inherent in P_3), being transported by a given mode and configuration (inherent in P_1 and P_2) over a given infrastructure (inherent in P_1). When R is combined with the number of in the surrounding population and/or nature and area of exposed environments, a determination of risk is obtained. This determination includes measures such as number of predicted fatalities, injuries, impacted facilities, and/ or type and amount of damage to an impacted area.

Estimates for general crash, or accident, rates are readily available in the literature. However, estimates of crash rates involving hazardous materials, probabilities of release, and estimates of resulting consequences are generally not available to support detailed estimates except where focused studies have been performed. Similarly, when using crash/accident/incident data, three principal difficulties exist in creating specific estimates (List et al., 1990):

(1) selecting data from the set of reported crashes/accidents/incidents that represent relevant events;
(2) determining an appropriate measure of exposure that is consistent with the problem being addressed; and
(3) recognizing the uncertainty in the estimates as a result of small numbers of reported crashes/accidents/incidents in specific categories.

Another commonly used method of assessing risk is the use of a risk profile. This is a plot of $1 - F(x)$ versus x, where x is some measure of consequences from a specified activity, such as the number of fatalities, and $F(x)$ is the cumulative distribution function of that consequence. This provides an exceedance probability, or the probability that consequences will exceed a given level. Kaplan and Garrick (1981) and Kaplan (1982) have provided an analytical procedure for constructing risk profiles.

3.2. Routing

For transporting hazardous materials, the level of risk can be managed by making appropriate routing and scheduling decisions. Techniques for routing and scheduling vehicles have been developed with improved methods for minimizing measures of risk and determining trade-offs between risk and other system performance measures.

The simplest routing algorithm solves the classical least "*cost*" route problem. This procedure finds the route from an origin to a destination that has a minimum (or, in some applications, maximum) measure or "*cost*." This measure must be additive across links, or segments, of the route. Adaptation of this procedure to hazardous materials routing has traditionally minimized a single metric defined by the user, such as *travel time* or *accident likelihood*. Another common metric of risk has been *exposure*, which is usually defined as the adjacent population within some prescribed bandwidth multiplied by link length. These link-based risks are summed to determine an overall risk for each route, and the route with the lowest total risk is selected. As a note of caution, most routes optimized on minimizing *exposure* result in much longer and circuitous trips with increased accident likelihoods because of this circuity.

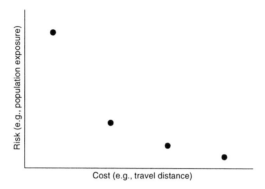

Figure 1. Results using an early multi-objective model approach to risk estimation.

With the expansion of desktop computing and geographic information systems, the single metric has been enhanced to include risk measures similar to that defined by equation (1), which incorporates accident rates by highway segment, type of material, and type of transport. This measure assumes that the entire population within the band is equally at risk and that, where bands around connecting links overlap, the population may be double counted, resulting in overestimation of the risk measure.

Because the use of a single risk measure for hazardous materials models fails to represent the competing needs of risk reduction – usually minimizing exposure – and cost optimization – usually minimizing distance – multiple objective formulations have been developed. When this occurs, the concept of an optimal solution is replaced with that of an efficient solution. The earliest effort to explicitly include multiple objectives were performed by Shobrys (1981) and others in the early 1980s. Shobrys used different weights to combine two objectives, distance and population cost to obtain several non-dominated or pareto-optimal solutions, which explicitly represent trade-offs between different objectives. An example of results from this approach is shown in Figure 1. This and other early multi-objective models are based, explicitly or implicitly, on identifying some weighting scheme to make the multiple objectives commensurable. A limitation with this approach is that only a subset of possible routes are identified, providing a partial set of solutions.

In contrast to the weighting scheme is the constraint method, which reduces the problem to a single-objective problem by converting all but one of the objectives into constraints. Each link is assigned a "cost" vector that characterizes that link in terms of constraining criteria. These constraints provide bounds on the converted cost vector. The decision maker examines different routing scenarios by changing the goal attainment levels and the priority for their attainment, thus having considerable flexibility. However, this is also its greatest limitation since inferior solutions can be selected if the goal attainment levels lie inside the defining curve.

When multiple shipments are necessary, a single population or region may be at greater risk if all shipments are transported over the same route. Therefore, risk equity, or spreading the risk across multiple paths, has been considered. However, the set of paths that most equitably distributes risk may be more hazardous overall then a single route with multiple shipments.

3.3. Routing and scheduling

Inclusion of scheduling in routing decisions is considered important in managing risk because risk estimates depend on underlying parameters that are time-of-day dependent. For example, accident rates vary by light conditions and by congestion that is dependent on time of day and day of week. Time-of-day dependency of

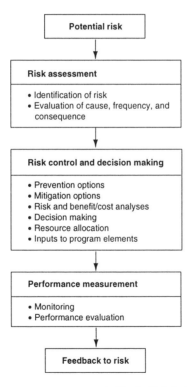

Figure 2. Risk management strategy recommended in the U.S.A. (source: U.S. Department of Transportation, 1995).

segment attributes implies that scheduling decisions must be made in conjunction with routing decisions and that management decisions must include this combination explicitly.

Much current research is focusing on developing algorithms that incorporate network segment attributes that vary with time into the optimization process. Sulijoadikusumo and Nozick (1998) propose an approach to find the set of non-dominated paths from a given origin to a given destination when waiting in the network is forbidden and the departure time is given. Their proposed heuristic is similar to the class of label-correcting shortest-path algorithms. Miller-Hooks and Mahmassani (1998) assume that both travel time and risk measures are not constant over time and are not known with certainty, and thus are incorporated into the solution procedure as random variables with probability distributions functions that vary over time.

Simultaneous routing and scheduling analyses require more information and more sophisticated analysis methods than simpler static routing solutions. However, current research indicates that incorporation of changes in trip departure times can reduce some of the path selection risk measures by orders of magnitude, thus justifying the increased cost and effort.

4. Risk management

Many responsibilities associated with the transport of hazardous materials have been placed with jurisdictional governments including (Abkowitz et al., 1991):

(1) community preparedness and emergency response,
(2) risk evaluation and communication,
(3) routing and siting considerations,
(4) data collection and information management, and
(5) inspection and enforcement.

As part of this, policy-making agencies have used risk management techniques to establish designated routes for vehicle use or, in limited areas, exclusion routes that prohibit vehicles carrying hazardous materials. In some cases, agencies have established time-of-day constraints and lane restrictions.

Risk management has been defined as the systematic application of policies, practices, and resources to the assessment and control of risk affecting human health and safety and the environment (U.S. Department of Transportation, 1998). Hazard, risk, and benefit/cost analysis are used to support development of risk reduction options, program objectives, and prioritization of issues and resources. Figure 2 provides a summary of a risk management strategy recommended in the U.S.A. States are not required to designate routes, but only to follow established rules if they do so (U.S. Department of Transportation,

Table 3
Route designation factors

Population density	Type of highway
Type and quantity of hazardous materials	Emergency response capability
Results of consultations	Terrain considerations
Route continuity	Alternate route availability
Effects on commerce	Climatic conditions
Delays in transportation	Congestion
Accident history	

1995). These rules are intended to permit individuals with limited knowledge of hazardous materials safety issues to analyze and compare risks and accordingly designate routes for transport of hazardous materials. Route designation factors are given in Table 3.

A current U.S. initiative is a study on the applicability of applying Hazard Analysis and Critical Control Points (HACCP) to the management of risks posed by hazardous materials transportation. Results of this study are anticipated to lead to voluntary, "best-practice" risk management techniques applicable to various parties involved in hazardous materials transportation, and may eventually identify a need for changes to the current regulatory system. The HACCP system, first developed by the Pillsbury Company in cooperation with the National Aeronautics and Space Administration, consists of the following steps:

(1) analyze hazards,
(2) identify critical control points,
(3) establish preventive measures with critical limits for each control point,
(4) establish procedures to monitor critical control points,
(5) establish corrective actions to take when monitoring shows that a critical limit has not been met,
(6) establish procedures to verify that the system is working properly, and
(7) establish effective record keeping.

While most recent experience in broadening the application of the concept occurs with respect to food safety, the HACCP approach may by relevant for other safety systems, including hazardous materials transportation.

5. Technology

Intelligent transportation systems (ITSs) include advanced technologies that can be used to enhance the safe transport of hazardous materials, including vehicle control and driver information systems, heavy vehicle detection systems, and driver/vehicle performance monitoring systems (see Chapter 34). While not specifically used for assessing risks or determining routes, these technologies can provide information for increasing the safety of hazardous materials transport.

Vehicle control systems assist drivers through tracking surrounding vehicles and notifying the driver of unsafe conditions to assist in collision avoidance. Although still in the developmental phase, these systems, when operational, should provide techniques to reduce crashes. Driver information systems improve the quality and real-time availability of traffic and route-availability information that can be provided to drivers and dispatchers. Automatic vehicle location (AVL) technologies provide real-time information about vehicle location. At any time, the dispatch office has the ability to locate vehicles geographically, which in combination with driver information systems can be used for routing decision support en route. Driver performance systems can monitor the performance of a driver or vehicle through recorded information such as time on the road, vehicle speed, and time between travel breaks. Weigh-in-motion (WIM) systems measure vehicle weight as a vehicle travels over the system. This allows enforcement personnel to efficiently identify vehicles operating at unsafe weights on the highway.

Advanced technologies complement efforts to reduce risk in the transport of hazardous goods. Enhanced communication afforded by these systems can also provide improvements in logistics and productivity.

References

Abkowitz, M., P. Alford, A. Boghani, J. Cashwell, E. Radwan and P. Rothberg (1991) "State and local issues in transportation of hazardous materials: toward a national strategy", in: *Freight transportation: truck, rail, water, and hazardous materials*, TRR No. 1313, pp. 49–54. Washington, DC: National Academy Press.

Kaplan, S. (1982) "Matrix theory formalism for event tree analysis: application to nuclear risk analysis", *Risk Analysis*, 2:9–18.

Kaplan, S. and T. Garrick (1981) "On the quantitative definition of risk", *Risk Analysis*, 1:11–27.

Keller and Associates (2000) *Hazardous materials compliance pocketbook*. Neenah: Keller and Associates.

List, G., P. Mirchandani and M. Turnquist (1990) *Logistics for hazardous materials transportation: scheduling, routing and siting*. Washington, DC: U.S. Department of Transportation.

Miller-Hooks, E. and H.S. Mahmassani, (1998) "Optimal routing of hazardous materials in stochastic, time-varying transportation networks", in: *Forecasting, travel behavior, and network modeling*, TRR No. 1645, pp. 143–151. Washington, DC: National Academy Press.

Shaver, D. K. and M. Kaiser (1998) *NCHRP synthesis 261, criteria for highway routing of hazardous material*. Washington, DC: National Academy Press.

Shobrys, D. (1981) "A model for the selection of shipping routes and storage locations for hazardous substance", Ph.D. dissertation, Johns Hopkins University, Baltimore.

Sulijoadikusumo, G. S. and L.K. Nozick (1998) "Multiobjective routing and scheduling of hazardous materials shipments", in: *Freight transportation planning, motor carrier size and weight issues, international trade, and hazardous materials transport*, TRR No. 1613, pp. 96–104. Washington, DC: National Academy Press.

United Nations (1999) *United Nations recommendations on the transport of dangerous goods: model regulations*, 11th revised edn. New York: United Nations.

U.S. Department of Transportation (1995) *Guidelines for applying criteria to designate routes for transporting hazardous materials*, FHWA-SA-94–083. Washington, D.C.: U.S. Department of Transportation, Federal Highway Administration, Office of highway Safety.

U.S. Department of Transportation (1998) *Risk based decision making in the hazardous materials safety program*. Washington, D.C.: U.S. Department of Transportation, Research and Special Programs Administration (available online: http://hazmat.dot.gov/riskprog.pdf [2000, March 13]).

U.S. Department of Transportation and U.S. Department of Commerce (1999) *United States hazardous materials, 1997 economic census, transportation, 1997 commodity flow survey*. Washington, DC: U.S. Department of Transportation and U.S. Department of Commerce.

Chapter 32

DYNAMICS AND SPATIAL PATTERNS OF INTERMODAL FREIGHT TRANSPORT NETWORKS

HUGO PRIEMUS
Delft University of Technology

ROB KONINGS
OTB Research Institute, Delft

1. Introduction

The spatial pattern of freight transport is determined not only by the location pattern of manufacturers and consumers, but also by the transport networks of transport and distribution companies and logistic service suppliers. This is explained by the rising trend of out-sourcing transport and logistic services to professional providers.

While it is the case that production and consumption are the basic sources for transport flows, these transport flows may have very diffuse patterns and involve different logistical requirements that demand different kinds of logistic and transport chains. Therefore, different modes of transport involving different infrastructural facilities are required (Figure 1). Alert producers, shippers, and carriers must weigh matters up again and again. As a result, incidental and more structural changes may occur in the pattern of transport flows. In other words, production, distribution, and logistics introduce dynamics into the networks and terminals for goods transport. This sets high performance requirements for a transport system, which needs to be characterized by intermodality to link all segments of the total transport chain.

This chapter relates the dynamics of transport flows to terminal and network configurations and evaluates the implications for the performance of intermodal transport. Tools are presented to integrate the intermodal transport system with the logistical requirements of transport flows.

The structure of this chapter is as follows. In Section 2 we briefly discuss some basic notions about the position and function of transport networks and terminals in the supply and distribution chain. This chain can be characterized by a flow of goods, with value added several times, until ultimately the goods arrive at the final

Handbook of Logistics and Supply Chain Management, Edited by A.M. Brewer et al.
© 2001, Elsevier Science Ltd

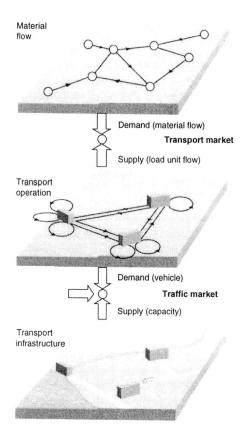

Figure 1. Basic relationships between flows of goods and transport flows (source: Ruijgrok., 1992).

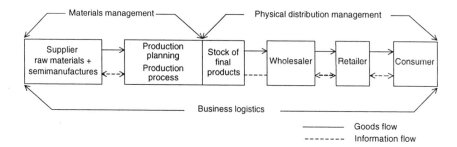

Figure 2. Flow of goods from the supplier of raw materials to the consumer (source: Van Goor, 1986).

consumer. This flow is directly related to a transport chain. The spatial patterns of goods transport are in a state of flux as a result of the constantly changing production, distribution, and transport environment. In Section 3 we discuss the driving forces behind these changes and the implications for intermodal transport. The basic requirements of intermodal transport networks are the subject of Section 4. In Section 5 the development of intermodal transport networks is illustrated with some empirical evidence from Europe. A more theoretical analysis of these network developments is given in Section 6, where tools for evaluating network performances are presented. The chapter ends with some findings about the major requirements of intermodal transport networks to anticipate the dynamics of goods transport (Section 7).

2. Basic patterns of transport flows: The transformation of the production chain into the transport chain

Several production chains can be distinguished within a supply chain. The production chain usually starts by extracting raw materials. Firms involved in these activities deliver their final products as production factors to firms active further on in the production chain; they transform the products into new products with higher added value. This process of delivering intermediate products and creating added value occurs several times within the production chain. At the end of this chain the final consumer is reached.

On the basis of these fundamental notions of its function, the production chain could be transformed into a flow of goods, which starts with the supplier of the raw materials and ends with the consumer (Figure 2). The flow of goods is focused on the client, with intermediate actors and consumers. At the same time, the needs of the client initiate an information flow that triggers intermediate actors upwards in the chain and, ultimately, the producers. All in all, the "material management" part is related to the flow of goods that enter firms, while "physical distribution management" refers to the flows that leave the firms.

If we transform these notions of flows of goods into a transport network, schemes such as those drawn by INRO-TNO (1993) and Vermunt (1993) are generated (Figures 3 and 4). The basic idea is that collection, distribution, and trunk networks can be distinguished within a transport chain. The trunk network connects collection and distribution nodes, which themselves provide the connections to the original origin (collection) or final destination (distribution) of the goods. Generally, the nodes that connect the collection or distribution networks with the trunk networks present themselves as the natural locations where exchange of load between (different) modalities takes place. This is the major function of terminals within the transport chain.

Transporting goods along the trunk network requires the consolidation of transport flows. Opportunities for the consolidation of transport flows depend primarily on the demand for transport, taking into account spatial, time, and product requirements. In other words, logistical requirements will strongly influence the organization of the transport chain. According to Vermunt (1993) the collection or distribution network is characterized far more by activities involving transformation and handling functions, while on the trunk network activities are concerned above all with the pure transport function.

In relation to the way collection, trunk movements, and distribution are organized, many different network structures are conceivable, varying from very simple to rather complex. As a result, the role of terminals in these structures may also differ. Although many variants are possible, they can be derived in some way

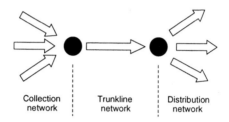

Figure 3. The structure of the transport chain according to INRO-TNO (source: INRO-TNO, 1993).

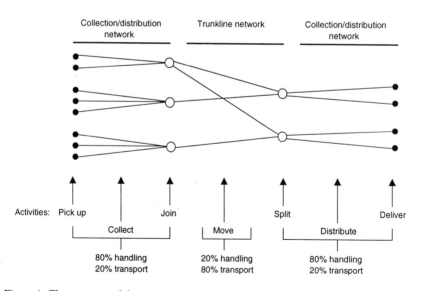

Figure 4. The structure of the transport chain according to Vermunt (source: Vermunt, 1993).

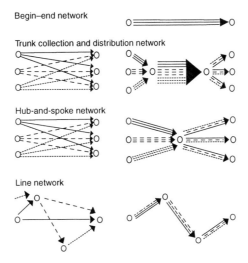

Figure 5. Basic freight bundling concepts (source: Kreutzberger, 1999).

or other from the following basic freight bundling network concepts (Figure 5) (Kreutzberger, 1999):

(1) begin–end network;
(2) trunk collection and distribution network;
(3) hub-and-spoke network; and
(4) line network.

The basic features of transport networks and the goods flow chains seem, however, to be insufficient to derive universal relationships that are valid for the total supply chain. Transporting raw materials or intermediate products is quite different from transporting final consumer goods. Therefore, differences in product or market characteristics will result in different logistic characteristics. Coopers and Lybrand (1989) have described the relationships between these characteristics. It transpires that network and node profiles will diverge for different segments of the supply chain. In other words, which kind of networks emerge and which node characteristics result will depend strongly on the place within the goods flow chain (Figure 6). Moreover, because the complexity of the goods flow chain increases at the end of the chain, even more complex and different chain structures are conceivable.

To summarize, it has been shown that transport nodes are closely related to the production chain, the supply chain, the logistical chain, and the transport chain connected with each other by network structures. The flow patterns of goods are the central variables, which relate production, transport, and distribution with each other.

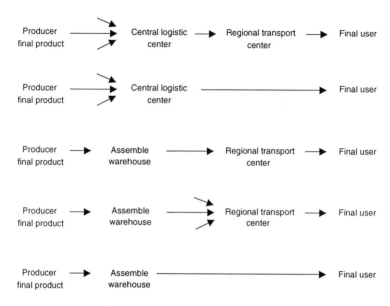

Figure 6. Differentiated goods flow chain (source: Vermunt, 1993).

3. Driving forces for changing patterns of goods flow chains: Starting-points for development of intermodal transport networks

In general terms, production and distribution chain cost reductions and quality enhancement are pursued. Sometimes production can become cheaper by increasing transport costs, so that on balance the costs are lower. Increasing the reliability of delivery can enhance quality. Concerns and logistic service providers try increasingly to arrive at chain integration. In doing so, improvement of the cost/quality ratio is always pursued. The scale on which alternatives are compared is increasingly the whole world – global sourcing, global production, global marketing. Through growing chain integration and the increasing computerization of society, the dynamics in the chain will be considerable in the years to come.

The world is in a state of flux. Business tries to anticipate changes by repeatedly making a trade-off between production costs and purchasing prices, on the one hand, and transport costs on the other. From the viewpoint of global sourcing, many firms increasingly see the whole world as a possible purchasing market for raw materials and semimanufactures. Firms try to purchase goods from anywhere in the world at the lowest integrated costs. Product assembly also takes place at locations where an advantage in terms of integrated costs can be gained. This approach calls for flexibility, so that changes in production, supply, and transport can be implemented without high costs. The withdrawal of firms to "core business" fits in with this. Parts of the firm that do not fit in with the core activities of the firm are disposed of, or farmed out to third parties. In this way fixed costs are converted into variable costs that can be better controlled.

From the perspective of global marketing, the whole world is seen as a possible market. Higher transport costs are compensated for by lower costs of production (e.g., lower labor costs) and/or by higher total returns. Changes in the international division of labor come about through changes in comparative advantages between countries. As production processes globalize, transport flows increase accordingly; these flows also become larger, but the need for flexibility still remains.

The economic integration of Europe gives a vigorous impulse to the internationalization of production processes inside and outside Europe. The growth of European distribution centers is a current phenomenon. Here we see obvious concentration processes in distribution. The development of a "post-Fordist" environment is characterized by more and temporary relationships between businesses and a need for just-in-time (JIT) deliveries. These circumstances tend to reduce warehousing and increase the integration between the elements of the production system in a complex network of relationships. The flexibility to adapt to constant fluctuations of origins and destinations is a major requirement in this business environment. In such an environment, the transport function is closely integrated to production and distribution, and is the main element in minimizing delays and warehousing (Rodrigue, 1999).

In the logistic chain we see the customer order disconnecting point coming to the fore, so that the influence of customers in production and logistics becomes noticeable at an earlier stage. This shift brings about concentration upwards in the production chain, and deconcentration in the distribution downwards chain. The shift of the customer order disconnecting point is accompanied by steadily decreasing order lead times. In 1987, the average order lead time, in industry and commercial trade still aggregated 27 days. Whereas the average lead time dropped to 18 days in 1993, it decreased to 12 days in 1998, and is expected to come down to only 9 days by 2003 (AT Kearney, 2000). At the same time, the requirements on the reliability of delivery have considerably increased. Both developments endorse the increasing requirements regarding transport. Since truck transport is increasingly confronted with congestion, these requirements will inevitably support developments in the direction of multimodality. Uncertainties about the development of transport costs (general and specific) also call for multimodality, so that shippers can always choose optimum combinations.

Finally, there is the, as yet hesitant, trend in the direction of reverse logistics, which could offer a counterbalance for increasingly long production and transport chains. Chains become closed cycles whereby the consumer market becomes an increasingly important supplier of secondary raw materials. Facts indicate that concentration and deconcentration trends in goods flows patterns occur simultaneously. Parts of the logistic chains will remain fine mesh, above all in the distribution phase. Here the truck remains an attractive means of transport, possibly with the interspersion of urban distribution centers, partly to be replaced by and supplemented with environmentally acceptable vehicles, such as electric cars. Environmental considerations will gain in importance in the years to come. The pursuit of the use of environmentally acceptable means of transport (e.g., pipeline, train, ship) will become more vigorous. The disadvantage of these modes of transport is that in certain respects they lack flexibility. The infrastructural provisions for pipeline transport and trains are expensive. Each of the environmentally acceptable means of transport requires large goods flows. This seems at odds with the need for flexibility. The way out of this dilemma is multimodality and strategic use of terminals with advanced trans-shipment techniques. In the logistic chain it is rare for only one means of transport to be utilized from raw material to customer, and vice versa. More often there will have to be a switch in transport modality. As a result, the market share of environmentally acceptable means of transport can be increased without having to forego the flexibility of certain less environmentally acceptable modes of transport.

From the perspectives outlined in this section, it is likely that, as the international economy grows and the information society further expands, the process of globalization will also further intensify. The globalization process will be accompanied by larger international transport flows. These flows are likely to concentrate increasingly on main ports from where more fine-meshed networks servicing large hinterlands should be developed (see also O'Kelly, 1998). The concentration of goods flows on an intercontinental and international scale is likely to go with a dispersion of flows on national and regional level. However, as a result of increasing goods flows to the main ports, congestion in and around these hubs will also increase (Fleming and Hayuth, 1994). These developments in themselves are a substrate for intermodal transport, since short sea shipping, inland shipping, and rail transport between these hubs will become more attractive.

It is evident that, within the patterns of goods flows, locations are found where goods come to a temporary, often inevitable, standstill. This halt might be due to purely transport considerations (e.g., a seaport where the maritime and continental section within the logistical chain are linked) or logistical reasons (e.g., warehouse and distribution policies). These locations will present themselves as sites where typical terminal activities may be expected. These observations are a major starting point for developing intermodal network structures that comply with the pattern of goods flows and their logistical requirements.

4. Basic requirements of intermodal transport networks

In order to provide a serious alternative to unimodal road transport, intermodal transport must offer competitive services in terms of costs and quality at the door-to-door level. Shippers will evaluate transport alternatives on the integral logistical costs. The quality criteria for intermodal transport refer to the following points (Kreutzberger, 1999):

(1) short door-to-door lead time;
(2) high frequency of transport services;
(3) high time reliability of transport services;
(4) easy access to a large number of destination terminals for each origin terminal;
(5) great flexibility of services;
(6) maximum volumes and weights of load units;
(7) opportunities for services for less than container load (LCL) shipments; and
(8) minimizing freight damage vulnerability.

Points (1)–(3) lead to lower interest costs for shippers for goods in circulation. For point (2) this reduction is the consequence of shorter intervals (i.e., waiting times) for the load which needs to be transported. Point (4) indicates opportunities to offer intermodal transport services at a large number of transport relations. In other words, this aspect is directly related to the size of the catchment area of intermodal transport services. Finally, points (5)–(8) indicate opportunities to conquer new transport markets; that is to say, to increase the degree of containerization in certain transport markets.

Obviously, a field of tension exists between the transport volumes needed to offer intermodal services meeting the required quality conditions, on the one hand, and the costs of these services on the other. This calls for strategies for bundling transport flows. As shown in Section 2, some kind of bundling seems inevitable in transport business, because evident advantages arise from it. As Figure 7 shows, the advantages of bundling may be threefold (Kreutzberger, 1999):

(1) the loading degrees of transport units or load units can be increased, and/or
(2) the transport frequency can be increased, and/or
(3) the number of destinations from one origin terminal can be increased.

Point (1) refers to costs, while points (2) and (3) refer to the quality of intermodal transport services. Ultimately, an increase in the degrees of loading of the transport units (trains and barges) may result in increasing the scale of operation to gain economies of scale.

However, Figure 7 also shows that bundling has some disadvantages. It causes:

(1) Additional handling at the node, such as:
    (i) exchanging load units (i.e., trans-shipment) between trains or barges; and

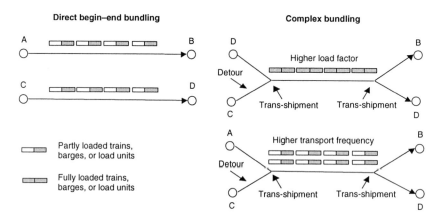

Figure 7. The impacts of bundling (source: Kreutzberger, 1999).

    (ii) exchanging train wagons (i.e., shunting) between trains or push barges or between push barge units – this additional handling involves time and costs, and may reduce the door-to-door reliability of the transport chain.

(2) Detour of routes for most transport services. The transport distances are larger than in the case of direct terminal–terminal services without intermediate bundling operations. The additional distance increases lead time and costs.

In the more complex bundling concepts, intermediate terminals are also used for bundling purposes.

    The best bundling concept for a given pattern of goods flows is that which creates the optimum between the advantages mentioned (loading degrees, economies of scale on the links, transport frequencies, number of destinations) and disadvantages (additional node costs, detour costs).

5. Some typical patterns of European intermodal network developments

Direct shuttle trains and barge services between a container port and various destinations in the hinterland comprise the most obvious formula for the transportation by other means than trucks of containers from a seaport to its hinterland (and vice versa). Several of the main European loading centers generate sufficiently thick freight flows to guarantee a sufficient loading level of trains and barges. For a number of places lying further away in the hinterland, freight flows would be too thin to make possible a system of frequent direct shuttle trains. This is even more the case for shipping connections; not all places can be reached by inland waterways and, even where they can be reached via a river or a canal, the freight flows may be too thin. Here the bundling of container flows can offer a future for inland hubs from which directed services by rail or inland waterway vessels could be provided. Freight trucks could again play an indispensable part in the fine-meshed collection and distribution networks.

Apparently, the development of shuttle train or barge services seems the most natural concept for the development of intermodal networks. Their uniform and straightforward operations on the infrastructural network and their uncomplicated operations at terminals could lead to the designation of these services as the simplest configuration of an intermodal network. However, their very simplicity leads to some drawbacks, especially with respect to more diffuse patterns of goods flows. Some kind of network development will therefore be required. A spatial model for the development of a port related network is presented in Figure 8. This model was put forward by Notteboom and Winkelmans (1998).

Notteboom and Winkelmans (1998) describe the four phases of this spatial model as follows:

"In the beginning period of containerization the railroad network is still underdeveloped. The market area for smaller container ports remains limited to the directly surrounding hinterland. The concentration movement in the port system resulting from the arrival of maritime hub and spoke systems then enables the larger loading centers to extend their rail connections to considerably more destinations (phase 2). The nearby smaller ports seek to establish connections via feeders with the extended hinterland network of the loading centers. The large loading centers compete in the more or less captive hinterland, or further outlying smaller container ports (in figure 8, the port on the right hand side) to a very restricted extent. Indeed, the larger loading centers generate too little rail volume to extend their hinterland further in that direction.

The third phase is characterized by the formation of hub and spoke systems in the outlying hinterland, so that the maritime container volumes destined for the regions concerned become totally concentrated in a few prominent corridors between the loading centers and the "master hubs" in the hinterland.

The intense traffic flows over the corridors make a system of frequent, cost effective shuttle trains feasible and thus improve the competitive position of the loading centers with respect to the smaller, more distant container ports. Along the corridors new logistic nodes develop which, in the beginning phase, actually make little profit from the through rail traffic. In the last phase (phase 4) the number of these new logistic nodes grows through the development of 'master hubs' in the nearby hinterland of the loading centers. This development leads neighboring smaller container ports to seek direct connections to the new master hubs, through which the extended hinterland network becomes accessible without having to remain dependent on the large maritime hubs. When the various master hubs have been interconnected via high frequency shuttle trains, the intermodal container network reaches the phase of complete integration."

Phase 1: Limited penetration of the hinterland

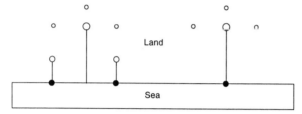

Phase 2: Concentration in the port system and strong hinterland penetration

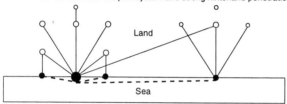

Phase 3: Formation of inland hubs in the more distant hinterland

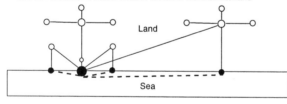

Phase 4: Formation of inland hubs in the less distant hinterland and the interconnections between inland hubs

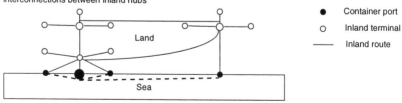

Figure 8. Spatial model of the development of main-port related rail networks (source: Notteboom and Winkelmans, 1998).

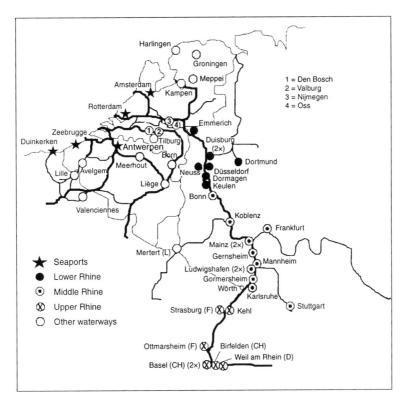

Figure 9. Overview of the container terminals for the inland waterways in north-west Europe (source: Notteboom and Winkelmans, 1998).

The European intermodal railroad network can be said to be in phase 3 of the proposed model. The hinterland network that connects with the ports of Rotterdam and Antwerp actually shows clear signs of a transformation to the fourth phase, although the interconnections between the inland hubs have not yet been completely put in place (Notteboom and Winkelmans, 1998).

Finally, a major observation is that the master hubs, notably the main ports, seem increasingly to transfer the second line terminal activities, such as storage and assembly of freight, to satellite facilities, notably at inland terminals. Removing these activities permits the main ports to improve their throughput performance and therefore their actual transfer function (see also Slack, 1999). As a consequence the hub function will be improved, which means that there is also more flexibility to adapt to changing transport flows. With respect to the development of barge networks in Europe, to some extent similar developments can be encountered. However, the geographical pattern of the most important waterways in north-west Europe is of critical importance here (Figure 9).

The Rhineland, with its integrated rivers and canals, comprises by far the most important waterway in Europe. The inland shipping of containers in north-west Europe is increasing. In 1997, the total European container traffic by barge amounted to about 2.2 million TEU (20-feet equivalent units), a figure that has risen considerably year by year. In 1997, Rotterdam achieved a container inland waterway traffic of an estimated 1.4 million TEU compared with 225 000 TEU in 1985 and 670 000 TEU in 1990. In 1997, about 1.1 million TEU was transported by barge in and out of the port of Antwerp, compared with 128 700 TEU in 1985 and 293 000 TEU in 1990. An important share of the traffic is accounted for by the Antwerp–Rotterdam connection. In 1996, the container flow transported by barge between these two ports was about 560 000 TEU, amounting to a 35% share of the modal split on this transport axis. In the other container flows the Rhine waterways dominate.

Following Notteboom and Winkelmans (1998), the following analysis of the spatial and organizational development of the barge networks can be given. Inland waterway shuttles connect the Rhine–Schelde Delta ports with groups of inland terminals within each of the separate waterway areas of the Rhine (upper Rhine, lower Rhine, middle Rhine) and the Danube. In order to raise the level of service and offset destructive competition, the inland shipping companies operate joint services on the various sections of waterway on the Rhine and the Danube through co-operative agreements. In the Rhineland, and in particular in Belgium and northern France, the principle of direct shuttle services between a port and a particular inland terminal operates. The more the terminal network outside the Rhine area grows, the greater the expectation becomes that here too the Rhine system of the grouped development of inland waterway terminals will become feasible.

In contrast with the rail network, not a single container terminal along the Rhine fulfills the function of master hub. A possible concentration of river associated container traffic has not resulted in the formation of hub-and-spoke systems in the terminal network of barge transport itself. However, we see that the hub function of inland waterway terminals is being developed by linking barge with rail services. In other words, inland waterway terminals are focusing to an increasing extent on the complementary relationship between rail and inland shipping. For example, some of the container flows between the Rhine–Schelde ports and Eastern Europe arrive by barge in Duisburg, where they are then transferred to shuttle trains with destinations in the Slovakian Republic, Czechia, or Poland. The same phenomenon occurs in the traffic relationships between the Benelux ports and northern Italy.

We see how the network of inland terminals for rail and barge transport are snowballed together via a number of trimodal terminals. The developments on the east–west axis, for example between Rotterdam or Antwerp and Eastern European countries such as Poland and Czechia and Italy, demand a combination of inland waterway transport and rail transport. In north-west Europe it would seem that the pattern of freight flows has reached the phase where a limited number of hub-and-spoke networks has been developed. Careful choice of location needs to be followed by good design of the corridors connecting the most important inland terminals to each other and to the gateways.

The development of bundling networks and the resolution of terminal problems depend in part on the development of European and intercontinental transport networks. Shippers are important players in the development of transport services and networks into the hinterland. We see a development in which people seek to move containers and other freight away from a seaport area as quickly as possible and look for consolidation at an inland terminal. Whoever is best able to resolve the associated logistic problems will have the most competitive edge with respect to traditional unimodal road transport services.

6. Tools for matching intermodal transport networks to patterns of goods flows

Transport volumes and frequencies of services are important determinants for bundling concepts, because there is a relationship between the number of origin and destination terminals in a network and the transport volumes required to offer service frequencies that conform with market requirements. In addition, distances to be bridged in the network by trains or barges are an important entity because, for their part, they are related to the quality and frequency of services. Transport volumes, frequencies, and distances within the network are major determinants for the costs of network operations, and therefore for network performances. These entities are the basic parameters of an intermodal transport cost model that has been developed within the framework of the European research project TERMINET (Kreutzberger, 1999).

The basic idea behind this model is to optimize the circulation time of trains and barges in a network in order to obtain the lowest costs for intermodal transport network operations. In this model the concept of "circulation time" proves to be a useful tool for defining a network structure with optimal cost conditions for intermodal transport. This includes the ability to point to optimal relative locations of terminals. On the other hand, given a pattern of goods flows, the model can match a transport network structure that fits this pattern best, for example in terms of costs. In the same way, the model can be used to adapt and redefine the structure in case patterns of goods flows change. Empirical results of this model, which is still under development, can be found in case studies of the TERMINET project (TERMINET, 2000a).

Another interesting model for analyzing intermodal networks for freight transport is a spatial model implemented in the GIS software NODUS (Jourquin, 1995; Jourquin and Beuthe, 1996). The framework of the model is based on constructing a virtual network where a particular link is associated with every distinct transport operation that takes place over the real geographic network, thus linking in a systematic way all the possible successive operations (loading, moving, unloading, transshipping, transiting) in the geographical space (Figure 10). Three modes, with various means, are incorporated in the model: railways, roads, inland waterways.

NODUS generates virtual links for each possible mode t and means m on each link, which should be adopted to the real network and node configuration and according to the requirements within a specific application of the model. Relevant cost functions of each particular operation can be attached to the virtual links, so that it is possible to search for a minimum-cost solution to a transportation task that must be performed on the network. Based on minimizing the shippers' generalized costs, the model provides as solutions assignments of the transportation task flows between modes, means, and routes. Given a pattern of goods flows (origins–destinations), the NODUS software tool will be able to provide information on how freight will flow between transport modes, transport units, and routes.

Since it is possible to analyze and compare various scenarios (e.g., changing patterns of goods flows, new network links, cost changes at network links or terminals), NODUS is a powerful tool for evaluating the competitiveness of intermodal transport networks (see also TERMINET, 1998).

7. Conclusions

The spatial patterns of goods transport will become more complex through fluxes in the strategies of producers, shippers, and carriers. The requirements imposed on the transport system will be further increased. Customers in a global supply chain will continue to demand higher quality delivery of their products, with great flexibility as the marketplaces shift and change. There is continuous pressure to reduce the total transit time, improve on-time performance, and lower transport costs. In this context, the quality of intermodal transport could become a decisive factor in the competition between supply chains, in particular in an environment of global competition. Much will depend on the ability of intermodal transport to satisfy the shippers' requirements and to respond to changes.

Adapting intermodal transport services to the needs of shippers first requires the ability to fit alternative services into the overall planning schedule of intermodal transport services. As far as rail transport is concerned, this requirement means that the desired train paths should be available. Therefore,

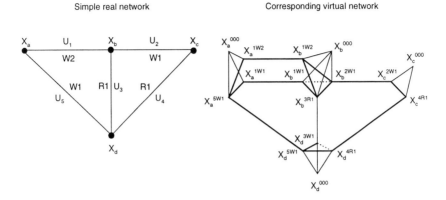

Figure 10. Representation of network analysis with NODUS (source: TERMINET, 2000b).

organizational as well as technical constraints (i.e., limited infrastructural capacity) may act as barriers.* Opportunities to change the services on the links (which means on the network itself) comprise a necessary, but not sufficient, condition. The terminal should also be able to accommodate alternative time schedules of train and/or barge services. This requires sufficient handling capacity in the first place, but also more sophisticated, intelligent operational processes at the terminal. To enable more complex bundling network structures, new terminal technologies may even be required. The significant importance of terminal performances is not only relevant for remaining flexible as transport flows shift and change, but even more for improving the total transit time. Evidence suggests that contemporary improvements of transport systems are mostly undertaken at terminals, notably through synchronization (Rodrigue, 1999). These findings underline the importance of improving terminal performance, while maintaining maximum flexibility with respect to different kinds of handling operations. This should ensure that terminals have enough flexibility and adaptability to changing patterns of goods flows. In a global environment in which the decision to build fixed and expensive infrastructure represents a considerable risk, these qualities of a terminal are critical (see also McCalla, 1999).

Terminal and network operations cannot be considered in isolation. The intermodal capability will have to be an integrated and seamless network structure, with good connections between modes at strategic locations. Network optimization models can prove useful tools for this design task. The merits of the

---

*At a higher level, one may even state that the intermodal infrastructural network of waterways, rail lines, and motorways itself may act as a barrier to the optimization of intermodal transport services because, for instance, important missing links within and between these networks exist.

transport circulation model as a tool for network analysis are found in the possibility of constructing and optimizing an intermodal network structure. Given the characteristics of the transport landscape, that is to say the pattern of goods flows, it is possible to analyze which bundling concept fits best.

On the other hand, the NODUS network model is a particularly useful tool in evaluating the performance of an intermodal network in relation to alternative transport solutions. Cost functions for different modes are incorporated in the model. It is therefore also a useful instrument for evaluating the effects of different transport policies on the network performances. In this perspective, the two types of models can be considered as complementary tools for developing intermodal transport networks that are the most competitive with unimodal road transport.

As both network models confirm, the location of multimodal terminals is of strategic importance. These terminals should be situated on, or close to, the main transport axes; that is, along well-navigable waterways and/or railways along which many goods are transported. In addition to the now familiar road–water terminals and road–rail terminals, increasing attention will have to be devoted to road–rail–water terminals integrating three modes of transport. The construction and further development of a number of advanced multimodal terminals allows improvements in the dynamics in production and logistic chains, and thus also in the dynamics in goods transport, to be anticipated.

References

AT Kearney (2000) "Insight to impact", European Logistics Association, Brussels.

Coopers and Lybrand (1989) "Uitbesteding logistieke functies" ("Outsourcing logistical functions"), Ministeries van Economische Zaken en Verkeer en Waterstaat, The Hague.

Fleming, D.K. and Y. Hayuth (1994) "Spatial characteristics of transportation hubs: Centrality and intermediacy", Journal of Transport Geography, 2(1):3–18.

INRO-TNO (1993) "De logistieke ketens gekraakt: Case studies ter verklaring van de ontwikkelingen in de bedrijfsmatige logistiek" ("The logistic chains broken: Case study to explain the developments in business logistics"), INRO, Delft.

Jourquin, B. 1995 "Un outil d'analyse économique des transports de marchandises sur des réseaux multi-modaux et multi-produits: Le réseau virtuel, concepts, méthodes et applications", Ph.D. thesis, Faculté Universitaires Catholiques de Mons.

Jourquin, B. and M. Beuthe (1996) "Transportation policy analysis with a geographic information system: The virtual network of freight transportation in Europe", Transportation Research C, 4(6):359–371.

Kreutzberger, E. (1999) "Promising innovative intermodal networks with new-generation terminals", project for the 4th Framework Programme of the EU DG VII, Brussels, OTB/TRAIL, Deliverable D7.

McCalla, R. (1999) "Global change, local pain: Intermodal seaport terminals and their service areas", Journal of Transport Geography, 7:247–254.

Notteboom, T. and W. Winkelmans (1998) "Bundeling van containerstromen in het European havensysteem en netwerkontwikkeling in het achterland" ("Bundling of container flows within the European port system and network developments in the hinterland"), Tijdschrift Vervoerwetenschap, 4:379–398.

O'Kelly, M.E. 1998 "A geographer's analysis of hub-and-spoke networks", *Journal of Transport Geography*, 6(3):171–186.

Rodrigue, J.P. (1999) "Globalization and the synchronization of transport terminals", *Journal of Transport Geography*, 7:255–261.

Ruijgrok, C.J. (1992) *Sustainable development and mobility*. Delft: INRO-TNO.

Slack, B. (1999) "Satellite terminals: A local solution to hub congestion?", *Journal of Transport Geography*, 7:241–246.

TERMINET (1998) "GIS-presentation of innovative bundling concepts", project for the 4th Framework Programme of the EU DG VII, Brussels, FUCAM, Deliverable D3.

TERMINET (2000a) "Performance analyses: 5 new-generation terminal case studies", project for the 4th Framework Programme of the EU DG VII, Brussels, OTB/TRAIL, Deliverable D10.

TERMINET (2000b) "New-generation terminals and innovative networks", project for the 4th Framework Programme of the EU DG VII, Brussels, OTB/TRAIL, Final Report.

Van Goor, A.R. (1986) "Fysieke distributie" ("Physical distribution"), Alphen aan den Rijn (Samsom).

Vermunt, A.J.M. (1993) "Wegen naar logistieke dienstverlening" ("Ways towards provision of logistical services"), University of Tilburg, Tilburg.

Chapter 33

SUPPLY CHAIN STATISTICS

MICHAEL S. BRONZINI
George Mason University, Fairfax

1. Introduction

One of the most revealing statistics about the importance of supply chains as an active area of concern comes from searching the internet for sources of information. At the time of writing, a search using a relatively sophisticated search engine, with the term "supply chain management," yielded approximately 627 000 world wide web references. Even the more restricted term "supply chain management transportation and logistics data" yielded 22 300 hits. Although comparable data from earlier years are not available, these large numbers of references suggest that interest in this topic is growing geometrically or exponentially. Hopefully, information and expertise are experiencing similar growth.

With this exploding interest as a backdrop, this chapter offers guidance on what it is important to measure about supply chains, particularly within the transport and logistics worlds, and provides examples of data that are available. For the most part the data and examples relate to the U.S.A., with some commentary on what might reasonably be expected to be found in the developed world. Data availability is very much more restricted in the developing world, so no attempt is made to extend the coverage to those countries. After some introductory comments on data availability, data from private firms and public agencies are discussed, including a brief note on military supply chains. The chapter concludes with a discussion of outstanding data needs and how they might be filled.

2. Current state of knowledge

Making authoritative statements about supply chains, and indeed about freight transport and logistics in general, is very difficult, as there is a profound lack of publicly available statistics. This lack of data is pervasive, extending throughout the world's developed and developing economies. Supply chains are made up

Handbook of Logistics and Supply Chain Management, Edited by A.M. Brewer et al.

predominantly of private firms, which is one of the primary factors leading to this lack of up-to-date and reliable data. Shippers have detailed knowledge about their outbound and inbound freight, and about the resources that they expend on supply chain activities. However, each firm adopts its own definitions, accounting systems, performance standards, and measurement processes, to name just a few of the internal decisions that affect data availability and data quality. Hence, when data are obtained from a number of firms, there can be significant and unknown differences in exactly what each firm reports. A complicating factor is the increasing reliance of firms on out-sourcing some or all of their supply chain management (SCM) operations. This can lead to differences across firms, depending on their degrees of out-sourcing, in what information can be obtained and in data comparability. Out-sourcing also adds to the complexity of constructing shipper surveys. In many cases the data needed will reside with the third-party logistics provider (3PL), but 3PLs are not usually included in the sample frame of shipper (and carrier) surveys. The responding firm may not always have access to the detailed data held by the 3PL, or worse may hold and report conflicting data. Finally, there is the issue of data confidentiality, a concern that exists within both shippers and logistics firms. In summary, most data of interest lie within private firms who collect the data for their own purposes, and the data do not often make their way into public data sets.

Even when data are collected by or otherwise provided by government agencies there are still many problems in using the data for research or planning. This is particularly true in the case of international freight movements, where data on some aspects of the shipment will be collected by each country (and in some cases by each state or province) transited. The single biggest impediment is lack of standardization in definitions of the data items, such as commodities, geography (e.g., port identification), modes, vehicles, tariffs, accidents (e.g., damage cost reporting thresholds), and units of measure. A concrete example of standardization problems is the treatment of international marine containers, where an attempt to introduce some degree of comparability to activity measurement led to the adoption of the 20-foot equivalent unit (TEU) as the preferred measurement unit. When 20-foot containers were the norm this made a lot of sense. Now, with the proliferation of container sizes and the increasing use of 40-foot (and longer) boxes, the 40-foot equivalent unit (FEU) is appearing in many reports. While TEUs and FEUs may be fine for marine reporting, even when mixed but clearly identified in the data, they are not as well suited for tracking rail, truck, and air container moves. A second important difference between countries lies with their data collection and reporting systems, including what data are collected, how often, with what sampling rates, and with what accuracy. While this litany of public sector data ills applies to virtually all developed countries, the situation is even worse in the developing countries, where there is an almost complete lack of hard data of any sort.

The preceding paragraphs treat the private and public sectors as though they operate in distinct and disjoint spheres. In some cases a portion of the supply chain has both private and public operators. For example, in many countries the state owns part of the freight system, such as a railroad, airline, or marine port. On the infrastructure side there are privately owned and operated toll facilities that exist alongside public highways and bridges. There is even out-sourcing of traditional government functions, such as highway maintenance. In all these cases, where both government agencies and private firms are engaged in providing similar supply chain facilities and services there will be differences in data availability and comparability. The public–private dichotomy tends to accentuate these differences.

Finally, there are two issues that permeate all collection schemes and are at the root of many of the problems noted above. First, it is universally true that comprehensive freight surveys, even when they exist, are conducted infrequently. In the U.S.A., for example, the Census of Transportation's Commodity Flow Survey (CFS) is scheduled only every 5 years, and the 1993 CFS was the first that had been conducted in over 20 years. Second, while supply chains are by their very nature integrative, all surveys, be they of shippers or carriers, cover only one link or a limited set of the links in the chain. Hence, attempts to reconstruct the operation, performance, or impact of entire supply chains are impeded by the different methodologies that were used within each of the surveys.

As noted above, supply chains involve both private sector firms and government agencies. Supply chain statistics, even though problematic, are available from both sectors, which are discussed in turn in the next two sections.

3. Private firms: supply chain performance

Two types of private firms are involved in supply chain performance: shippers and transportation carriers.

3.1. Shippers

The private firms that are most actively engaged in SCM are the shippers; that is, those industrial and commercial companies trying to improve their financial and market performance through optimizing operations along their entire supply chains. These firms have myriad needs for information on how well their SCM efforts are succeeding. Some of their most important measurement needs relate to (University of Tennessee, 1999):

Table 1
Vehicle-miles (millions) of travel in the U.S.A.

Mode	1985	1990	1995	1997
Single-unit truck	45 400	51 900	62 700	66 800
Combination truck	78 100	94 300	115 500	124 500
Rail class I freight (car-miles)	24 920	26 159	30 383	31 660

Source: Bureau of Transportation Statistics (1999b).

(1) the share of the overall cost attributable to logistics,
(2) inventory turnover,
(3) the capital tied up in inventory,
(4) customer service requirements,
(5) the service quality provided to customers,
(6) the inbound logistics performance of suppliers, and
(7) the relationship of logistics results to shareholder value.

By their very nature, data such as these reside within the individual firms and are closely held. Hence, other than in case studies that appear in the literature, these data are not readily available to the public.

There are, however, some revealing data from various industry surveys and studies on the pay-offs that can be achieved through SCM initiatives. Anderson and Lee (1999) quote a Council of Logistics Management study showing that in North America the total logistics cost as a percentage of sales dropped from nearly 50% in 1983 to about 10% in 1990, and remained near that level through 1997. Here, the total logistics cost includes finished goods transportation, warehousing, inventory carrying cost, and customer service and administrative costs. Some other bottom-line results have been reported by Quinn (1999). Summarizing an integrated supply chain benchmarking study that was conducted by a consulting firm, he noted that those companies employing best-practice SCM spend only about 55% as much on supply chain activities as do their average cost competitors. In the computer industry, for example, the average percentage of company revenue spent on supply chain activities is 13.3%, while the best-practice companies spend 7.3%. Quinn also reported that companies studied in the MIT Integrated Supply Chain Management Program had recorded some impressive bottom-line results, including: 50% inventory reduction, 40% increase in on-time deliveries, 27% decrease in cumulative cycle time, and 17% increase in revenue.

One imponderable is how the emergence of e-commerce might produce further gains. To some extent, the supply chain performance improvements noted above have resulted from applications of information technology (IT) to reducing inventory, transportation, and customer service costs. It has been reported, for example, that tracking a package via telephone costs, on average, U.S. $2.50 per inquiry, while obtaining the same information via the internet costs only U.S. $0.10. Most knowledgeable observers seem to believe that the e-commerce explosion is in its infancy, and that supply chain efficiency gains from IT will soon spread throughout industry.

3.2. Carriers

The second set of private sector firms that are affected by SCM initiatives are the transportation companies, particularly trucking companies, railroads, airlines, and ocean carriers. To the extent that these companies are privately owned, as is the case in the U.S.A., there are few data available on their supply chain activities. However, some insights can be gained by considering recent trends in industry-wide statistics, which mostly relate to SCM.

Overall, from 1993 to 1997 the total value of freight moved in the U.S.A. increased by 17% and freight tonnage increased by 14%, while during that same period gross domestic product (GDP) increased by 13% and resident population by only 4% (U.S. Bureau of Transportation Statistics, 1996). Some added detail is provided in Table 1, which shows recent trends in truck and rail vehicle-miles of travel in the U.S.A. Truckers are perhaps the most affected by supply chains, as they participate heavily in both single-mode moves of high-priority freight, and in the pick-up and delivery legs of most high-value freight that moves by rail, water, and air. As shown, single-unit truck travel grew by 47% and combination truck travel by 59% from 1985 to 1997, at rates outpacing the general economic and freight traffic growths cited above. Rail freight car-miles increased by 27% over the same time period. Much of the growth in rail traffic has been fueled by intermodal (trailer or container on flatcar) traffic, which amounted to 8.7 million car-loadings accounting for 28% of rail car-loads in 1997 (Association of American Railroads, 1998).

The air and marine modes have also participated in the expedited freight movements that characterize much of SCM strategy. Air freight increased by a factor of 2.6, as measured by ton-miles, from 1985 to 1997, and airports held three of the top six spots (including the top spot) in the 1997 ranking of U.S. foreign trade gateways by value of shipments (U.S. Bureau of Transportation Statistics, 1999b). Finally, U.S. international waterborne container traffic has also seen spectacular growth. The number of TEUs handled by U.S. ports increased from 3.1 million in the first quarter of 1996 to 4.2 million in the first quarter of 2000. In fact, traffic grew by 9.5% from 1999 to 2000 (U.S. Bureau of Transportation Statistics, 2000b). Similar growth in air freight and marine container traffic can be observed worldwide.

Table 2
Extent of the U.S. transportation system (a)

Mode	Components
Highway	46 344 miles of Interstate highway 113 757 miles of other National Highway System roads
Rail	119 800 miles of class 1 freight railroads 21 360 miles of regional freight railroads
Air	5354 public-use airports
Water	26 000 miles of navigable waterways 276 navigation locks
Pipeline	87 660 miles of crude petroleum lines 90 990 miles of petroleum product lines

Source: U.S. Bureau of Transportation Statistics (2000b).
Note: (a) An incomplete list of system elements.

4. Public sector performance

Public sector agencies are involved in supply chains primarily as the provider and operator of the infrastructure for many of the modes. In the U.S.A., for example, only the railroads and pipelines are privately owned and operated. The other modes all have substantial government involvement. In other countries, even the railroads are public-sector entities.

Since the public sector makes decisions about infrastructure investments, and is involved in safety and economic regulation, public agency needs for data about supply chains have quite a different focus than do the needs of private firms. Public agencies need to know how the private sector is using the transportation system, and how developing trends in SCM might affect usage levels and patterns. A list of their typical data requirements concerning freight transportation would include the following:

(1) freight quantities, by commodity, origin, and destination;
(2) commodity characteristics, including value, density, and packaging;
(3) the modes used and intermodal linkages;
(4) routes, travel times, and trip-length distributions;
(5) movements of empty vehicles;
(6) transport and logistics costs;
(7) safety, product loss, and damage;
(8) public sector costs of providing freight services; and
(9) environmental impacts of freight traffic.

This is a fairly long list, with some elements more closely related to supply chains than others. For a general treatment of data availability the reader is referred to appropriate government agencies, such as the U.S. Bureau of Transportation Statistics in the U.S.A., and to recent research reports (e.g., Cambridge Systematics, 1995). A sample of what is available is presented below.

4.1. System infrastructure

The most basic data about a nation's transportation system is the inventory of right-of-way miles and public-use facilities such as ports and airports. These data are usually readily available from national and state or provincial authorities. As an example, an extract from the Department of Transportation's data on the extent of the U.S. transportation system is presented in Table 2. Data are available in much more modal and geographic detail, and usually include system maps. See, for example, the National Transportation Atlas Database (NTAD), which provides transportation network data in a geographic information system (GIS) format (U.S. Bureau of Transportation Statistics, 2000a). Data such as these on transport system components and their extent are useful to policy-makers for what they reveal about modal availability and importance, and the potential for intermodal service. The NTAD includes a data file of intermodal terminals to aid such analyses.

Transportation vehicles are also an essential part of the system, but these are typically privately owned and operated, so authoritative public sector data on vehicle fleets are sometimes lacking. Nonetheless, data on vehicles by mode can be indicative of how transportation services are evolving. For example, the U.S. Department of Transportation (2000) estimates that the number of container vessels in operation worldwide increased by 25% in just 2 years (from 1996 to 1998), and now exceeds 2500. This provides one strong indicator of the rapid increase in cargo shipments requiring premium service.

4.2. System usage

Most data on freight traffic flows is obtained from modal traffic counts and surveys, which produce vehicle-mile estimates of the type presented above. Unfortunately, these data do not include the commodity and geography detail specified in the data requirements list. The reason for this is that counting vehicles is relatively simple and inexpensive, while data on commodity origin–destination movements must be obtained using expensive and time-consuming shipper survey methods. Even carrier-based surveys do not provide the requisite detail, since they cover only one mode and thus miss the intermodal connections of shipments using

Table 3
Domestic and export U.S. freight shipments, 1993 and 1997

Mode	Value (U.S. $ million)			Tons (thousands)		
	1993	1997	Change (%)	1993	1997	Change (%)
Parcel, postal, and courier services	616 839	865 561	40.3	18 892	24 677	30.6
Truck, for-hire, and private	4 822 222	5 518 716	14.4	6 385 915	7 992 437	25.2
Rail (including truck and rail)	361 901	383 222	5.9	1 584 772	1 538 538	−2.9
Water	197 370	195 461	−1.0	1 465 966	1 522 756	3.9
Air (including truck and air)	152 213	213 405	40.1	3 139	5 047	60.8
Pipeline	340 664	330 176	−3.1	1 870 496	1 881 209	0.6
Other and unknown	283 106	447 908	58.2	706 690	753 848	6.7
Total	6 774 414	7 954 549	17.4	12 035 870	13 718 512	14.0

Source: U.S. Bureau of Transportation Statistics (1999a).

more than one mode, and often also miss the true origins and destinations of the traffic. For example, U.S. waterborne commerce data are obtained from carrier reports, and track shipments only from the port terminal where the commodity is loaded into a vessel to the terminal where it is unloaded (or, in the case of imports and exports, from where the vessel enters and leaves U.S. waters). This totally misses any overland hauls to or from actual commodity production or use locations. Similar statements can be made about all carrier-based commodity traffic data.

There have been some efforts to remedy this situation. In the U.S.A., for example, the U.S. Census Bureau and the U.S. Department of Transportation conducted comprehensive commodity flow surveys in 1993 and 1997. In these surveys a statistically selected sample of shippers was asked to sample their outbound freight documents and report, for each sampled shipment, the commodity, origin postal code, destination postal code, and modes used. Similar efforts have been mounted by consulting firms in Australia (Rockliffe et al., 1998) and elsewhere.

Table 3 shows summary data from the U.S. commodity flow surveys. These summary data immediately show the impact that aggressive SCM is having on the U.S. transportation system. Truck shipments, whether measured by value or tons, dominated by a wide margin, and also exhibited considerable growth. The largest growth occurred in parcel and air freight shipments, both of which grew by 40%

between the two survey years. World containerized trade moves also increased by 40% over those years, from 30 million to 42 million TEUs (U.S. Department of Transportation, 2000). The picture that emerges is one of industry's increasing reliance on moving small shipments of high-value goods quickly and efficiently, using the higher cost transport modes. Also, intermodal shipments are increasing faster than are other types of traffic. The increases in e-commerce transactions are likely to reinforce these trends. The summary data sketched here only scratch the surface of what is available. The data sources cited provide considerably more detail on the current status of freight transportation.

4.3. Military supply chains

One aspect of public sector performance that is often overlooked is the use of transportation infrastructure and other assets by the military. In the U.S.A., military shipments and vehicles are excluded from most published transportation data. Military needs for rapid deployment and sustainment of forces is of concern, and the armed services are increasingly looking to SCM strategies as ways to increase both responsiveness and efficiency.

The travails encountered by the U.S. military deployment to the Persian Gulf in 1990–1991, known as Operation Desert Storm, are by now well known within logistics circles. According to Muller (1999), in that operation the military moved roughly the equivalent of the entire population of Atlanta, Georgia, together with their cars, food, clothing, and other belongings halfway around the world. Problems developed because of inadequate container documentation, weak port operations, and the failure of many shippers to comply with standard military transportation procedures. As a result, about 40% of the containers landed in Saudi Arabia had to be opened to determine their contents and destination. These delays caused field commanders to re-order the required materiel, further overburdening the transport system.

IT is one of the keys to solving the military's modern deployment problems. An analysis of the Desert Storm logistics operation indicated that the deployment time could have been shortened by 100 days and the materiel shipped could have been reduced by 1 million ton with better information, producing a savings of U.S. $650 million in transportation costs alone (Lynn, 1997). These kinds of savings are comparable to the gains achieved by private firms through improved SCM.

5. Data needs

Comparing the list of desired freight and logistics data with what is available, there are still some unmet needs, as follows:

(1) Detailed shipment level data by commodity, origin, destination, and mode are not available in a timely fashion. The U.S. commodity flow surveys are done only every 5 years, and data are released about 2 years after the collection period. Furthermore, there is no guarantee that this survey will continue, as it is very expensive and is subject to the vagaries of federal data collection budget priorities. The lack of similar public datasets worldwide is a good indication that cost and budget concerns are widespread.

(2) Data on shipment time and cost are lacking. These data are not routinely collected in current surveys, and so must be obtained in special surveys or estimated through modeling. Cost data, including transportation rates paid to for-hire carriers and other costs, are part of the commercial transactions between shippers and carriers. Transportation performance data may also be part of the transaction, or may be used internally by firms for competitive strategy development. Thus, there are many reasons for firms to keep these data private, and little in the way of incentives for the data to be released to others.

(3) True total logistics cost data are not available, except through case studies of individual firms or shipments. 3PLs have access to these data, but again have reasons and obligations to protect them. This type of data goes to the heart of supply chain statistics, and the lack of such statistics makes it difficult to track the effectiveness of SCM efforts in a comprehensive fashion.

The U.S. Bureau of Transportation Statistics (1999a) suggests that new data collection mechanisms, such as unobtrusive data collection using administrative records and remote sensing rather than relying on surveys, may contribute to meeting these needs.

References

Anderson, D.L., and H. Lee (1999) "Synchronized supply chains: The new frontier", in: D.L. Anderson, ed., *Achieving supply chain excellence through technology*. San Francisco: Montgomery Research, Inc. (http://www/ascet.com/ascet/docs).

Association of American Railroads (1998). *Railroad facts*. Washington, DC: Association of American Railroads

Cambridge Systematics (1995) "Characteristics and changes in freight transportation demand: A guidebook for planners and policy analysts", Transportation Research Board, Washington, DC, NCHRP Project 8-30.

Lynn, L. (1997) "DARPA's advanced logistics program", in: *National conference on setting an intermodal transportation research framework, Conference proceedings 12*, pp. 13–22. Washington, DC: Transportation Research Board.

Muller, G. (1999) *Intermodal freight transportation*, 4th edn. Washington, DC: Eno Transportation Foundation, Inc.

Quinn, F.J. (1999) "The payoff potential in supply chain management", in: D.L. Anderson, ed., *Achieving supply chain excellence through technology*. San Francisco: Montgomery Research, Inc. (http://www/ascet.com/ascet/docs).

Rockliffe, N., M. Wigan and H. Quinlan (1998) "Developing a database of nationwide freight flows for Australia", *Transportation Research Record*, 1625:147–155.

University of Tennessee (1999) *Keeping score: Measuring the business value of logistics in the supply chain*. Oak Brook, IL: Council of Logistics Management.

U.S. Bureau of Transportation Statistics (1999a) "Transportation statistics annual report", U.S. Department of Transportation, Washington, DC, BTS99-03.

U.S. Bureau of Transportation Statistics (1999b) "National transportation statistics 1999", U.S. Department of Transportation, Washington, DC, BTS99-04.

U.S. Bureau of Transportation Statistics (2000a) "National transportation atlas database 2000", U.S. Department of Transportation, Washington, DC.

U.S. Bureau of Transportation Statistics (2000b) "Transportation indicators, September 2000", U.S. Department of Transportation, Washington, DC.

U.S. Department of Transportation (2000) *The changing face of transportation*. Washington, DC: U.S. Department of Transportation.

Chapter 34

NEW TECHNOLOGIES IN LOGISTICS MANAGEMENT

ROGER R. STOUGH

George Mason University, Fairfax

1. Introduction

With the unfolding of the knowledge age, new technology has induced multiple waves of change in the logistics industry. For example, by the mid-1990s, increasing adoption, application, and maturing of technologies such as barcodes, radiofrequency identification, warehouse management systems, and global positioning systems had significantly reduced transaction and transport costs. This contributed to the belief that industry consolidation would occur in logistics management. Yet, a few years later, the industry was decentralizing (Cooke, 2000b). The internet, the supporting technology for e-commerce, has made a new partnering business model, sometimes called "collaborative commerce" or "collaborative fulfillment," possible for a more integrated provision of logistics services up and down the value delivery chain. Thus, in logistics, as in all other industries, the evolution of generic knowledge-age technologies (information technology (IT), computers, telecommunications) have impacted and continue to impact organizational and industry structure and operations. In this chapter the technologies that are inducing this change and the changes they have induced in structure and operations are examined.

The chapter begins with an examination of the definition of logistics. This is important because of the turbulent industry dynamics. Next, the core technologies supporting the logistics industry are identified and described. A discussion of the way the internet and related technologies are redefining the industry is provided in the next section. This is followed by a discussion of a set of technologies called "intelligent transportation systems," which hold promise for increasing the capacity and productivity of existing surface transport infrastructure. The final part of the chapter provides a summary and conclusions, and speculates on possible future technologies, and how they may impact the nature and provision of logistics services.

Handbook of Logistics and Supply Chain Management, Edited by A. Brewer et al.
© 2001, Elsevier Science Ltd

2. The logistics industry

The Council of Logistics Management defines logistics as: "[T]he process of planning, implementing, and controlling the efficient, effective flow and storage of goods, services, and related information from point of origin to point of consumption for the purpose of conforming to customer requirements." As such, logistics is about supply chain management, or the integration of management processes from the customer to the provision and acquisition of raw material inputs to the production process and to the suppliers of products, services, and information that add value for customers. All the supply chain processes, including planning, buying, raw materials management, production (materials management), stock management, distribution and recycling (reverse logistics), and related management integration, have been impacted by several waves of technology over the past decade or so.

3. Base technologies impacting logistics management

The list of technologies that are relevant to logistical management and distribution is potentially very large. For example, physical distribution involves vehicles that operate on land, sea, and in the air. In recent years there have been many new vehicle technologies, such as sensors, fuel injection systems, and safety devices, not to mention technologies that have improved components such as tires, batteries, paint, and materials. It is not possible to examine all these here. Nor is it possible to consider important emergent technologies in air and ship propulsion and design, and in pipelines, which could provide major cost and time savings in sourcing and distribution. Consequently, the focus in this chapter is on technologies that have or will directly affect the management of logistics. These include barcodes, RFID, GPS, and computer hardware and software.

Barcodes are electronic readable labels that are attached to goods or containers of goods, and provide information such as origin, destination, type of goods, and billing information. In logistics this technology facilitates the identification, tracking, processing, and delivery of raw materials and goods. Barcodes use the thickness and separation between bars to code information. An electronic scan across one dimension of the barcode label provides full information. However, because of the unidimensionality of these labels, they have limited capacity to store information. More recently two-dimensional barcodes have been developed that extend the concept to two dimensions, thus increasing the capacity to store information. The two-dimensional codes (Moore, 1999) must be scanned in two dimensions to read all the information. United Parcel Service (UPS) now uses two-dimensional barcodes.

Radiofrequency identification (RFID) (Cooke, 1999) technology enables a wide range of products and objects to be tracked and managed. RFID uses small radio tags or transponders and readers (encoders) that are linked to an information system, thus enabling the reading of the information contained in the tags at a distance. The tag, with a chip and integrated antenna, is activated by a radio signal and sends a return signal. Encoders can read information from the chip (read-only tags) and also write information on the chip (if the tag is read–write). RFID technology is highly versatile, and is applied in a wide range of industrial and commercial sectors. It enables information to be exchanged at a distance, can be read even if not seen (unlike barcodes), can be used in hostile environments (e.g., toxic, high temperature), can be programmed and reprogrammed many times, and can be made secure. This technology facilitates automation across the value delivery chain, lowers operation cost, and determines the location of tagged objects in real time. For many uses, RFID is a replacement technology for barcodes (Moore, 1999).

Global positioning systems (GPS) (Kennedy, 1996) enable the location of objects, vehicles, and individuals by using satellites to determine the co-ordinates via triangulation (i.e., by measuring the location from the positions of three satellites). Location information obtained via GPS is real time, and when connected to a communication system can be used to track shipments and specific parcels in a shipment, identify destinations, and estimate delivery times. As such, GPS is also a base technology for logistics management.

Computer hardware and software underlie the management and communication systems that make it possible to use barcode, RFID, and GPS technologies to streamline the logistics management process, whether it is for specific components of the process (e.g., warehouse management) or for total supply chain management. Warehouse management (Brockman, 1999) is used to illustrate how the base technologies have reduced time in the value delivery chain, transaction costs, and real costs, and improved efficiency (see also Chapter 14).

Warehousing of goods as one component in the supply chain process includes the storage of resources while being moved to production facilities, and the storage of goods while en route to customers. Minimizing the time that goods are in storage and handling through just-in-time (JIT) inventory processes (Claycomb et al., 1999) is fundamental to competitiveness, profitability, and survival. Most warehousing is underlain by a computer-based information system that is used to track warehoused goods as they enter, are handled, and removed from storage. Recently, by linking barcode and RFID, and in some cases GPS, through computer information systems, advanced WMS have been created. With the appropriate software communication between warehoused goods and encoders located on vehicles and/or hand-held by workers, inventory counts can be updated in real time via radiofrequency transmissions. As barcodes or RFID devices are scanned on items placed into or removed from storage, workers can update inventory counts and update information on available stock. In the past, warehouse management was labor intensive, with items added to and removed from storage actually counted and recorded by workers. Much of this work has been reduced due to the base technologies linked to computer-based WMS and inventory management systems (see Chapter 12).

However, warehouse management is only one component in the supply chain process. Similar integrated applications of the base technologies are occurring up and down the value delivery chain in areas such as sourcing (i.e., buying processes, production processes, distribution processes, customer relations, reverse logistics) (Andel, 1997). More importantly, great interest has been focused on the development and application of integrated software systems (Fingar, 2000) to address the total logistics management problem. Moves are underway to integrate these supply chain logistics systems more fully into enterprise-wide information systems (e.g., Oracle, SAP, People Soft). The combination of computer technology coupled with IT and telecommunications technology is making this higher level of integration increasingly realistic.

4. The internet and e-commerce

e-Commerce is having major logistical knock-on effects that are rapidly unfolding (Schwartz, 1999; Collins, 2000; Seidenman, 2000). e-Commerce is expected to become a nearly U.S. $6 trillion business by 2004 (Anon., 2000b). e-Commerce and its underlying technology, the internet, provide a completely new infrastructure for doing business in a new way, where customers interact with the total business system or value web, which is made up of all companies and industries not just the individual company (Fingar, 2000). In this context all business processes are expected to be organized around the customer, leading to a reversal of the traditional linear supply chain concept that begins with sourcing. Moreover, a vision of business processes for the future is refocusing to include suppliers and trading partners and to ensure that when customers come into contact with any of the resources of the corporation they also rub up against the full value or supply chain.

While time compression in the value delivery chain has been the hinge variable in profitability for more than a decade (Kash, 1990), it is even more so in the e-commerce age. Today, as over the past decade, the organization that responds to customer demands fastest will likely achieve greater profitability and survive. Thus, large decreases in cycle time are required. This is more likely to be achieved in consumer pull-oriented systems, where goods are warehoused on the basis of usage, not estimated on the basis of expected consumption. In this environment the "virtual" supply chain (Fingar, 2000) (i.e., the envisioned web of organizations that source, produce, and distribute) will become the central institutional infrastructure as competition will be across the total value chain, with supply chain of company A competing against that of company B.]

The internet and e-commerce make increased value chain competition and optimization driven by customer demand a reality. This emerging outline of the future of the environment for logistics management suggests the unfolding of a revolution in fulfilling such traditional parts of the logistics management process as sourcing, production, and distribution, as well as non-traditional ones such as customer and supplier relations. An e-commerce platform makes it possible for an enterprise to extend supply chain automation to its suppliers and its suppliers' suppliers, and its customers and their customers' customers, thereby forming dynamic trading networks. For the optimization of these networks, shared real-time logistical support systems and data warehouses of information will be base requirements. These will make it possible for all agents, including the increased participation by small- and medium-sized enterprises, in the extended supply chain to improve their competitiveness, performance, and profits.

The vision of competition around an extended supply chain is what led Cooke (2000b) to invent the concept of "supply chain communities"; that is, trading partners which will conduct business electronically in real time. In this view, supply chain communities will operate as a dynamic hub-and-spoke model for rapid customer-demand-driven inventory renewal.] This model is in contradistinction to the more linear and sequential logistics model. To date, supply chain communities have focused mostly on linking trading partners with carriers, with third-party logistics providers not yet participating (Cooke, 2000b). Given the vision of widespread participation precipitating a major change in the business relationship between shippers and carriers, transport companies may end up working for the "community" rather than for a specific producer or shipper. Alternatively, competition for freight among carriers might be increased by the formation of bidding exchanges, one of the newest and most dynamic developments in logistics management (Ansberry, 2000; Cooke, 2000c; Anon., 2000a; Rosen, 2000), where bidding for shipments occurs in real time.

While the internet may be the new technology that is most affecting logistics management, there are some specific technical bottlenecks that are constraining the rate at which it will impact the industry. One of these is order fulfillment. e-Commerce must be supported by a solid distribution system, as the explosive growth in on-line sales has not been accompanied by a growth in distribution capacity (Jedd, 2000). The process innovation that seems most promising and is being widely explored is disintermediation, or the bypassing of intermediaries between buyer and seller by introducing a middle man. Furthermore, technologies (physical and process) to improve the reverse logistics process are sorely needed.] Finally, supply chain communities will need to be able to seamlessly exchange information at low cost to ensure that small- and medium-sized enterprises can participate. Electronic data interchange (EDI) technology has been used to send, read, and process purchase orders, bills of lading, and advance shipment notices. However, utilization has been limited to larger companies with the resources to invest in the necessary supporting computer hardware and software (Cooke, 2000c). A new technology called "extensible mark-up language" (XML) (Cooke, 2000a), recently developed and now being tested by Lucent Technologies, Inc., suggests that a low-cost alternative may be available in the near future.

5. Intelligent transportation systems

Intelligent transportation systems (ITS) are a complex of interrelated IT and telecommunication technologies that are applied to transportation infrastructure and vehicles (Cambridge Systematics, 1999). ITS technologies include, but are not limited to, advanced traffic/transportation management systems (ATMS), advanced transportation information systems (ATIS), and commercial vehicle operations (CVO). These technologies have a variety of base components, including collision avoidance, in-pavement sensors, video cameras to support traffic management, variable message signs (VMS), electronic guidance systems (EGS), electronic toll collection (ETC), and integrated computer information systems that support improved provision of emergency medical, police and fire services, and information for travelers. More specifically, the individual technologies include electronic, sensors, wire and wireless communication devices, computer hardware and software, automatic vehicle location (AVL), supported by GPS and geographic information systems (GIS). Barcodes and RFID are also technologies that support ITS. All of these ITS component technologies and systems are together envisioned to greatly increase the productivity and capacity of existing transport infrastructure. ITS is also expected to contribute to meeting other transportation goals, including multimodal integration, systems integration, and improved safety and environmental quality.

The vision of ITS is that it will provide potential capacity improvements as high as 20% (Johnson, 1997). However, it is still in the early stages of deployment. Existing for the most part for little more than 10 years, ITS have been deployed on a demonstration basis, and only in the past few years has deployment on a non-demonstration basis begun to occur. At the same time, in metropolitan areas, where most early deployment is expected, plans to support the full deployment of ATMS and ATIS systems are in a generally nascent stage of development. As a consequence, the extent to which these technologies will increase capacity and productivity, as well as other transport goals, is uncertain. Institutional barriers embedded in intergovernmental relations, traditional roles of the public and private sectors, and social equity issues related to access and privacy are larger barriers to deployment than are technical barriers.

The contribution of ITS, especially ATMS, ATIS, VMS, and CVO, to logistics management is expected to reduce uncertainty in sourcing and physical distribution systems. In this way it will contribute to reduced transaction and transportation costs. Intelligent transportation technologies, unlike the logistics-enhancing technologies discussed above (barcodes, RFID, GPS, the internet, computer hardware and software in support of enterprise or the value chain community logistics management), are mostly being developed by the public sector, although sizeable amounts are being out-sourced. Here the public good from these investments is expected to result in improved infrastructure, thereby creating knock-on effects that will enhance the business climate through reduced transaction and transport costs.

6. Summary and conclusions

The analysis shows that base technologies such as barcodes, RFID, and GPS, are being coupled with computer information systems and, in turn, contributing to an ever more integrated approach to logistics management. Yet, while this integration is unfolding, internet technology has created the opportunity for a new business model, but one that grows out of a community or web of agents all organized around managing the supply chain. The customer, who in the internet age has gained new power in the buyer–seller relationship, drives this model. Supply chain community webs are envisioned to be the primary model providing logistics services in the future.

At the same time that technologies are driving change in supply chain management and its organization, ITS has emerged with a potential to increase the capacity and productivity of existing transport infrastructure and, therefore, to greatly reduce transport time and cost. While ITS includes vehicle-specific technologies, such as collision avoidance, automatic vehicle location systems (GPS linked to RFID), and smart information devices supported primarily by the private sector, major components such as ATMS, VMS, and ATIS are being developed through a public sector lead.

As noted in the beginning of this chapter, the range of technologies considered is limited. There are, however, technologies that have not been discussed here, which are either in the early stages of development or are not yet well developed, that may affect logistics management in a significant way. One of these, XML, is expected to reduce the cost of supporting the communication processes necessary for supply chain management within the community model of logistics management. Another technology with the ability to greatly enhance computer hardware and software communication compatibility, Enterprise JavaBeans (Johnson, 1998), promises to greatly increase the ability to rapidly and easily interface logistics client–server systems. Large-scale vacuum pipeline transport systems and new high-speed propulsion systems for shipping could significantly reduce the cost and time in sourcing and distribution. Much further away is the vision of the mechanical extension of microelectronics (MEMS) and nanotechnology (Shipbaugh, 2000), which could revolutionize identification, communication, distribution, and production processes, while at the same time reducing the demand for logistics.

References

Andel, T. (1997) "Reverse logistics: A second chance to profit", *Transportation and Distribution*, 38(7), 61–66.

Anon. (2000a) "The brave new world of online marketplaces", *Logistics Management & Distribution Report*, April.

Anon. (2000b) "e-Commerce projected to reach $5.7 trillion in four years", *Logistics Management & Distribution Report*, 6.

Ansberry, C. (2000) "Let's build and online supply network!", *Wall Street Journal*, April, B1.

Brockmann, T. (1999) "21 warehousing trends in the 21st century", *IIE Solutions*, July, 36–40.

Cambridge Systematics (1999) *Challenges and opportunities for an ITS/intermodal freight program*. Washington, DC: U.S. Department of Transportation,.

Claycomb, C., C. Droge and R. Germain. (1999) "The effect of just-in-time with customers on organizational design and performance", *International Journal of Logistics Management*, 10(1):37–58.

Collins, P. (2000) "e-Logistics 2000: Re-thinking the supply chain", *Management Services*, June, 6–10.

Cooke, J.B. (1999) "Auto ID, software drive the supply chain", *Logistics Management & Distribution Report*, 38(7):101–102.

Cooke, J.B. (2000a) "ZML may represent the way of the future for exchanging supply chain messages over the internet", *Logistics Management & Distribution Report*, January.

Cooke, J.B. (2000b) "The dawn of supply chain communities", *Logistics and Management & Distribution Report*, February

Cooke, J.B. (2000c) "Technology: Good news, bad news", *Logistics Management & Distribution Report*, July.

Fingar, P. (2000) "e-Commerce: Transforming the supply chain", *Logistics Management & Distribution Report*, April, E7–E10.

Jedd, M. (2000) "Sizing up home delivery", *Logistics Management & Distribution Report*, February.

Johnson, C.M. (1997) *The national ITS program: Where we've been and where we're going*. Washington, DC: U.S. Department of Transportation, Federal Highway Administration.

Johnson, M. (1998) "A beginner's guide to Enterprise JavaBeans", *JavaWorld*, October.

Kash, D.E. (1990) *Perpetual innovation: The new world of competition*. New York: Basic Books.

Kennedy, M. (1996) *The global positioning system and GIS: An introduction*. Ann Arbor, MI: Ann Arbor Press.

Moore, B. (1999) "Bar code or RFID: Which will win the high-speed sortation race?", *Automatic ID News*, June, 29–36.

Rosen, C. (2000) "Logistics: The next step for online marketplaces", *Information Week*, May.

Schwartz, B.M. (1999). *Information Integration*, December, 43–46.

Seiderman, T. (2000) "Weapons for a new world", *Logistics Management & Distribution Report*, April (Suppl.).

Shipbaugh, C. (2000) "Thinking small: Technologies that can reduce logistics demand", *Army Logistician*, 32:20–25.

Chapter 35

FREIGHT AND LOGISTICS MODELING

GLEN D'ESTE
PPK Environment & Infrastructure Pty Ltd, Sydney

1. Introduction

Modeling is an extremely powerful tool for the analysis of logistics systems. When faced with the task of planning and managing logistics systems, most practitioners start by constructing a mental picture (or model) of the situation. This model is then used to develop and test possible courses of action. However, as supply chains and associated logistical systems become more and more complex, the mental map required to accurately represent the real world situation becomes so large and complex that it is virtually impossible to rely solely on intuitive methods. An alternative is to build a mathematical or computer-based model that reliably mimics the behavior of the real world system. This model can then take care of managing the data and performing the required calculations while the practitioner concentrates on developing scenarios, interpreting results, and making decisions.

Modeling techniques can be applied to all aspects of the supply chain and to strategic, tactical, and operational planning and management. This chapter focuses on systems modeling, in particular on modeling the transport-related aspects of freight logistics. It provides an overview of the role of freight and logistics modeling in an organization; the types of problems that are amenable to modeling; an introduction to modeling techniques; and a discussion of issues involved with development and implementation of freight and logistics models.

2. The role of modeling

As part of an integrated approach to logistics management, modeling can serve several important functions:

(1) *Testing theories* – testing ideas and theories is expensive and time-consuming, and in many cases it is impossible or impractical to experiment on the actual system, especially if the experiments may cause disruptions whose implications are not fully understood. Modeling also provides a

Handbook of Logistics and Supply Chain Management, Edited by A.M. Brewer et al.
© *2001, Elsevier Science Ltd*

mechanism for isolating the influence of specific factors affecting the logistics system and for identifying cause and effect relationships. Modeling is generally cheaper, safer, and more flexible than full-scale testing on an operational logistics system. In many cases, modeling is the only option.

(2) *Prioritizing options* – when faced with the problem of refining existing logistics systems or planning new systems, there may be many possible designs and management strategies. Modeling can be used to test, screen, and rank the options, and identify the most promising options for further investigation.

(3) *Minimizing risk* – models can be used to anticipate and diagnose the effects of different operating conditions and external factors on the functioning and robustness of a logistics system. For example, modeling can be used to anticipate the effects of possible changes in ordering patterns, and to identify potential bottlenecks in a supply chain. Therefore, modeling can be an integral part of a continual monitoring and improvement process.

(4) *Explaining ideas* – last, but by no means least, modeling can play an important role in understanding and visualizing complex logistics systems and the consequences of alternative scenarios and management strategies. Having decided on a course of action, modeling can also be a powerful tool to sell the idea.

Therefore in a competitive business environment, modeling allows a business to test and prioritize alternative responses to new service requirements, changes in market conditions, and identified weaknesses in current logistics systems and procedures.

Modeling can play a key role at all stages of the planning and implementation cycle: strategic, tactical and operational. At the strategic planning level, modeling can assist with long-term resource planning for design and re-engineering of supply chain and logistics systems. Modeling is especially valuable where the system is highly integrated with complex relationships between system components and there are many options but limited availability of financial, human, and infrastructure resources. While strategic planning takes a global and long-term view of the supply chain and logistics arrangements, tactical planning takes an intermediate view. It involves incremental adjustments to transport arrangements, inventories, and storage capacity over a period of days or months. In this case, modeling will help the business to anticipate, plan for, and respond to changing market conditions. Operational planning focuses on the daily or real-time management of individual activities in the logistics chain. Modeling can assist with optimization of operational procedures and day-to-day allocation of resources. At all stages of the planning hierarchy, modeling can help a business to be competitive by helping it to identify available opportunities and take advantage of them by adopting the best response.

The sort of strategic planning issues and questions that fall within the scope of freight and logistics modeling include:

(1) developing and testing alternative supply chain and logistics strategies and concepts, such as alternative transportation and inventory strategies;
(2) optimizing supply and distribution arrangements;
(3) optimizing the number and location and size of distribution points to meet delivery deadlines and cost requirements;
(4) testing the robustness of logistical arrangements to changes in scale and pattern of demand, such as changes in the ordering habits of customers;
(5) testing system velocity and reliability;
(6) finding the lowest-cost approach to meeting required service levels;
(7) supply chain capacity analysis and finding the bottlenecks, weak links, and system limits;
(8) evaluating the impact of investing to remove capacity and service constraints;
(9) evaluating cost trade-offs, such as cheaper overseas manufacturing versus shipping costs, or cheaper warehousing and primary transport cost compared with higher secondary transport costs;
(10) testing synergies between products or companies to determine whether overall system efficiency could be improved by combining logistics operations; and
(11) resource allocation.

3. Approaches to modeling logistics systems

Modeling can be applied to individual components of a logistics systems or the system as a whole. There are a range of techniques available for analyzing and modeling specific aspects of logistics systems (e.g., inventory models) and for representing particular processes and relationships in the logistics chain. These techniques are discussed in other chapters in this handbook. For modeling complex freight and logistics systems, there are three major approaches:

(1) optimization;
(2) simulation; and
(3) network modeling.

3.1. Optimization

Optimization is a mathematical modeling technique designed to find the best course of action subject to constraints which set the bounds of what is feasible or necessary under the prevailing conditions. In general, this means finding the best allocation of limited resources to specific activities. For example, consider the problem of finding the cheapest way of supplying two distribution centers from three factories with known transport costs between each factory and distribution center, known maximum production levels from each factor, and known requirements at each distribution center. Optimization can be used to find the least-cost supply strategy that meets all of these constraints.

When applied to real-world problems in business management and operations, the optimization approach is often referred to as *operations research*. The operations research approach involves formulating the problem as a set of mathematical expressions, comprising an objective function and a set of constraints. The objective function is a statement of the measure of system performance that is used to evaluate and rank alternatives. In most cases this will be the total cost of running the system, but it could be the total time taken for goods to move through the system, or any other factor or combination of factors that need to be maximized or minimized. The constraints represent limits on available resources or levels of performance that the logistical system must meet, such as

(1) available resources in terms of personnel, materials, and infrastructure,
(2) production capacity and demand patterns,
(3) capacities for transport links, warehouses, etc.,
(4) incompatibilities between activities, and
(5) minimum service levels.

The main purpose of the constraints is to impose realistic conditions on the generation of possible solutions and exclude unfeasible options. The objective function and constraints are expressed mathematically in terms of decision variables and parameters. The decision variables represent specific activities that can be varied at the discretion of the decision maker. Alternative values of these decision variables correspond to different scenarios or strategies. The parameters represent known characteristics of the systems. In most cases, the parameters are measures of the costs associated with performing these activities. In terms of the example of supplying distribution centers:

(1) the decision variables are the quantities of product dispatched from each supplier to each distribution center;
(2) the parameters are the known unit transport costs;
(3) the objective function is the total transport cost;
(4) the constraints stipulate that total shipments from a factory cannot exceed its production capacity and that total shipments to each distribution center must satisfy its demand.

The best course of action corresponds to the optimum values of the decision variables. In this example, it will be expressed in terms of the optimum quantity of product to ship between each combination of factory and distribution center.

The technique used to find the solution to the optimization problem will to some extent depend on the type of problem and the way that it is formulated. There are a range of general techniques, such as linear, integer, and dynamic programming, that can be applied to a wide range of optimization problems. There are also many special-purpose methods that have been developed to efficiently solve particular types of linear and non-linear optimization problems. Examples of optimization problems for which efficient special-purpose algorithms have been developed include transportation problems with or without trans-shipment (such as the factory to distribution center example), and the "travelling salesman problem" with or without constraints on delivery within specified time windows. The "travelling salesman problem" involves visiting a set of locations in the most efficient way, and has obvious applications to routing and scheduling of delivery vehicles. For a comprehensive but mathematically oriented introduction to optimization models, techniques and algorithms, see Taha (1992), and for a more application-oriented approach with examples drawn from logistics and supply chain issues, see Heizer and Render (1996).

Optimization techniques are especially useful for large complex strategic planning problems. The example of supplying two distribution centers from three factorizes is trivial, but if extended to a global logistics network with multiple suppliers of multiple products, many transport options including trans-shipment opportunities, and several possible locations for distribution centers, then the value of systematic techniques for system optimization becomes evident. Another strength of optimization is its flexibility. Although the requirement to formulate the situation into an objective function and constraints may appear to be restrictive, in practice it is possible to represent an extremely wide range of strategic, tactical, and operational planning problems in this way. All that is required is to formulate an optimization model is to address to following questions:

(1) What is the desired outcome (objective)?
(2) What aspects of the system can be controlled or varied (decisions variables)?
(3) What limits are there on what can be done (constraints)?

Optimization is a well-proven method that is backed up by a large corpus of research and practice and many standard algorithms and software packages that simplify the process of constructing and solving the model and allow the practitioner to focus on model formulation and interpretation of results. However, like all modeling techniques, optimization is based an idealized representation in which certain features of the real-world operation of the system are omitted to achieve a tractable model. In particular:

- Most optimization models do not take account of the inherent variation that is major feature of most logistics activities. For instance, the time that it takes for a truck to load and travel to its destination will vary, but for the purposes of optimization modeling it is usually assumed that this time and other model parameters are fixed. Therefore, most optimization models represent the long-run steady-state behavior of the system.

- All aspects of the problem must be reduced to mathematical expressions. As a result, subtle constraints and decision factors that are not amenable to mathematical representation tend to be omitted from the model. The results are simplified constraints and an objective function. In particular, the use of a simple linear objective function does not capture the richness of the actual decision-making process.

In summary, optimization is a powerful and flexible technique for finding the best feasible solution for problems in logistics and supply chain management. It is particularly well suited to large complex strategic planning studies, involving logistics issues such as transport strategy, facility location, inventory control, routing and scheduling, and supplier selection.

3.2. Simulation

Simulation modeling takes a different approach. Whereas the goal of optimization is to find the best course of action, simulation is a way of mimicking the behavior of the system and the testing and comparing alternatives. Therefore, it is essentially a "what if" approach to modeling.

Simulation models use mathematical and logical relationships to represent interactions between system components and the sequence of logistical activities. The model starts with a description of the system in terms of systems components, the way that they interact and their performance characteristics, and then progressively updates the state of the system according to well-defined operating rules. Therefore, a simulation model is essentially a laboratory version of the real-world system. The process of constructing a simulation model involves identifying:

(1) the items that are being processed through the system, such as consignments – these items then become the basic units of the simulation model;
(2) all the steps in the process and the possible paths through the system, such as processing, transport and warehousing – the degree of detail of the breakdown of steps will largely determine the complexity of the model;
(3) the rules that govern the way that items are processed, such as first-come first-served queuing discipline at loading and unloading stations, or limits on the number of items that can be loaded onto a vehicle or stored in a warehouse; and
(4) the performance of system components, such as the time taken to load a vehicle and the variability in loading time.

The last step highlights an important feature of simulation modeling. The models are inherently stochastic, which means that the performance of system components and the model as a whole is not fixed or predictable. For instance, in the real world, the time take to load a vehicle is not the same each time. In a simulation model this behavior is modeled by sampling from a prescribed probability distribution. Therefore, simulation models capture the variability that is a key feature of most real-world logistics systems. It also means that running a simulation model should be viewed in the same way as a statistical experiment. The output from the model is a sample from the set of all possible outcomes of the model. By running the model several times, it is then possible to get an idea of both the typical operation of the system and unusual behavior corresponding to exceptional events. Therefore, simulation can provide valuable insight into the robustness of a system to possible variations in inputs and the performance of specific system components.

Simulation models are implemented by converting the set of system processes and rules into instructions that can be processed by a computer. There are two basic types of simulation modeling: discrete and continuous. Discrete modeling processes the items as individual units, and progressively moves them through the model by updating their state each time some process takes place. Since most goods are moved in discrete packages or consignments, this is the approach most commonly used for modeling logistics systems, especially at the tactical and operational levels. The continuous approach looks at flows of materials, such as water in pipelines and reservoirs. For a technical discussion of simulation methods, see Taha (1992) and for a more application-oriented approach with examples drawn from logistics and supply chain issues, see Heizer and Render (1996).

There are a large number of general-purpose and application-specific software packages that simplify the process of constructing and running a simulation model. Many of these simulation software packages have powerful visualization tools that allow the user to draw the model as a process flow chart on the computer screen and provide an animated view of the model as it runs. The ability to watch the modeled system evolve is a powerful tool for understanding the dynamics of the system; diagnosing and correcting possible problems with the model; and explaining the results. Because of the level of detail involved in simulation modeling, models of logistics system of realistic scale and practical interest tend to have large demands in terms of data and computing power.

Like the optimization approach, simulation has its strengths and weaknesses. The strengths are:

(1) the ability to model complex dynamic systems involving feedback, non-linearity, and probabilistic variation;
(2) the break down of the logistics system into its component processes, from which complex system dynamics emerge naturally as these processes interact; and

(3) sensitivity to the timing, sequencing, and interaction between events
    because the simulation model tracks the evolution of the system as it
    evolves through time.

However, a simulation model does not inherently optimize. Simulation tests the
performance of a scenario designed by the modeler but does not find the best
scenario. A variety of techniques have been proposed to find optimum states of
simulation models. Most are essentially intelligent search techniques that vary
parameters in a logical way and converge toward the optimum. In other words, the
simulation is run over and over with different values of key parameters until the
best combination of parameters values is found.

In summary, simulation models are especially well suited to "what if" modeling
and testing of logistics systems that involve complex system dynamics and inherent
variation. This makes the simulation approach well suited to tactical and
operational modeling and for testing the robustness of systems to variations in
inputs and the performance of individual components of the system, and for
identifying potential bottlenecks and weaknesses in logistics systems.

3.3. Network modeling

The third approach to modeling freight flows and logistics is based on
representing the logistics system as a network of linked activities. For example, a
network of transport links naturally has this structure. Most network models can
be converted to an equivalent set of equations in an operations research format, so
the network approach could be considered to be a subtype of the optimization
category. However, the network approach is particularly well suited to modeling
complex freight transport and handling systems, and has enough special features
to be considered a category in its own right.

The network approach involves building a mathematical representation of the
pattern of activity linkages and investigating the characteristics of freight flows
through the network model. The model comprises nodes, each of which
corresponds to a particular location or activity center (supplier, storage,
distribution center, customer) in the logistics system, and links which represent
the movement and handling of goods. Therefore, the network model can be seen
as an idealized representation of the pattern of transport linkages between
suppliers, distribution centers, and customers. Having mapped the real-world
pattern of transport linkages onto a network model, the optimization process
involves finding the best path or combination of paths through the network.
A network model can be used to optimize the movement of individual
consignments through a network of transport options or to identify a system-wide
optimum for a complex pattern of freight demand. The approach is general

enough to cover multiple modes, trans-shipment, and multiple products, and can also be used to investigate bottlenecks and other network performance characteristics.

The network approach builds on mathematical graph theory and practical experience gained from modeling urban road traffic and transit systems. For instance, optimization of a logistics network model is equivalent to the network assignment phase of urban traffic and transit modeling. There are a large number of algorithms that have been developed to efficiently solve large complex network models. Solution algorithms and methods for formulating transport network models are discussed in detail in Volume 1 of this series (Hensher and Button, 2000).

The strength of the network approach is that it is particularly well-suited to modeling of transport systems and can efficiently represent and optimize a large, complex transport network. In addition, the formulation of the logistics system as a network model is easy to understand and visualize because there is a direct correspondence between model components (nodes and links) and real-world features. However, like other approaches, a network model is a highly idealized representation of the logistics system. In particular, standard network formulations and algorithms are not proficient at modeling the discontinuity and "lumpiness" that is a feature of logistics systems. For instance, demand for a product may be expressed in terms of individual items while consignments are in units such as shipping containers or cartons, which are then consolidated for transport and supplied in units such as ship or truck loads. There is thus a mismatch between the units used at different points in the logistics chain. This is difficult to model using the network approach. For a full critique of the network approach, see D'Este (1996).

In summary, the network approach provides a general framework that is well suited to strategic planning of freight transport systems. Its main features are that it is easy to formulate and visualize the models, and the availability of efficient special-purpose solution algorithms.

3.4. Other approaches

In addition to the three major approaches, there are a large number of techniques that have been proposed to complement and extend their capabilities, and to address their weaknesses. These techniques include:

(1) heuristic methods;
(2) expert systems;
(3) genetic algorithms and related optimization techniques; and
(4) the event-based approach.

Heuristic methods

Heuristics are "rules of thumb" that have been observed to provide an effective approach to solving a particular problem. They are based on human judgement and experience rather than mathematical theory. In general, heuristics do not guarantee an optimal solution but simply seek to produce a good solution in an efficient way. Heuristics may also used as a way of speeding up other methods, such as optimization algorithms, without sacrificing the overall validity of the method. They are especially important for certain types of optimization problems for which a formal solution algorithm is not available.

Heuristic methods are widely used across all types of freight and logistics modeling. They can play an important role in developing efficient and tractable models; however, some care should be taken because the use of heuristics can exclude possible solutions by constraining the model to comply with "conventional wisdom."

Expert systems

An expert system is a computer program that mimics the analytical processes of an expert in the field. An expert system typically comprises a large set of decision rules, data, and general "rules of thumb" that have been determined by an analysis of the system and the thought processes followed by experts. This information is then combined using conventional logic or inferential techniques such as neural networks and fuzzy logic. A neural network is an inference model based on an analogy with the functioning of neurons in the human brain. Fuzzy logic extends conventional logic by allowing fuzzy and qualitative definitions of inputs. The user presents the expert system with a description of a situation, and the expert system automatically draws conclusions and makes recommendations based on its embedded knowledge and rules. Therefore, an expert system does not model the situation directly; instead, it models the process of analyzing a situation and reaching conclusions.

In terms of logistics modeling, an expert system can be linked to an optimization, simulation, or network model to extend the overall capabilities of the model and provide an "expert" interface to the technical aspects of the modeling. For more details on the use of expert systems in business applications, see Watkins and Elliot (1993).

Genetic algorithms and related approaches

Nature is a very efficient optimizer. Over time, most natural systems adapt to prevailing conditions and constraints by settling down to an optimum state that minimizes requirements for energy and resources. This phenomenon has prompted researchers to look to nature for novel optimization techniques. One of the most promising of these techniques is genetic algorithms. The genetic algorithm approach imitates the evolution of species by using mathematical structures equivalent to chromosomes to represent possible solutions to the problem being modeled. The modeling process starts by randomly generating an initial population of chromosomes, then these chromosomes are evolved over several generations through processes of selection, cross-breeding, and mutation. The probability that each chromosome will survive to breed the next generation depends on how well it is adapted to the required task – survival of the fittest. Over successive generations, the process will "breed" new and better solutions. Using the terminology of the optimization approach, a chromosome is equivalent to a particular set of values of the decision variables, and the fitness function reflects the objective function and system constraints. For a detailed discussion of techniques and applications of genetic algorithms and related approaches, see Davis (1991).

The genetic algorithm and related approaches have several advantages. First, it is possible to model highly complex dynamic systems that would be difficult to model by other methods. Genetic algorithm models are not limited to linear or quasi-linear relationships or to continuous functions. They can handle a wide range of highly non-linear systems and mixed non-continuous and qualitative data. Secondly, the performance measure that guides the evolution of solutions can be very general, which means that genetic algorithms techniques are not limited to a simple objective function, such as minimizing total system cost. Thirdly, by breeding a large number of alternatives and simultaneously adapting all of these options, genetic algorithms often find multiple feasible solutions and can discover non-intuitive management strategies. This can be an extremely valuable outcome because adopting a novel strategy can potentially provide a market edge or differentiate a business from its competitors. Genetic algorithms and related approaches have proven to be an effective approach for a variety of logistics problems which are not amenable to other optimization and problem-solving strategies, in particular complex scheduling problems. They are likely to have an increasing role in freight and logistics modeling.

Event-based approach

Most freight and logistics models focus on the spatial dimension of goods movement and handling, and on the costs associated with physical transfers of goods. As a result, they do not reflect the growing importance of service and information factors in logistics management and competitiveness. The event-based approach (D'Este, 1996) is an attempt to unify the technical, information, and service aspects of logistics planning in a single framework.

This approach views the passage of a consignment through a freight transport and handling system as equivalent to a sequence of logistical events. Each event takes the goods from one logistical state to another by performing a logistical activity. The range of possible activities encompasses all aspects of the logistics chain, from physical movement of goods to transfer of information. The conceptual framework of logistical states, events, and activities closely mirrors the operation of the real-world system and provides a great deal of generality in modeling logistical systems. It also directs attention away from transport, technology, and spatial concerns and toward services, transitions, and outcomes, in the broadest sense. By abstracting from transport links to logistic events, the event-based approach integrates the full range of relevant logistical activities into a single coherent framework. In many respects, it shares more in common with project management concepts than with traditional freight and logistics models. The event-based approach provides an alternative paradigm for formulating models, and can be implemented using optimization, simulation, and network techniques.

4. Designing and using logistics models

In all modeling applications there are design decisions and trade-offs. This section provides a brief commentary on practical aspects of formulating, developing, and using freight and logistics models.

The first step in developing a model is to decide on the approach. This involves matching the type of model and its degree of detail and complexity to the objective and purpose of the model. A strategic planning model must reproduce the overall system dynamics but can take a reasonably broad-brush approach as long as there is sufficient detail to enable a valid comparison of options. Modeling of the timing of events is less important than modeling the overall pattern and sequence of activities. The main purpose of strategic modeling is to evaluate all reasonable options and find the optimum. Tactical modeling needs greater detail with more attention to the time dimension and reproducing the behavior of a particular system scenario, while an operational model must be an extremely faithful representation of the real-world process. However, it must be recognized that greater detail usually means that the model will be more complex and have greater demands on data and be more difficult to implement and consume more corporate resources.

Optimization and network approaches are well suited to finding the best management strategy in large, complex freight and logistics systems with many options. However, the level of detail is generally low compared with a simulation model, which is better suited to evaluating and refining a particular option. An extremely large amount of data and model complexity and computing power is required to find optimum solutions for large complex systems using simulation techniques. This suggests that, as a general "rule of thumb," optimization and network techniques work well for strategic and tactical planning, while simulation is a valuable tool for tactical and operational planning. However, there are always exceptions. For instance, optimization techniques (generally in combination with heuristics) can play a valuable role in routing and scheduling for complex distribution tasks.

Another design decision relates to the way that the model is implemented. In almost all logistics models of practical value, the model will be implemented on a computer using some sort of general-purpose or specific software tools. Spreadsheets are a very popular tool for financial modeling, but they have limitations for modeling freight logistics. A spreadsheet becomes unwieldy for large, complex models, and a spreadsheet approach is not well suited to modeling situations that involve time phasing of activities or to iterative optimization. The alternative is to use specialized software packages that are available for implementing optimization and network models and for developing simulation models. Off-the-shelf packages are a cost-effective way to implement and customize a model to the required situation but may be restricted in their capability to model particular aspects of the system. On the other hand, developing a bespoke model using low-level tools is a time-consuming and expensive approach that requires considerable specialist expertise. However, it will deliver a model that is highly tuned to the required task.

Quality models and decisions come from quality inputs and a detailed understanding of the linkages and cause and affect relationships that drive the logistics system. The model developer needs to understand the system being model and the limitations of the model. All modeling involves compromise, and in all cases a model should not be given credit for more accuracy than it has. A model of a logistics system is a management tool whose results need to be considered along with other inputs to the decision-making process. The model should be seen as an extension of the logistics manager's capability, not as an alternative. For further discussion of designing and using models, see Powers et al. (1983).

In summary, the best approach is to keep it simple, and avoid building unnecessarily complicated models. The model should only be as complex as it needs to be to provide the required answers – not all the answers. The complexity of the model will to some extent be dependent on the complexity of the supply chain, but the model developer must always be mindful of the purpose for which the model is being developed and match the model with the objective and customize it to the particular situation. An effective strategy for planning logistics systems is to use a combination of an optimization model for strategic planning and a simulation model for testing and verification. The first step is to develop an optimization model at a sufficient level of detail to allow valid comparison of options, and use the model to identify a preferred option. Then the operational performance and robustness of the preferred option under a range of operating scenarios can then be tested using a simulation model.

5. Concluding remarks

Modeling can be an extremely powerful tool for planning and managing logistics and supply chains. It provides a mechanism for testing and evaluating ideas about how to improve the operation of a logistics system. Modeling is generally cheaper, safer, and more flexible than full-scale testing on an operational logistics system. Modeling can help a business to see potential problems and opportunities faster and adapt to them, and enhance its competitive edge. This chapter has presented several approaches to modeling logistics systems that are robust, transferable, and suitable for use under a wide range of circumstances. However, when developing models it must be remembered that every situation is unique and demands a model that is customized to match the particular situation and purpose.

References

Davis, L. (1991) *Handbook of genetic algorithms*. New York: Van Nostrand Reinhold.

D'Este, G.M. (1996) "An event-based approach to modeling intermodal freight systems", *International Journal of Physical Distribution and Logistics Management*, 26:4–15.

Heizer, J. and B. Render (1996) *Production and operations management*. Englewood Cliffs: Prentice Hall.

Hensher, D.A. and K.J. Button (2000) *Handbooks in transport 1. Handbook of transport modelling*. Oxford: Elsevier.

Powers, R.F., J.J. Karrenbauer and G. DoLittle (1983) "The myth of the simple model", *Interfaces*, 13(6):84–91.

Watkins, P.R. and L.B. Elliot (1993) *Expert systems in business and finance: issues and applications*. Chichester: Wiley.

Taha, H.A. (1992) *Operations research: an introduction*. New York: Macmillan.

AUTHOR INDEX

Abdulaal, M., 499, 505
Abkowitz, M.P., 476
Aldag, R.J., 134
Andan, O., 303
Andel, T., 516
Andel, T.J., 340
Anderson, D.L., 301, 504
Anderson, J.C., 136
Ansberry, C., 517
Arentze, T.A., 306, 307
Argyris, C., 361
Arntzen, B.C., 100
Ashford, N., 325
Atkinson, A.A., 15

Ballou, R.H., 66
Banister, D., 340
Bardi, E.J., 219
Barnett, W.P., 133
Barney, J., 128, 137
Beaumont H., 63, 65
Bechtel, C., 100
Beesley, A., 180, 181
Belohradsky, E., 456
Ben-Akiva, M., 294, 299
Bennis, W., 253
Berkley, B.J., 287
Berry, L.L., 282, 283
Beuthe, M., 251, 496
Bhatt, C.R., 299
Billington, C., 100
Birou, L.M., 214
Bitran, G., 285
Bleijenberg, A., 347
Blood, M.R., 134
Bohnhoff, A., 457, 458, 463, 464, 465
Bonney, J., 18
Bontekoning, Y.M., 247
Borts, G.H., 322
Bowen, J., 279
Bowersox, D.J., 84, 100, 159, 160, 165, 166, 263, 407
Bowman, R.J., 370
Boyson, S., 47, 48, 52, 54
Braeutigam, R.R., 321
Braithwaite, A., 38
Brandenburger, A.M., 128, 133
Brax, B., 456, 458, 464
Brewer, A.M., 253

Brewer, P.C., 131
Brief, A.P., 134
Bröcker, J., 43
Brockmann, T., 515
Brooks, M.R., 422, 425
Brown, S., 408
Bucklin, L.P., 109
Burbidge, J., 187
Burgelman, R.A., 133
Burkhalter, L.A., 372, 373
Burton, B.K., 133
Button, K.J.152, 250, 294, 340, 395, 432, 439, 529
Byrne, P., 340

Camp, R.C., 100
Carman, J.M., 285
Carron, A.S., 437, 433
Caves, R.E., 128
Cecchini, P., 29
Champy, J., 116, 118, 120, 159, 172
Chase, R.B., 276, 279, 286, 288
Chow, G., 21
Christopher, M., 38, 82, 99, 129, 131, 159, 160, 163, 354, 379, 416
Claes, F., 281, 283
Clarke-Hill, C.M., 389
Claycomb, C., 515
Clendein, J.A., 111
Clinger, J., 142, 154
Clinton, S.R., 166
Closs, D.C., 407
Closs, D.J., 100, 166
Clutterbuck, D., 216
Colbert, D.N., 100
Coleman, J.L., 121
Colin, J., 159
Collins, P., 516
Congram, C., 287
Cook, D.P., 275
Cooke, J.B., 513, 515, 517, 518
Coone, T., 373
Cooper, D., 334
Cooper, J.C., 161, 162, 163, 167, 343, 340, 347
Cooper, M.C., 99, 101, 102, 104, 107
Cooper, R., 92
Corsi, M., 47, 48,
Cox, J., 89
Coyle, J., 366, 407

SUBJECT INDEX

3M, 110
advanced planning and scheduling, 414–15
advanced traveler information systems, 221
Ahold, 386, 388, 389
air freight. 2, 38, 41, 69, 148, 152, 153, 229, 241,
    243, 342–46 *passim*, 381, 431–39, 458,
    459–61 *passim*, 503, 508, 514
  belly hold, 38
Airborne Express, 433
Airports Council International, 434
airports, 2, 38, 277, 325, 326, 329–37 *passim*,
    404
Albert Heijn, 386, 390
alliances, 26, 75, 94, 131, 148–9, 152, 219–20,
    251, 263, 419
  consortia, 427
  cross-equity, 90–1
Amazon.com, 277, 390
American consumer satisfaction index, 281
American President Lines, 145
APL Logistics, 423–4, 426
applications service providers, 54–5, 58
Arcadia, 390
Arrow Air, 432
Asda, 390
Association of American Railroads, 501
AT Kearney, 164, 264, 371, 487
Australian Bureau of Industry Economics,
Australian Bureau of Statistics, 218
Australian Bureau of Transport and
    Communications Economics, 445, 447
Australian Productivity Commission, 18
automatic vehicle location, 478
  see also tracking and tracing
Avis, 274,

back-loads, 398
Baltic International Futures Exchange, 449–50
banks, 130
bar codes, 329
barges, *see* inland waterways
benchmarking, 118, 136, 160, 325–37, 353, 422,
    477, 504
Benetton, 190, 379, 380
best practice, *see* benchmarking
billing, 2, 16, 329, 411
BMW, 340
Boeing, 38
border controls, 31

break bulk, 36, 245, 257, 434, 484, 490
Broken Hill Propriety Ltd, 451
Burlington Northern, 368
business process re-engineering, 159–60, 167,
    172

Cable and Wireless, 72
Cambridge Systematics, 507
Canadian National Railway, 368
Canon, 51
car codes, *see* tracking and tracing
Cargolux, 432
Carrefour/Promodes, 389
CELO, 39, 40
Census of Transportation Commodity Flows,
    503
Challenge Air Cargo, 433
Chrysler Corporation, 123
Cisco Systems, 47, 50, 51–2, 57
Citibank, 274
city logistics, *see* urban logistics
Civil Aeronautics Board, 437
Coca-Cola, 382
combined transportation, *see* intermodal
    transportation
Competition Policy Agreement, 1995
    (Australia), 446
competition, *see* competitiveness
competitiveness, 20 23, 43, 45, 76, 99, 133, 134,
    162–3, 163, 240, 254, 416, 422, 495, 518,
    522
computer reservation systems, 432
congestion, 22, 34, 221, 247, 393, 394, 395, 399,
    400, 408, 439, 466, 487
Conrail, 368
consolidation, 40, 73, 219, 236, 239–251, 402–4,
    433
containers, 45, 143–9, 151, 152, 229–30, 240,
    247, 427–28, 489, 491–2
  terminals, 493, 494
  see also intermodal transportation
Contship Containerlines, Ltd., 427–28
Cooper and Lybrand, 382, 383, 387
cost centers, 320
costing, 314–23
Council of Logistics Management, 100, 101,
    162, 341, 371, 504, 514
courier services, *see* express deliveries *and*
    parcels